CHEMISTRY RESEARCH AND APPLICATIONS

COMBUSTION

TYPES OF REACTIONS, FUNDAMENTAL PROCESSES AND ADVANCED TECHNOLOGIES

CHEMISTRY RESEARCH AND APPLICATIONS

Additional books in this series can be found on Nova's website under the Series tab.

Additional e-books in this series can be found on Nova's website under the e-book tab.

CHEMISTRY RESEARCH AND APPLICATIONS

COMBUSTION

TYPES OF REACTIONS, FUNDAMENTAL PROCESSES AND ADVANCED TECHNOLOGIES

JOSEPH M. GRIER
EDITOR

nova publishers
New York

For permission to use material from this book please contact us:
Telephone 631-231-7269; Fax 631-231-8175
Web Site: http://www.novapublishers.com

NOTICE TO THE READER

Library of Congress Cataloging-in-Publication Data

Combustion: types of reactions, fundamental processes and advanced technologies / [edited by] Joseph M. Grier.
pages cm
Includes bibliographical references and index.
ISBN: 978-1-62948-967-4 (hardcover)
1. Combustion. I. Grier, Joseph M., editor of compilation.
QD516.C6175 2014
541'.361--dc23

LCCN: 2013048325

Published by Nova Science Publishers, Inc. † New York

CONTENTS

PREFACE

In this book the authors present current research in the study of different types of reactions, fundamental processes and advanced technologies of combustion. Topics discussed in this compilation include the Lagrangian formulation to treating the turbulent reacting flows; g-equation in white-noise in time turbulent velocity field; deposition of thin functional coatings at atmospheric pressure using combustion chemical vapor deposition; fundamentals of oxy-fuel carbon capture technology for pulverized fuel boilers; combustion of lignocellulosic biomass and marine biomass by means of thermal analyses; methane combustion by electrochemical promotion of catalysis phenomenon; novel nitrate-free acetate-H_2O_2 combustion synthesis; and solution combustion method.

Chapter 1 - In this chapter the authors outline a Lagrangian approach to treating problems of turbulent reacting flows. In order to avoid unnecessary complications, and to highlight the basic issues they limit ourselves to the particular case of a constant density, homogeneous isotropic turbulent flow. The case a passive reactive scalar is considered, i.e. it is assumed that the turbulence is not affected by the combustion.

Chapter 2 - A promising approach to simulating premixed turbulent combustion in gas turbine and spark ignition engines consists of tracking a flame surface in the turbulent velocity field in the framework of so called $G-$equation, associating the evolution of the flame surface with the evolution of zero (or G_0) level set of a function $G(\mathbf{x},t)$. The goal of the present chapter is to derive the equation for the probability density function of G straightforwardly from the well-known G-equation for tracking an infinitely thin flame sheet (flamelet) in a white-noise in time statistically homogeneous isotropic constant density turbulent flow.

Chapter 3 - The deposition of thin functional layers under atmospheric pressure conditions using flame-based techniques is a cost-effective solution for several applications in comparison with the well known vacuum techniques. The basic principles of the flame-based deposition techniques are reviewed and the technical realization is shown by means of selected examples of the current research and development of Innovent e.V.

In the case of combustion chemical vapour deposition (CCVD) a combustible gas-air mixture is ignited in a burner. The temperature of the flame depends on the gas-air ratio and strongly on the zone of the flame. In the simplest case oxydizing and reducing flame are differed. In dependence of both of these main flame-properties, the chemistry of precursor reactions and film-deposition is influenced. The film morphology can also be adjusted for example by the distance of the substrate to the flame. Additionally a specialized remote CCVD system is developed in order to deposit hydrolyzing precursers.

The formation of silicone oxide layers with a wide range of properties is investigated in detail. The most famous applications like adhesive, optical effective, barrier and matrix layers for embedding particles are reviewed. Fundamental investigations were performed to deposit zinc oxide layers with the objective of transparent and conducting coatings. Further coating systems like photocatalytic active (TiO_2), electrochromic (wolfram oxide), silver, aluminium oxide and rare-earth layers are also introduced.

Chapter 4 - Coal is expected to continue to contribute to electricity production in the near future. However, it generates more CO_2 per unit of produced energy than other fossil fuels and its use is associated with the formation of contaminants such as NO_x, SO_x, Hg, etc. In recent years, oxy-fuel combustion has attracted great interest as a promising Carbon Capture and Storage (CCS) technology, since it can be easily adapted to both existing and newly built coal-fired power plants. During oxy-fuel combustion coal is burnt in a mixture of oxygen and recycled flue gas (RFG) (mainly CO_2 and water vapor), that is used to lower the flame temperature and yield a rich stream of CO_2. However, the successful implementation of oxy-fuel combustion technology depends on having a full understanding of the difficulties that arise as a result of replacing inert N_2 by reactive CO_2 in the oxidizer stream.

Biomass is a renewable fuel which can be used to reduce CO_2 emissions, as it is considered carbon-neutral. The combination of oxy-coal combustion with biomass co-firing could provide an effective method of CO_2 disposal which has only partially been studied. Furthermore, biomass co-firing is presented as an option for increasing capture efficiency, as the costs and efficiency penalties would be reduced. This chapter presents a technical review of the oxy-fuel combustion process, covering the most recent experimental studies at both laboratory and pilot scale. The impact of replacing N_2 by CO_2 upon heat transfer, devolatilization, ignition, combustion kinetics, pollutant formation (mainly NO_x and SO_x) and fly ash and slag formation is also evaluated. Furthermore, the effect of blending coal and biomass under oxy-fuel conditions on combustion properties is assessed.

Chapter 5 - Biomass fuels are gaining particular attention as a potential alternative to increase energy independence of fossil fuels and reduce environmental pollution. The main ones refer to lignocellulosic biomass and marine biomass (especially microalgae). There is a growing interest in the cultivation of lignocellulosic biomass for energy production as they typically have less capital-intensive conversion technologies, attractive opportunity for local and regional self-sufficiency, reduction in greenhouse gas emissions and viable alternative to fossil fuel use. On the other hand, microalgae have received increasing attention due to the fact that they can be cultured in ponds or photobioreactors with supply of nutrients or waste water. Moreover, the production of microalgae does not require of high quality arable land and therefore it does not compete with food crops.

Direct combustion is the oldest and still the most commonly used route for converting biomass to heat, power and combined heat and power. Combustion can be defined as the conversion of biomass fuels to several forms of useful energy in the presence of air or oxygen. Despite its inherent potential as a biofuel resource, the commercial viability of biomass combustion technology has not been achieved yet. In this regard, thermogravimetric analysis is one of the most common techniques used to rapidly investigate and compare thermal events and kinetics during combustion of biomass. TGA provides useful information concerning the temperatures at which combustion or decomposition reactions in the sample start. Biomass characteristics and kinetics of biomass combustion are essential for modeling the combustion in industrial processes. Therefore, from TGA data suitable mathematical models can be derived for a better comprehension of the oxidation behavior of these complex feedstock that allow to perform economic analysis and develop technology for a more efficient biomass conversion.

Additionally, during the combustion of biomass, the composition of the gas emissions should be determined before industrial application. In spite of the environmental advantages, some aspects concerning the release of contaminants during biomass combustion must be taken into account. In this regard, NO_x and SO_x emissions depend on raw biomass composition, which usually is variable. Furthermore, the chloride amount in biomass might turn into operational problems such as corrosion. Other organic compounds such as benzene and toluene are considered to be part of the most dangerous emissions from biomass combustion causing diseases as lung infection and leukemia. Therefore, the knowledge of pollutant release during biomass conversion is truly important in order to reassure the use of biomass from the environmental point of view. In this regard, the coupling of thermogravimetric analysis with mass spectrometry (TGA-MS) stands out as the only experimental technique able to afford real-time and sensitive detection of evolved gases from the thermochemical conversion of a very small sample.

In this chapter, the combustion behavior and the gas evolution analysis of different biomass feedstock (lignocellulosic and microalgae) by means of TGA-MS will be addressed. Finally, the kinetic analysis of the combustion process will be presented.

Chapter 6 - Methane is the main component of natural gas and is probably the most abundant organic compound on Earth. In view of the fact that the world petroleum reserves are on the decline, a great deal of emphasis has been placed on developing alternatives for energy production. Significant efforts have been focused on finding effective processes and technology for the optimum utilization of the abundant natural gas inenergy production. During the last decades, the catalytic combustion of methane has been extensively studied as an alternative to conventional thermal combustion. This method allows converting methane at relatively low temperatures and with very high combustion efficiency, thus avoidingemissions. Hence, it is considered an effective route to produce power with low environmental impact and then suitable for gas turbine applications.

On the other hand, one of the lastest advances in catalysis is the phenomenon of Electrochemical Promotion Of Catalysis (EPOC), or Non-faradaic Electrochemical Modification of the Catalytic Activity (NEMCA), that has had a strong impact in the fields of electrochemistry, heterogeneous catalysis and surface science. This phenomenon is an important tool to improve the catalytic performance of a metal catalyst interfaced with a solid electrolyte by the application of low current or potentials over the catalytic film. During the last years, it has been demostrated that itis based on the change of the work catalytic function due to the elctrochemical pumping of ions from the solid electrolyte to the catalyst.

In this chapter, the catalytic combustion of methane by this new phenomenon will be addressed. Finally, the application of this tecnhology to the methane combustion over palladium based catalyst-electrodes will be presented.

Chapter 7 - One of the key challenges in materials science research is the synthesis of ceramic materials with the desired phase purity, morphology, particle size and expected properties for defined applications, while controlling cost. The selection of the synthesis method is vital to regulate the composition, structure, and morphology of a preferred material. Ceramic oxide materials are typically prepared by solid-state reactions, for example simple mechanical mixing of ceramic oxide powders, or by precipitation from solution and following decomposition. Nonetheless, the solid-state route can often lead to the appearance of non-homogeneity in the final products and low phase purity that can deteriorate the desired properties.

A novel nitrate-free acetate-H_2O_2 combustion synthesis method has recently been developed that has been demonstrated for the successful preparation of Ni-$BaZr_{0.85}Y_{0.15}O_{3-\delta}$ cermet anodes and $Ba(Ce,Zr,Y)O_{3-\delta}$ electrolytes for Proton Ceramic fuel cell (PCFC) applications. Metal acetates (M-$(OCOCH_3)x$ yH2O) and 30% H_2O_2 are employed as starting precursors for this combustion reaction, using microwave heating. The new route contrasts dramatically to traditional nitrate-based combustion processes in that the metal precursors now constitute the fuel rather than the oxidant in the combustion reaction. Another significant distinction of the novel route is that the acetate precursor solution is neutral in contrast to typical nitrate-based combustion routes that involve acidic solutions. This factor has been shown to be highly important for the successful preparation of cermet anodes such as Ni-$BaZr_{0.85}Y_{0.15}O_{3-\delta}$ that contain alkaline earth metals. The novel combustion route produces powders with a fine crystallite size in the nanometric scale. The presence of an intermediate peroxy (-OOH) bond has been identified and is suggested to aid the combustion reaction by harnessing the required oxidant in the form of labile peroxide complexes. This soft chemical method offers homogeneous distribution and stabilization of the metal ions in the solution state and can be applied for the formation of further multi-element ceramic oxide materials, offering economical advantages over more classical nitrate-based combustion routes, as well as significant environmental benefits due to the avoidance of releasing NO_x gases.

Chapter 8 - The development of advanced catalyst synthesis methods to meet the stringent environmental regulations in the production of ultraclean transportation fuels represents a major challenge for catalyst suppliers. In this chapter, the positive impact of solution combustion synthesis (SCS)-based preparation approach in the synthesis of alumina-supported NiMoP and NiMoW hydrotreating catalysts has been examined. The results show that the use of urea as fuel can tune the metal oxide-support interaction, altering the distribution of Mo-oxo species onto alumina support. Furthermore, the hydrodesulphurisation (HDS) activity of alumina-supported NiMoP sulphide catalysts achieved a maximum at Φ ratios between 0.6 and 1.2. This maximum HDS activity is attributed to a higher surface density of polymolybdate relative to molybdate species in the catalyst precursor and the effective interaction of nickel with the edges of MoS_2 that facilitates the formation of NiMoS structure. On the other hand, the sequential deposition of NiMo-urea solution on metal oxide-modified alumina surface produces a reduction of the catalyst bed temperature rate and an increase of the decomposition period of the redox mixture. The addition of Ni and Mo to WO_x-modified alumina assists the segregation of MoO_x and WO_x species on the alumina surface and promotes their reducibility. Furthermore, the NiMoW catalyst prepared impregnation-solution based combustion method showed higher hydrotreating activity than a commercial NiMo catalyst.

In: Combustion
Editor: Joseph M. Grier
ISBN: 978-1-62948-967-4

Chapter 1

LAGRANGIAN FORMULATION TO TREATING THE TURBULENT REACTING FLOWS

V. A. Sabelnikov*
ONERA - The French Aerospace Lab., Palaiseau, France

ABSTRACT

In this chapter we outline a Lagrangian approach to treating problems of turbulent reacting flows. In order to avoid unnecessary complications, and to highlight the basic issues we limit ourselves to the particular case of a constant density, homogeneous isotropic turbulent flow. The case a passive reactive scalar is considered, i.e. it is assumed that the turbulence is not affected by the combustion.

Keywords: Advection-reaction equation, Eulerian frame, Lagrangian frame, Lagrangian characteristics, Eulerian characteristics, mixed Lagrangian-Eulerian quantities, Premixed turbulent combustion, combustion progress variable, Taylor theory, turbulent diffusion coefficient, G -equation

* E-mail: vladimir.sabelnikov@onera.fr.

1. Introduction

The impetus for this study stems from two recent publications [1], [2] devoted to the analysis of the Lagrangian properties of turbulent diffusion that, as the authors of [1] wrote, "even if they are ignored in the Eulerian approach used for the description of heat and mass transfer, are important in several applications, in particular in turbulent premixed combustion". The authors [1] derived the equations which illustrate, in particular, the nonequilibrium behavior of the turbulent diffusion coefficient. This effect, as it is well-known, is important for modeling the initial stage of combustion for example, in the spark ignition engine. Unfortunately some conclusions of the paper [1], as will be demonstrated in the present paper, are incorrect.

The main objective of the present paper is to develop a systematic approach to cope with the Lagrangian formulation to treating the turbulent reacting flows. The case a passive reactive scalar is considered, i.e. it is assumed that the turbulence is not affected by the combustion. We revise also the incorrect results in [1]. We begin with, in the section 2, the formulation of an instantaneous advection-reaction equation for reactive scalar. In the section 3 the relations connecting the instantaneous and expected Lagrangian and mixed Lagrangian-Eulerian quantities are derived. In the section 4 the relations connecting the instantaneous Eulerian and Lagrangian quantities, and expected Eulerian and mixed Lagrangian-Eulerian quantities are derived. In the section 5 the equations for the expected mixed Lagrangian-Eulerian quantities are derived. In the section 6 we apply the Taylor theory of the turbulent diffusion to transform the equation for the expected mixed Lagrangian-Eulerian reactive scalar derived in the section 5. The section 7 is devoted to consideration of the turbulent premixed flame employing the equations derived in the section 6. The discussion addresses an asymptotic case of an infinitely thin flame front (flamelet) that separates the reactants and products in a turbulent flow and self-propagates at a speed S_L with respect to the products. When the flame front hits a reactant particle it turns into a product particle. An exact analytical expression for the density probability function of the flame front hitting the reactant particle is derived. To this end we applied the well-known G-equation (the level set method for tracking interfaces). The section 8 continues the analysis of the premixed combustion. An approximate expression connecting expected Eulerian progress variable and the Lagrangian transition probability density function is derived. Then an approximate equation for the mean progress variable is derived. This equation coincides with well-known in the theory of the premixed combustion Prudnikov equation. In section 9 we summarize the principal results and discuss the relation and an advance to what was previously published. In section 10, the principal results are summarized.

2. THE ADVECTION-REACTION EQUATION

The derivation of the Lagrangian and expected mixed Lagrangian-Eulerian equations follows closely to the approach presented in the book [3]. The main part of the notations is taken also from this book. Boldface notations refer to the three-dimensional (3D), otherwise to one-dimensional (1D) quantities.

Let us consider the initial boundary value problem (IVBP) for an advection-reaction equation

(i.e. the case of zero molecular diffusion)

$$\frac{\partial c}{\partial t}+\mathbf{u}\cdot\nabla_{\mathbf{X}}c=W(c,\mathbf{X},t)\,,\nabla_{\mathbf{X}}=\frac{\partial}{\partial\mathbf{X}}, \tag{1}$$

$$c(\mathbf{X},t=0)=c_0(\mathbf{X}), \tag{2}$$

where $c(\mathbf{X},t)$ is a reactive scalar, $\nabla_{\mathbf{X}}$ is a nabla symbol (i.e. $\nabla_{\mathbf{X}}c$ and $\nabla_{\mathbf{X}}\cdot\mathbf{u}$ stand for gradient of $c(\mathbf{X},t)$ and divergence of velocity $\mathbf{u}(\mathbf{X},t)$, respectively), $\mathbf{u}\cdot\nabla_{\mathbf{X}}c$ stands for the scalar product, $\mathbf{u}(\mathbf{X},t)$ is an isotropic homogeneous incompressible turbulent (i.e. random) flow field, which is assumed not affected by chemical reaction (passive reactive scalar), i.e.

$$\nabla_{\mathbf{X}}\cdot\mathbf{u}=0\,, \tag{3}$$

with zero mean value $\langle\mathbf{u}\rangle=0$. The density is assumed constant $\rho=const$. Without loss of generality we assume that an initial scalar field $c_0(\mathbf{X})$ is deterministic. The case of a random initial scalar field $c_0(\mathbf{X})$ is considered in the Annex A. Eq. (1) is a stochastic partial differential equation (SPDE) with multiplicative noise term $\mathbf{u}\cdot\nabla_{\mathbf{X}}c$.

Eq. (1) can be rewritten in the conservative form

$$\frac{\partial c}{\partial t}+\nabla_{\mathbf{X}}\cdot(\mathbf{u}c)=W(c,\mathbf{X},t), \tag{4}$$

by virtue of the incompressibility Eq. (3).

In the section 7 the particular case of premixed combustion is studied. In this case the reactive scalar c is called the combustion progress variable. Analysis is conducted in the frame of Bray-Moss-Libby (BML) model [4] (detailed discussion of the BML model can be found, e.g. in [5] and [6]). BML approximation for the turbulent premixed combustion is based on the assumption that the flame front thickness is negligibly small in comparison with the length scales of the turbulent velocity field. In such a limit it is assumed that the infinitely thin flame front separates the products and reactants, i.e. the progress variable c takes only two values $c=0$, in reactants and $c=1$, in products. In other words, we neglect the probability of observation of intermediate values $0<c<1$ which is proportional to the flame front thickness. The flame front propagates with the normal velocity of flame propagation S_L, supposed below to be constant, without loss of generality. The combustion signifies jump of the progress variable c from $c=0$ (in the reactants) to $c=1$ (in the products). This jump takes place when the flame front hits some fluid particle in the reactants.

Accordingly, $\nabla_{\mathbf{X}}c=0$ everywhere with the exception of the flamelet surface, where $\nabla_{\mathbf{X}}c$ is the Dirac δ function. The source in the frame of BML model reads, e.g. [5], [6]

$$W(c,\mathbf{X},t)=S_L\left|\nabla_{\mathbf{X}}c(\mathbf{X},t)\right|. \tag{5}$$

A particular case of the source $W(c,\mathbf{X},t)$ that does not depend on the scalar, i.e. $W(c,\mathbf{X},t)=q(\mathbf{X},t)$ is considered in the Annex B.

Averaging (4) yields

$$\frac{\partial \langle c \rangle}{\partial t} + \nabla_{\mathbf{X}} \cdot \langle \mathbf{u}c \rangle = \langle W \rangle, \tag{6}$$

where $\langle \mathbf{u}c \rangle$ is the turbulent flux. The unconditional ensemble averages over the random velocity field $\mathbf{u}(\mathbf{X},t)$ will be denoted by angle brackets.

Eqs (1), (4), (6) are written in the Eulerian frame. Let us transform these equations into Lagrangian counterpart.

3. LAGRANGIAN FRAME

3.1. Instantaneous Lagrangian Equations

First order partial differential equation (PDE) (1) can be equally presented in a characteristic ordinary differential equation (ODE) form

$$\frac{d\mathbf{X}}{dt} = \mathbf{u}(\mathbf{X},t)\,, \mathbf{X} = \mathbf{x}_0\,, t = 0\,, \tag{7}$$

$$\frac{dc}{dt} = W(c,\mathbf{X},t)\,, c = c_0(\mathbf{x}_0)\,, t = 0\,, \tag{8}$$

where $c_0(\mathbf{x}_0)$ is the initial scalar field (2).

In what follows, for simplicity of writing, the initial time t_0 is assumed equal to zero, i.e. $t_0 = 0$. Hence the elapsed time $t - t_0$ is simply t. The solution of (7) is a particle Lagrangian trajectory

$$\mathbf{X}(t|\mathbf{x}_0,0)\,, \tag{9}$$

where $\mathbf{x}_0$ stands for an initial position of a fluid particle. Initial position is the inverse function

$$\mathbf{x}_0 = \mathbf{x}_0(\mathbf{X},t)\,, \tag{10}$$

i.e. plugging the expression (10) for the initial position $\mathbf{x}_0$ into (9) yields

$$\mathbf{X}(t|\mathbf{x}_0(\mathbf{X},t),0) = \mathbf{X}. \tag{11}$$

The Jacobian of the mapping $\mathbf{X}(t|\mathbf{x}_0,0)$ is equal to one

$$j = \left|\frac{\partial \mathbf{X}}{\partial \mathbf{x}_0}\right| = \frac{\partial(X_1, X_2, X_3)}{\partial(x_{01}, x_{02}, x_{03})} = 1, \tag{12}$$

by virtue of the incompressibility condition (3). The Cartesian components of the vectors $\mathbf{X} = (X_1, X_2, X_3)$ and $\mathbf{x}_0 = (x_{01}, x_{02}, x_{03})$ are used in (12).

The substitution of (9) into Eulerian velocity field $\mathbf{u}(\mathbf{X},t)$ yields a Lagrangian velocity

$$\mathbf{v}(t|\mathbf{x}_0,0) = \mathbf{u}(\mathbf{X}(t|\mathbf{x}_0,0),t). \tag{13}$$

Eq. (8) can be rewritten as follows

$$\frac{dc}{dt} = W(c, \mathbf{X}(t|\mathbf{x}_0,0),t), c = c_0(\mathbf{x}_0), t = 0. \tag{14}$$

The solution of (14) is a functional of the particle trajectory $\mathbf{X}(t|\mathbf{x}_0,0)$ and can be symbolically written as

$$c(t|\mathbf{x}_0,0; R_c[\mathbf{X}(\tau|\mathbf{x}_0),\tau]), 0 \le \tau \le t, c = c_0(\mathbf{x}_0), t = 0, \tag{15}$$

where $R_c[\mathbf{X}(\tau|\mathbf{x}_0),\tau]$ stands for the functional. Hereinafter shorthand writing is used for the scalar in the Lagrangian frame $c(t|\mathbf{x}_0,0)$, suppressing the dependence on the functional of the particle trajectory.

Remark. It is worth noting that the particle trajectory $\mathbf{X}(t|\mathbf{x}_0,0)$ is, in its turn, a functional of the velocity field $\mathbf{u}(\mathbf{X},t)$, as follows from Eq. (7), i.e.

$$\mathbf{X}(t|\mathbf{x}_0,0) = R_{\mathbf{X}}[\mathbf{u}(\mathbf{X},\tau)],\ 0 \le \tau \le t.$$

Eqs (7), (8) (Lagrangian equations for the fluid particle) become

$$\frac{\partial \mathbf{X}(t|\mathbf{x}_0,0)}{\partial t} = \mathbf{u}(\mathbf{X}(t|\mathbf{x}_0,0),t) = \mathbf{v}(t|\mathbf{x}_0,0),\ \mathbf{X}(0|\mathbf{x}_0,0) = \mathbf{x}_0, \tag{16}$$

$$\frac{\partial c(t|\mathbf{x}_0,0)}{\partial t} = W(t|\mathbf{x}_0,0),\ c(t=0|\mathbf{x}_0,0) = c_0(\mathbf{x}_0), \tag{17}$$

taking into consideration Eqs (9), (13), (15). The shorthand notation for the source in the Lagrangian frame is used in Eq. (17)

$$W(t|\mathbf{x}_0,0) = W(c(t|\mathbf{x}_0,0),\mathbf{X}(t|\mathbf{x}_0,0),t).$$

In the particular case of an inert scalar, i.e. $W(c,\mathbf{X},t) = 0$, the scalar value does not change in the fluid particle, i.e.

$$c(t|\mathbf{x}_0,0) = c(t=0|\mathbf{x}_0,0) = c_0(\mathbf{x}_0), \tag{18}$$

as follows from (17).

Let the chemical source does not depend on the space coordinates, i.e.

$$W(c,\mathbf{X},t) = W(c,t). \tag{19}$$

Then Eq. (17) reduces to ODE

$$\frac{\partial c(t|\mathbf{x}_0,0)}{\partial t} = W(c(t|\mathbf{x}_0,0),t),\ c(t=0|\mathbf{x}_0,0) = c_0(\mathbf{x}_0). \tag{20}$$

Eq. (20) is a deterministic ODE, because, contrary to (17), it does not contain the functional of the particle trajectory $\mathbf{X}(t|\mathbf{x}_0)$, therefore its solution does not depend on the particle trajectory $\mathbf{X}(t|\mathbf{x}_0)$. Contrary to (15), the solution of Eq. (20), i.e. the scalar along the particle trajectory, $c(t|\mathbf{x}_0,0)$ is a deterministic function that depends only from the elapsed time t and the initial coordinates $\mathbf{x}_0$ (we remind that $c_0(\mathbf{x}_0)$ is a deterministic field, random field is considered in the Annex A).

3.2. Expected Lagrangian and Mixed Lagrangian-Eulerian Quantities

Let us introduce in addition to unconditional ensemble averages two kinds of expected conditional quantities, Lagrangian and mixed Lagrangian-Eulerian quantities (e.g. [3]). The expected Lagrangian quantities will be denoted by bar, e.g.

$$\overline{\mathbf{v}}(t|\mathbf{x}_0,0),\ \overline{c}(t|\mathbf{x}_0,0),\ \overline{W}(t|\mathbf{x}_0,0). \tag{21}$$

The expected Lagrangian quantities are the conditional averages with the selection of the initial positions of the particles. These conditional averages are calculated using the corresponding conditional Lagrangian probability density functions, e.g.

$$\overline{\mathbf{v}}(t|\mathbf{x}_0,0) = \int_{\mathbf{v}} \mathbf{v} P_L(\mathbf{v},t|\mathbf{x}_0,0) d^3\mathbf{v}\,. \tag{22}$$

It should be stressed that the expected Lagrangian quantities (21), by definition of the Lagrangian quantities, do not depend on the spatial coordinates $\mathbf{X}$. Averaging Eqs (16), (17) results in

$$\frac{\partial \overline{\mathbf{X}}(t|\mathbf{x}_0,0)}{\partial t} = \overline{\mathbf{u}}(\mathbf{X}(t|\mathbf{x}_0,0),t) = \overline{\mathbf{v}}(t|\mathbf{x}_0,0)\,,\ \overline{\mathbf{X}}(0|\mathbf{x}_0,0) = \mathbf{x}_0\,, \tag{23}$$

$$\frac{\partial \bar{c}(t|\mathbf{x}_0,0)}{\partial t} = W(t|\mathbf{x}_0,0), \bar{c}(t=0|\mathbf{x}_0,0) = c_0(\mathbf{x}_0). \tag{24}$$

It is worth noting that, in general case, the conditional averaging and the derivatives with respect to time and to space do not commute. The expected Lagrangian quantities, they do, e.g. [7].

The expected mixed Lagrangian-Eulerian quantities are defined with the aid of the conditional (transition) probability density function $P_L(\mathbf{X},t|\mathbf{x}_0,0)$

$$P_L(\mathbf{X},t|\mathbf{x}_0,0) = \langle \delta(\mathbf{X}(t|\mathbf{x}_0,0) - \mathbf{X}) \rangle, \int_{\mathbf{X}} P_L(\mathbf{X},t|\mathbf{x}_0,0) d^3\mathbf{X} = 1, \tag{25}$$

$$P_L(\mathbf{X},t=0|\mathbf{x}_0,0) = \delta(\mathbf{X} - \mathbf{x}_0). \tag{26}$$

The second expression in (25) is the normalisation condition of the transition function, (26) is the initial condition. The product $P_L(\mathbf{X},t|\mathbf{x}_0,0)d^3\mathbf{X}$ implies the probability as the random variable $\mathbf{X}(t|\mathbf{x}_0,0)$ lies among $\mathbf{X}$ and $\mathbf{X}+\mathbf{dX}$ at time t specified that this started out with value $\mathbf{x}_0$ at time 0.

Transition probability density function $P_L(\mathbf{X},t|\mathbf{x}_0,0)$ depends on the difference $\mathbf{X} - \mathbf{x}_0$, i.e.

$$P_L(\mathbf{X},t|\mathbf{x}_0,0) = P_L(\mathbf{X} - \mathbf{x}_0,t|,0),$$

by virtue of the assumption of homogeneity of the turbulent velocity field. For generality we will write below $P_L(\mathbf{X},t|\mathbf{x}_0,0)$. We note that

$$\delta(\mathbf{X}(t|\mathbf{x}_0,0) - \mathbf{X}) = \frac{1}{j}\delta(\mathbf{x}_0(\mathbf{X},t) - \mathbf{x}_0) = \delta(\mathbf{x}_0(\mathbf{X},t) - \mathbf{x}_0), \tag{27}$$

by virtue of the incompressibility condition (12), $j=1$. Hence averaging (27) yields

$$P_L(\mathbf{X},t|\mathbf{x}_0,0) = P_L(\mathbf{x}_0,0|\mathbf{X},t),\ \int_{\mathbf{x}_0} P_L(\mathbf{X},t|\mathbf{x}_0,0)d^3\mathbf{x}_0 = 1. \tag{28}$$

Following [3], let us introduce another type of statistical characteristics: so called expected mixed Lagrangian-Eulerian quantities, which are defined by

$$\langle Q(\mathbf{X},t)\rangle = \langle Q(t|\mathbf{x}_0,0)\delta(\mathbf{X}(t|\mathbf{x}_0,0)-\mathbf{X})\rangle = \overline{\overline{Q}}(t|\mathbf{X},\mathbf{x}_0,0)\langle\delta(\mathbf{X}(t|\mathbf{x}_0,0)-\mathbf{X})\rangle \tag{29}$$

$$\langle Q(\mathbf{X},t)\rangle = \langle Q(t|\mathbf{x}_0,0)\delta(\mathbf{X}(t|\mathbf{x}_0,0)-\mathbf{X})\rangle = \overline{\overline{Q}}(t|\mathbf{X},\mathbf{x}_0,0)P_L(\mathbf{X},t|\mathbf{x}_0,0), \tag{30}$$

where $Q(\mathbf{X},t)$ and $Q(t|\mathbf{x}_0,0)$ are arbitrary scalar or vector Eulerian and Lagrangian quantities, respectively. To elucidate the sense of $\overline{\overline{Q}}(t|\mathbf{X},\mathbf{x}_0,0)$, let us, departing from the Lagrangian instantaneous quantity $Q(t|\mathbf{x}_0,0)$, introduce a new random quantity $Q(t|\mathbf{X},\mathbf{x}_0,0)$ the sample values of which are the values of the Lagrangian quantity $Q(t|\mathbf{x}_0,0)$ at the fixed point $\mathbf{X}$. Thus the sample space of the random quantity $Q(t|\mathbf{X},\mathbf{x}_0,0)$ is a subset of the Lagrangian quantity $Q(t|\mathbf{x}_0,0)$. The expected mixed Lagrangian-Eulerian quantities will be denoted by two bars. They, as was just explained, are the conditional averages with the selection of the initial position of the particle and its position $\mathbf{X}$ at time t. Eq. (30) follows from (29) taking into consideration the definition of $P_L(\mathbf{X},t|\mathbf{x}_0,0)$ introduced by (25). The expected mixed Lagrangian-Eulerian conditional velocity, scalar and the chemical source, as follows from (30), are determined by the expressions

$$\overline{\overline{\mathbf{v}}}(t|\mathbf{X},\mathbf{x}_0,0) = \frac{\langle \mathbf{v}(t|\mathbf{x}_0,0)\delta(\mathbf{X}(t|\mathbf{x}_0,0)-\mathbf{X})\rangle}{P_L(\mathbf{X},t|\mathbf{x}_0,0)}, \tag{31}$$

$$\overline{\overline{c}}(t|\mathbf{X},\mathbf{x}_0,0) = \frac{\langle c(t|\mathbf{x}_0,0)\delta(\mathbf{X}(t|\mathbf{x}_0,0)-\mathbf{X})\rangle}{P_L(\mathbf{X},t|\mathbf{x}_0,0)}, \tag{32}$$

$$\overline{\overline{W}}(t|\mathbf{X},\mathbf{x}_0,0)=\frac{\langle W(t|\mathbf{x}_0,0)\delta(\mathbf{X}(t|\mathbf{x}_0,0)-\mathbf{X})\rangle}{P_L(\mathbf{X},t|\mathbf{x}_0,0)}, \tag{33}$$

respectively.

The expected mixed Lagrangian-Eulerian quantities $\overline{\overline{\mathbf{v}}}(t|\mathbf{X},\mathbf{x}_0,0)$, $\overline{\overline{c}}(t|\mathbf{X},\mathbf{x}_0,0)$, $\overline{\overline{W}}(t|\mathbf{X},\mathbf{x}_0,0)$, contrary to the expected Lagrangian quantities (21), as follows from (20), (32) and (33), depend on the space coordinates $\mathbf{X}$.

Let the chemical source does not depend on the space coordinates, Eqs (19), (20). Then, as was shown above, the reactive scalar $c(t|\mathbf{x}_0,0)$ is a deterministic function, the solution of Eq. (20). Scalar $c(t|\mathbf{x}_0,0)$ depends only from the elapsed time t and the initial coordinates $\mathbf{x}_0$. Therefore the expected Lagrangian scalar $\overline{c}(t|\mathbf{x}_0,0)$ coincides with its instantaneous value

$$\overline{c}(t|\mathbf{x}_0,0)=c(t|\mathbf{x}_0,0). \tag{34}$$

Eq. (24) reduces to (20), because

$$\overline{W}(t|\mathbf{x}_0,0)=\overline{W}(c(t|\mathbf{x}_0,0),t)=W(\overline{c}(t|\mathbf{x}_0,0),t)=W(c(t|\mathbf{x}_0,0),t). \tag{35}$$

Eqs (32) and (33) become

$$\overline{\overline{c}}(t|\mathbf{X},\mathbf{x}_0,0)=\frac{c(t|\mathbf{x}_0,0)P_L(\mathbf{X},t|\mathbf{x}_0,0)}{P_L(\mathbf{X},t|\mathbf{x}_0,0)}=c(t|\mathbf{x}_0,0), \tag{36}$$

$$\overline{\overline{W}}(t|\mathbf{X},\mathbf{x}_0,0)=\frac{W(c(t|\mathbf{x}_0,0),t)P_L(\mathbf{X},t|\mathbf{x}_0,0)}{P_L(\mathbf{X},t|\mathbf{x}_0,0)}=W(c(t|\mathbf{x}_0,0),t), \tag{37}$$

taking into account that $c(t|\mathbf{x}_0,0)$ is the deterministic function, (20), (34). We conclude that the expected Lagrangian reactive scalar $\overline{c}(t|\mathbf{x}_0,0)$ and mixed

Lagrangian-Eulerian scalar $\overline{\overline{c}}(t|\mathbf{X},\mathbf{x}_0,0)$ coincide with its instantaneous value $c(t|\mathbf{x}_0,0)$

$$\overline{\overline{c}}(t|\mathbf{X},\mathbf{x}_0,0)=\overline{c}(t|\mathbf{x}_0,0)=c(t|\mathbf{x}_0,0), \tag{38}$$

and, consequently $\overline{c}(t|\mathbf{x}_0,0)$ and $\overline{\overline{c}}(t|\mathbf{X},\mathbf{x}_0,0)$ do not depend on the space coordinates $\mathbf{X}$.

4. Eulerian Frame

4.1. Instantaneous Eulerian Quantities

Instantaneous Eulerian fields are found using the following expressions, e.g. [3]

$$c(\mathbf{X},t)=c\big(t|\mathbf{x}_0(\mathbf{X},t),0\big)=\int_{\mathbf{x}_0} c\big(t|\mathbf{x}_0,0\big)\delta(\mathbf{x}_0(\mathbf{X},t)-\mathbf{x}_0)d^3\mathbf{x}_0\,, \tag{39}$$

$$\mathbf{u}(\mathbf{X},t)=\mathbf{v}(t|\mathbf{x}_0(\mathbf{X},t),0)=\int_{\mathbf{x}_0} \mathbf{v}(t|\mathbf{x}_0,0)\delta(\mathbf{x}_0(\mathbf{X},t)-\mathbf{x}_0)d^3\mathbf{x}_0\,, \tag{40}$$

$$W(\mathbf{X},t)=W\big(t|\mathbf{x}_0(\mathbf{X},t),0\big)=\int_{\mathbf{x}_0} W(t|\mathbf{X},\mathbf{x}_0,0)\delta(\mathbf{x}_0(\mathbf{X},t)-\mathbf{x}_0)d^3\mathbf{x}_0\,. \tag{41}$$

Eqs (39)-(41) can be recast to the form

$$c(\mathbf{X},t)=\int_{\mathbf{x}_0} c\big(t|\mathbf{x}_0,0\big)\delta(\mathbf{X}(t|\mathbf{x}_0,0)-\mathbf{X})d^3\mathbf{x}_0\,, \tag{42}$$

$$\mathbf{u}(\mathbf{X},t)=\int_{\mathbf{x}_0} \mathbf{v}(t|\mathbf{x}_0,0)\delta(\mathbf{X}(t|\mathbf{x}_0,0)-\mathbf{X})d^3\mathbf{x}_0\,, \tag{43}$$

$$W(\mathbf{X},t) = \int\limits_{\mathbf{x}_0} W(t|\mathbf{X},\mathbf{x}_0,0)\delta(\mathbf{X}(t|\mathbf{x}_0,0) - \mathbf{X})d^3\mathbf{x}_0\,, \tag{44}$$

by virtue of incompressibility condition $j = 1$, (12), (27).

4.2. Expected Eulerian Quantities

Ensemble averaging (42)-(44) yields

$$\langle c(\mathbf{X},t)\rangle = \int\limits_{\mathbf{x}_0} \overline{\overline{c}}(t|\mathbf{X},\mathbf{x}_0,0)P_L(\mathbf{X},t|\mathbf{x}_0,0)d^3\mathbf{x}_0\,, \tag{45}$$

$$\langle \mathbf{u}(\mathbf{X},t)\rangle = \int\limits_{\mathbf{x}_0} \overline{\overline{\mathbf{v}}}(t|\mathbf{X},\mathbf{x}_0,0)P_L(\mathbf{X},t|\mathbf{x}_0,0)d^3\mathbf{x}_0\,, \tag{46}$$

$$\langle W(\mathbf{X},t)\rangle = \int\limits_{\mathbf{x}_0} \overline{\overline{W}}(t|\mathbf{X},\mathbf{x}_0,0)P_L(\mathbf{X},t|\mathbf{x}_0,0)d^3\mathbf{x}_0\,. \tag{47}$$

The turbulent flux $\langle \mathbf{u}c\rangle$ is presented by the following expression

$$\langle \mathbf{u}c\rangle = \int\limits_{\mathbf{x}_0} \overline{\overline{\mathbf{v}c}}(t|\mathbf{X},\mathbf{x}_0,0)P_L(\mathbf{X},t|\mathbf{x}_0,0)d^3\mathbf{x}_0\,, \tag{48}$$

where $\overline{\overline{\mathbf{v}c}}(t|\mathbf{X},\mathbf{x}_0,0)$ is the mixed Lagrangian-Eulerian turbulent flux. The shorthand notation

$$\overline{\overline{\mathbf{v}c}}(t|\mathbf{X},\mathbf{x}_0,0) = \overline{\overline{\mathbf{v}(t|\mathbf{x}_0,0)c(t|\mathbf{x}_0,0)}} \tag{49}$$

is used.

5. Governing Equations for Mean Mixed Lagrangian-Eulerian Quantities

Let us multiply (17) by $\delta(\mathbf{X}(t|\mathbf{x}_0,0)-\mathbf{X})$

$$\delta(\mathbf{X}(t|\mathbf{x}_0,0)-\mathbf{X})\frac{\partial c\left(t|\mathbf{x}_0,0\right)}{\partial t}=W\left(t|\mathbf{x}_0,0\right)\delta(\mathbf{X}(t|\mathbf{x}_0,0)-\mathbf{X}). \tag{50}$$

Eq. (50) transforms to

$$\frac{\partial c\left(t|\mathbf{x}_0,0\right)\delta(\mathbf{X}(t|\mathbf{x}_0,0)-\mathbf{X})}{\partial t}-c\left(t|\mathbf{x}_0,0\right)\frac{\partial\delta(\mathbf{X}(t|\mathbf{x}_0,0)-\mathbf{X})}{\partial t}=W\left(t|\mathbf{x}_0,0\right)\delta(\mathbf{X}(t|\mathbf{x}_0,0)-\mathbf{X}) \tag{51}$$

The time derivative $\dfrac{\partial\delta(\mathbf{X}(t|\mathbf{x}_0,0)-\mathbf{X})}{\partial t}$ reads

$$\frac{\partial\delta(\mathbf{X}(t|\mathbf{x}_0,0)-\mathbf{X})}{\partial t}=-\frac{\partial\delta(\mathbf{X}(t|\mathbf{x}_0,0)-\mathbf{X})}{\partial\mathbf{X}}\frac{\partial\mathbf{X}(t|\mathbf{x}_0,0)}{\partial t}=-\frac{\partial\delta(\mathbf{X}(t|\mathbf{x}_0,0)-\mathbf{X})}{\partial\mathbf{X}}\mathbf{v}(t|\mathbf{x}_0,0),$$

and then it transforms to

$$\frac{\partial\delta(\mathbf{X}(t|\mathbf{x}_0,0)-\mathbf{X})}{\partial t}=-\frac{\partial\mathbf{v}(t|\mathbf{x}_0,0)\delta(\mathbf{X}(t|\mathbf{x}_0,0)-\mathbf{X})}{\partial\mathbf{X}}, \tag{52}$$

by virtue of Eq. (16). Finally Eq. (52) becomes

$$\frac{\partial\delta(\mathbf{X}(t|\mathbf{x}_0,0)-\mathbf{X})}{\partial t}+\frac{\partial\mathbf{v}(t|\mathbf{x}_0,0)\delta(\mathbf{X}(t|\mathbf{x}_0,0)-\mathbf{X})}{\partial\mathbf{X}}=0. \tag{53}$$

Eq. (51) transforms to

$$\left(\frac{\partial}{\partial t}+\frac{\partial}{\partial\mathbf{X}}\cdot\mathbf{v}(t|\mathbf{x}_0,0)\right)c\left(t|\mathbf{x}_0,0\right)\delta(\mathbf{X}(t|\mathbf{x}_0,0)-\mathbf{X})=W\left(t|\mathbf{x}_0,0\right)\delta(\mathbf{X}(t|\mathbf{x}_0,0)-\mathbf{X}), \tag{54}$$

by taking into account (53). Ensemble averaging (53), (54) yields

$$\frac{\partial P_L(\mathbf{X},t|\mathbf{x}_0,0)}{\partial t}+\frac{\partial \overline{\overline{\mathbf{v}}}(t|\mathbf{X},\mathbf{x}_0,0)P_L(\mathbf{X},t|\mathbf{x}_0,0)}{\partial \mathbf{X}}=0\,, \tag{55}$$

$$\frac{\partial \overline{\overline{c}}(t|\mathbf{X},\mathbf{x}_0,0)P_L(\mathbf{X},t|\mathbf{x}_0,0)}{\partial t}+\frac{\partial \overline{\overline{\mathbf{v}c}}(t|\mathbf{X},\mathbf{x}_0,0)P_L(\mathbf{X},t|\mathbf{x}_0,0)}{\partial \mathbf{X}}=\overline{\overline{W}}(t|\mathbf{X},\mathbf{x}_0,0)P_L(\mathbf{X},t|\mathbf{x}_0,0)\,, \tag{56}$$

where the shorthand notation (49) is used.

Integrating (55), (56) over whole values of $\mathbf{x}_0$ results in

$$\nabla_{\mathbf{X}}\cdot\langle\mathbf{u}\rangle=0\,, \tag{57}$$

$$\frac{\partial\langle c\rangle}{\partial t}+\nabla_{\mathbf{X}}\cdot\langle\mathbf{u}c\rangle=\langle W\rangle\,, \tag{58}$$

taking into consideration the normalisation condition of $P_L(\mathbf{X},t|\mathbf{x}_0,0)$, (25), and Eqs (45)-(48).

Eqs (57) and (58) are the expected incompressibility (3) and the advection-reaction (1) equations, respectively.

Let us derive from (55), (56) the equation for the expected mixed Lagrangian-Eulerian scalar $\overline{\overline{c}}(t|\mathbf{X},\mathbf{x}_0,0)$. To this end, let us decompose the Lagrangian velocity and the scalar fields as follows

$$\mathbf{v}(t|\mathbf{x}_0,0)=\overline{\overline{\mathbf{v}}}(t|\mathbf{X},\mathbf{x}_0,0)+\mathbf{v}'(t|\mathbf{X},\mathbf{x}_0,0)\,, \tag{59}$$

$$c(t|\mathbf{x}_0,0)=\overline{\overline{c}}(t|\mathbf{X},\mathbf{x}_0,0)+c'(t|\mathbf{X},\mathbf{x}_0,0)\,, \tag{60}$$

where $\mathbf{v}'$ and c' are the fluctuations relative to the expected mixed Lagrangian-Eulerian velocity and scalar, respectively. Therefore the expected mixed Lagrangian-Eulerian fluctuations $\mathbf{v}'$ and c', introduced by (59), (60), are identically equal to zero

$$\overline{\overline{\mathbf{v}'}}(t|\mathbf{X},\mathbf{x}_0,0)=\frac{\left\langle\left(\mathbf{v}(t|\mathbf{x}_0,0)-\overline{\overline{\mathbf{v}}}(t|\mathbf{X},\mathbf{x}_0,0)\right)\delta(\mathbf{X}(t|\mathbf{x}_0,0)-\mathbf{X})\right\rangle}{P_L(\mathbf{X},t|\mathbf{x}_0,0)}=0\,, \tag{61}$$

$$\overline{\overline{c'}}(t|\mathbf{X},\mathbf{x}_0,0)=\frac{\left\langle\left(c\left(t|\mathbf{x}_0,0\right)-\overline{\overline{c}}(t|\mathbf{X},\mathbf{x}_0,0)\right)\delta(\mathbf{X}(t|\mathbf{x}_0,0)-\mathbf{X})\right\rangle}{P_L(\mathbf{X},t|\mathbf{x}_0,0)}=0\,, \quad (62)$$

by virtue of (31), (32).

The mean mixed Lagrangian-Eulerian turbulent flux $\overline{\overline{\mathbf{v}c}}(t|\mathbf{X},\mathbf{x}_0,0)$ becomes

$$\overline{\overline{\mathbf{v}c}}(t|\mathbf{X},\mathbf{x}_0,0)=\overline{\overline{\mathbf{v}}}(t|\mathbf{X},\mathbf{x}_0,0)\overline{\overline{c}}(t|\mathbf{X},\mathbf{x}_0,0)+\overline{\overline{\mathbf{v}'c'}}(t|\mathbf{X},\mathbf{x}_0,0)\,, \quad (63)$$

using (59)-(62).

Inserting expression (63) into Eq. (56) yields

$$\frac{\partial\overline{\overline{c}}(t|\mathbf{X},\mathbf{x}_0,0)P_L(\mathbf{X},t|\mathbf{x}_0,0)}{\partial t}+\frac{\partial\overline{\overline{\mathbf{v}}}\overline{\overline{c}}P_L}{\partial\mathbf{X}}+\frac{\partial\overline{\overline{\mathbf{v}'c'}}(t|\mathbf{X},\mathbf{x}_0,0)P_L}{\partial\mathbf{X}}=\overline{\overline{W}}(t|\mathbf{X},\mathbf{x}_0,0)P_L(\mathbf{X},t|\mathbf{x}_0,0) \quad (64)$$

Developing two first terms in the LHS of (64) results in

$$\overline{\overline{c}}\left(\frac{\partial P_L}{\partial t}+\frac{\partial\overline{\overline{\mathbf{v}}}P_L}{\partial\mathbf{X}}\right)+P_L\frac{\partial\overline{\overline{c}}}{\partial t}+\overline{\overline{\mathbf{v}}}P_L\frac{\partial\overline{\overline{c}}}{\partial\mathbf{X}}+\frac{\partial\overline{\overline{\mathbf{v}'c'}}(t|\mathbf{X},\mathbf{x}_0,0)P_L}{\partial\mathbf{X}}=\overline{\overline{W}}(t|\mathbf{X},\mathbf{x}_0,0)P_L(\mathbf{X},t|\mathbf{x}_0,0) \quad (65)$$

Finally (65) transforms to

$$\frac{\partial\overline{\overline{c}}(t|\mathbf{X},\mathbf{x}_0,0)}{\partial t}+\overline{\overline{\mathbf{v}}}(t|\mathbf{X},\mathbf{x}_0,0)\frac{\partial\overline{\overline{c}}(t|\mathbf{X},\mathbf{x}_0,0)}{\partial\mathbf{X}}=-\frac{1}{P_L}\frac{\partial\overline{\overline{\mathbf{v}'c'}}(t|\mathbf{X},\mathbf{x}_0,0)P_L}{\partial\mathbf{X}}+\overline{\overline{W}}(t|\mathbf{X},\mathbf{x}_0,0), \quad (66)$$

taking into account (55).

Eq. (66) for the expected mixed Lagrangian-Eulerian scalar $\overline{\overline{c}}(t|\mathbf{X},\mathbf{x}_0,0)$, contrary to Eq. (24) for the expected Lagrangian particle scalar $\overline{c}\left(t|\mathbf{x}_0,0\right)$, contains the convective terms: second term in the left-hand side (LHS) and first in the right-hand side (RHS) of Eq. (66), similar to (58).

Eq. (66) shows that in general case, the time derivative of the expected mixed Lagrangian-Eulerian scalar $\overline{\overline{c}}(t|\mathbf{X},\mathbf{x}_0,0)$ is not equal to the expected mixed Lagrangian-Eulerian source $\overline{\overline{W}}(t|\mathbf{X},\mathbf{x}_0,0)$, i.e.

$$\frac{\partial \overline{\overline{c}}(t|\mathbf{X},\mathbf{x}_0,0)}{\partial t} \neq \overline{\overline{W}}(t|\mathbf{X},\mathbf{x}_0,0). \tag{67}$$

The discussion of the equations for the expected mixed Lagrangian-Eulerian quantities will be continued in the section 6, invoking the celebrated Taylor theory of turbulent diffusion [8], (detailed discussion can be found in [3]). But first Eq. (56) is applied to the case when the chemical source does not depend explicitly on the space coordinates.

Then Eq. (55) becomes

$$\frac{\partial c(t|\mathbf{x}_0,0)P_L(\mathbf{X},t|\mathbf{x}_0,0)}{\partial t} + c(t|\mathbf{x}_0,0)\frac{\partial \overline{\overline{\mathbf{v}}}(t|\mathbf{X},\mathbf{x}_0,0)P_L(\mathbf{X},t|\mathbf{x}_0,0)}{\partial \mathbf{X}} = W(t|\mathbf{x}_0,0)P_L(\mathbf{X},t|\mathbf{x}_0,0), \tag{68}$$

by virtue of (36), (37). Eq. (68) transforms to

$$P_L\frac{\partial c(t|\mathbf{x}_0,0)}{\partial t} + c(t|\mathbf{x}_0,0)\left(\frac{\partial P_L}{\partial t} + \frac{\partial \overline{\overline{\mathbf{v}}}(t|\mathbf{X},\mathbf{x}_0,0)P_L}{\partial \mathbf{X}}\right) = W(t|\mathbf{x}_0,0)P_L. \tag{69}$$

We suppressed the variables in $P_L(\mathbf{X},t|\mathbf{x}_0,0)$. The second term in the LHS of Eq. (69) is equal to zero by virtue of Eq. (55), and Eq. (69) becomes

$$\frac{\partial c(t|\mathbf{x}_0,0)}{\partial t} = W(c(t|\mathbf{x}_0,0),t). \tag{70}$$

Eq. (70) is nothing but (20).

6. Application of Taylor Theory of the Turbulent Diffusion

Taylor proved [8] that for the random Gaussian Lagrangian velocity field $\mathbf{v}(t|\mathbf{x}_0,0)$ the following formula is valid

$$\bar{\bar{\mathbf{v}}}(t|\mathbf{X},\mathbf{x}_0,0)P_L(\mathbf{X},t|\mathbf{x}_0,0)=-D_T(t)\nabla_{\mathbf{X}}P_L(\mathbf{X},t|\mathbf{x}_0,0)=-D_T(t)\frac{\partial P_L(\mathbf{X},t|\mathbf{x}_0,0)}{\partial \mathbf{X}}, \quad (71)$$

where the turbulent diffusion coefficient $D_T(t)$ is determined by the Lagrangian velocity autocorrelation function $B_L(\tau)$ (e.g. [3])

$$D_T(t)=\int_0^t B_L(\tau)d\tau=\langle u^2\rangle\int_0^t R_L(\tau)d\tau,$$

$$B_L(\tau)=\frac{1}{3}\langle \mathbf{v}(\tau|\mathbf{x}_0,0)\cdot\mathbf{v}(0|\mathbf{x}_0,0)\rangle,\ R_L(\tau)=\frac{B_L(\tau)}{\langle u^2\rangle}. \quad (72)$$

The turbulent diffusion coefficient $D_T(t)$ depends on the elapsed time t (it is assumed that the initial instant $t_0=0$). Hence Eq. (55) becomes

$$\frac{\partial P_L(\mathbf{X},t|\mathbf{x}_0,0)}{\partial t}=D_T(t)\frac{\partial^2 P_L(\mathbf{X},t|\mathbf{x}_0,0)}{\partial \mathbf{X}^2}=D_T(t)\Delta P_L(\mathbf{X},t|\mathbf{x}_0,0), \quad (73)$$

where $\Delta=\nabla\cdot\nabla=\nabla^2$ is the Laplace operator (Laplacian). If elapsed time is much larger than Lagrangian integral timescale of turbulence T_L (defined by (74)), $t>>T_L$, the expression (72) reduces to

$$D_T(t)\approx D_T(\infty)=\langle u^2\rangle T_L,\ T_L=\int_0^\infty R_L(\tau)d\tau, \quad (74)$$

where $D_T(\infty) = \langle u^2 \rangle T_L$ is the asymptotic value of the turbulent diffusion coefficient.

It is readily verified that the solution of Eq. (73) with the initial condition (26) is a Gaussian function

$$P_L(\mathbf{X},t|\mathbf{x}_0,0) = \frac{1}{\left(2\pi\sigma(t)\right)^{3/2}} \exp\left(-\frac{(\mathbf{X}-\mathbf{x}_0)^2}{2\sigma^2(t)}\right),$$

$$\sigma^2(t) = 2\int_0^t (t-\tau) B_L(\tau) d\tau . \tag{75}$$

Eqs (64) and (66) reduce to

$$\frac{\partial \overline{\overline{c}} P_L}{\partial t} = D_T \frac{\partial}{\partial \mathbf{X}}\left(\frac{\partial P_L}{\partial \mathbf{X}} \overline{\overline{c}}\right) - \frac{\partial \overline{\overline{\mathbf{v}'c'P_L}}}{\partial \mathbf{X}} + \overline{\overline{W}} P_L, \tag{76}$$

$$\frac{\partial \overline{\overline{c}}}{\partial t} = D_T \frac{1}{P_L} \frac{\partial P_L}{\partial \mathbf{X}} \frac{\partial \overline{\overline{c}}}{\partial \mathbf{X}} - \frac{1}{P_L} \frac{\partial \overline{\overline{\mathbf{v}'c'P_L}}}{\partial \mathbf{X}} + \overline{\overline{W}} , \tag{77}$$

taking into consideration Eq. (73).

Let us transform the first term in the RHS of Eq. (76) with the aid of identity

$$\frac{\partial P_L}{\partial \mathbf{X}} \overline{\overline{c}} = \frac{\partial P_L \overline{\overline{c}}}{\partial \mathbf{X}} - P_L \frac{\partial \overline{\overline{c}}}{\partial \mathbf{X}} . \tag{78}$$

Eq. (76) becomes

$$\frac{\partial \overline{\overline{c}} P_L}{\partial t} = \left\{ D_T \frac{\partial^2 \overline{\overline{c}} P_L}{\partial \mathbf{X}^2} - D_T \frac{\partial}{\partial \mathbf{X}}\left(P_L \frac{\partial \overline{\overline{c}}}{\partial \mathbf{X}}\right) - \frac{\partial \overline{\overline{\mathbf{v}'c'P_L}}}{\partial \mathbf{X}} \right\} + \overline{\overline{W}} P_L . \tag{79}$$

Developing the second term in the braces in the RHS of Eq. (79)

$$\frac{\partial}{\partial \mathbf{X}}\left(P_L \frac{\partial \overline{\overline{c}}}{\partial \mathbf{X}}\right)=\frac{\partial P_L}{\partial \mathbf{X}}\frac{\partial \overline{\overline{c}}}{\partial \mathbf{X}}+P_L \frac{\partial^2 \overline{\overline{c}}}{\partial \mathbf{X}^2} \tag{80}$$

yields the equation equivalent to (79)

$$\frac{\partial \overline{\overline{c}} P_L}{\partial t}=\left\{D_T \frac{\partial^2 \overline{\overline{c}} P_L}{\partial \mathbf{X}^2}-D_T \frac{\partial P_L}{\partial \mathbf{X}}\frac{\partial \overline{\overline{c}}}{\partial \mathbf{X}}-P_L D_T \frac{\partial^2 \overline{\overline{c}}}{\partial \mathbf{X}^2}-\frac{\partial \overline{\overline{\mathbf{v}'c'P_L}}}{\partial \mathbf{X}}\right\}+\overline{\overline{W}}P_L . \tag{81}$$

Eq. (81) (or its equivalent Eq. (79)) is the expected mixed Lagrangian-Eulerian counterpart of expected Eulerian Eq. (6). The chemical source $\overline{\overline{W}}$ enters into the RHS of Eqs. (79), (81) explicitly. The terms in the braces in the RHS of Eqs (79), (81), are due to the turbulent velocity field (through D_T and P_L) and the chemistry (through $\overline{\overline{c}}$). One can guess that the terms in the braces in the RHS of Eqs (79), (81), are linked with the turbulent flux $\langle \mathbf{u}c \rangle$. It is indeed the case, as it will be seen soon.

The terms in the braces in the RHS of Eqs (79), (81) only redistributes the product $\overline{\overline{c}}P_L$ in the coordinate space $\mathbf{X}$. This conclusion follows from the fact that the integral of the quantity $\overline{\overline{c}}P_L$ over the whole space $\mathbf{X}$, $\int_{\mathbf{X}} \overline{\overline{c}}P_L d\mathbf{X}$ is conserved, because the integral of the braces in the RHS of (79), (81) over $\mathbf{X}$ is identically equal to zero

$$\int_{\mathbf{X}}\left\{D_T \frac{\partial^2 \overline{\overline{c}} P_L}{\partial \mathbf{X}^2}-D_T \frac{\partial}{\partial \mathbf{X}}\left(P_L \frac{\partial \overline{\overline{c}}}{\partial \mathbf{X}}\right)-\frac{\partial \overline{\overline{\mathbf{v}'c'P_L}}}{\partial \mathbf{X}}\right\}d^3\mathbf{X}=0 . \tag{82}$$

Indeed, the integrand of Eq. (82) is the divergence of the following flux vector

$$\mathbf{F}_T = D_T\left(\nabla_{\mathbf{X}}\left(P_L\overline{\overline{c}}\right)-P_L\nabla_{\mathbf{X}}\overline{\overline{c}}\right)-\overline{\overline{\mathbf{v}'c'P_L}} . \tag{83}$$

The integral (82) over the whole space can be considered as a limit of the integral over bounded space region V_R, e.g. a sphere with a radius R large, as desired. Then the volume integral (82) can be transformed to the surface integral using Gauss-Ostrogradsky theorem

$$\int_{V_R} \nabla \cdot \mathbf{F}_T d^3\mathbf{X} = \int_{S_R} \mathbf{F}_T \cdot \mathbf{n}_R dS_R = 0, \quad R \to \infty,$$

where $\mathbf{n}_R$ is the outward unit normal to the surface S_R of the sphere. The surface integral in the last expression tends to zero letting $R \to \infty$. Indeed, the flux vector $\mathbf{F}_T$ includes the transition probability density function $P_L(\mathbf{X},t|\mathbf{x}_0,0)$. For arbitrary, but fixed value of $\mathbf{x}_0$ we can chose the radius of sphere R large as desired, that $\mathbf{F}_T \to 0$, taking into account that $P_L(\mathbf{X},t|\mathbf{x}_0,0) \to 0$ in the limit $|\mathbf{X}-\mathbf{x}_0| \to \infty$.

To elucidate the result (82), let us present the turbulent flux $\langle \mathbf{u}c \rangle$ using the expected mixed Eulerian-Lagrangian quantity $\overline{\overline{\mathbf{v}c(t|\mathbf{X},\mathbf{x}_0,0)}}$ (48)

$$\langle \mathbf{u}c \rangle = \int_{\mathbf{x}_0} \overline{\overline{\mathbf{v}c}}(t|\mathbf{X},\mathbf{x}_0,0) P_L(\mathbf{X},t|\mathbf{x}_0,0) d^3\mathbf{x}_0 = \int_{\mathbf{x}_0} \overline{\overline{\mathbf{v}}}\overline{\overline{c}}_L P_L d^3\mathbf{x}_0 + \int_{\mathbf{x}_0} \overline{\overline{\mathbf{v}'c'}} P_L d^3\mathbf{x}_0 \quad , (84)$$

where the correlation $\overline{\overline{\mathbf{v}c}}$ is decomposed according to Eq. (63). The first term in the RHS of the second writing (84) transforms to

$$\int_{\mathbf{x}_0} \overline{\overline{\mathbf{v}}}\overline{\overline{c}} P_L d^3\mathbf{x}_0 = -\int_{\mathbf{x}_0} D_T \frac{\partial P_L}{\partial \mathbf{X}} \overline{\overline{c}} d^3\mathbf{x}_0, \tag{85}$$

by virtue of the Taylor formula (71).

Thus the expression (84) for the turbulent flux becomes

$$\langle \mathbf{u}c \rangle = \int_{\mathbf{x}_0} \left(-D_T \frac{\partial P_L}{\partial \mathbf{X}} \overline{\overline{c}} + \overline{\overline{\mathbf{v}'c'}} P_L \right) d^3\mathbf{x}_0 = \int_{\mathbf{x}_0} \left(-D_T \frac{\partial P_L \overline{\overline{c}}}{\partial \mathbf{X}} + D_T P_L \frac{\partial \overline{\overline{c}}}{\partial \mathbf{X}} + \overline{\overline{\mathbf{v}'c'}} P_L \right) d^3\mathbf{x}_0,$$

$$\langle \mathbf{u}c \rangle = -D_T \frac{\partial \langle c \rangle}{\partial \mathbf{X}} + \int_{\mathbf{x}_0} \left(D_T \frac{\partial \overline{\overline{c}}}{\partial \mathbf{X}} + \overline{\overline{\mathbf{v}'c'}} \right) P_L(\mathbf{X},t|\mathbf{x}_0,0) d^3\mathbf{x}_0, \tag{86}$$

because $\langle c(\mathbf{X},t)\rangle = \int\limits_{\mathbf{x}_0} \overline{\overline{c}} P_L d^3\mathbf{x}_0$ by virtue of Eq. (45). We remind that D_T and P_L depend only on the turbulent flow, by virtue of the assumption of a constant density (it is assumed that the turbulence is not affected by the combustion, passive reactive scalar). Chemistry enters into Eq. (86) implicitly through $\overline{\overline{c}}$ and the correlation $\overline{\overline{\mathbf{v}'c'}}$.

The spatial divergence of the turbulent flux $\langle \mathbf{u}c\rangle$ given by Eq. (86) then reads

$$\nabla_{\mathbf{X}} \cdot \langle \mathbf{u}c\rangle = \int\limits_{\mathbf{x}_0} \left\{ - D_T \frac{\partial^2 \overline{\overline{c}} P_L}{\partial \mathbf{X}^2} + D_T \frac{\partial}{\partial \mathbf{X}} \left(P_L \frac{\partial \overline{\overline{c}}}{\partial \mathbf{X}} \right) + \frac{\partial \overline{\overline{\mathbf{v}'c'}} P_L}{\partial \mathbf{X}} \right\} d^3\mathbf{x}_0 ,$$

$$\nabla_{\mathbf{X}} \cdot \langle \mathbf{u}c\rangle = -D_T \frac{\partial^2 \langle c\rangle}{\partial \mathbf{X}^2} + \int\limits_{\mathbf{x}_0} \left\{ D_T \frac{\partial}{\partial \mathbf{X}} \left(P_L \frac{\partial \overline{\overline{c}}}{\partial \mathbf{X}} \right) + \frac{\partial \overline{\overline{\mathbf{v}'c'}} P_L}{\partial \mathbf{X}} \right\} d^3\mathbf{x}_0 . \quad (87)$$

Comparing Eqs (87) and (79) we conclude, that the integrand in the Eq. (87) and the terms in the RHS of Eq. (79) in the braces differ only by the sign. Hence the terms in the braces in RHS of Eq. (79) with the sign minus is nothing but the integrand of the divergence $\nabla_{\mathbf{X}} \cdot \langle \mathbf{u}c\rangle$.

The equation for the mean reactive scalar in the Eulerian frame follows from Eq. (79) after integration over $\mathbf{x}_0$

$$\frac{\partial \langle c\rangle}{\partial t} = -\nabla_{\mathbf{X}} \cdot \langle \mathbf{u}c\rangle + \langle W\rangle , \quad (88)$$

by virtue of Eqs (45)-(47). The divergence of the turbulent flux $\nabla_{\mathbf{X}} \cdot \langle \mathbf{u}c\rangle$ is given by the relation (87).

To conclude, we proved that the turbulent flux is presented by the relation (85)

$$\langle \mathbf{u}c\rangle = -D_T \frac{\partial\langle c\rangle}{\partial \mathbf{X}} + \int_{\mathbf{x}_0} \left(D_T \frac{\partial \overline{\overline{c}}}{\partial \mathbf{X}} + \overline{\overline{\mathbf{v}'c'}} \right) P_L(\mathbf{X},t|\mathbf{x}_0,0)d^3\mathbf{x}_0 ,$$

where the integral term in the RHS is not equal to zero in the general case. We can introduce formally the turbulent diffusion coefficient D_T^c for the case of the reactive scalar c by the expression

$$\langle \mathbf{u}c\rangle = -D_T \frac{\partial\langle c\rangle}{\partial \mathbf{X}} + \int_{\mathbf{x}_0} \left(D_T \frac{\partial \overline{\overline{c}}}{\partial \mathbf{X}} + \overline{\overline{\mathbf{v}'c'}} \right) P_L(\mathbf{X},t|\mathbf{x}_0,0)d^3\mathbf{x}_0 = -D_T^c \frac{\partial\langle c\rangle}{\partial \mathbf{X}} \quad (89)$$

The integral term in the RHS of Eq. (89) is not equal to zero in the general case, therefore the turbulent diffusion coefficient D_T^c of the reactive scalar c is not equal to Taylor turbulent diffusion coefficient $D_T(t)$

$$D_T^c \neq D_T .$$

Let the chemical source does not depend explicitly the space coordinates. Then Eq. (48) transforms to

$$\langle \mathbf{u}c\rangle = \int_{\mathbf{x}_0} \overline{\overline{\mathbf{v}c}}(t|\mathbf{X},\mathbf{x}_0,0)P_L(\mathbf{X},t|\mathbf{x}_0,0)d^3\mathbf{x}_0 = \int_{\mathbf{x}_0} \overline{\overline{\mathbf{v}}}(t|\mathbf{X},\mathbf{x}_0,0)c(t|\mathbf{x}_0,0)P_L d^3\mathbf{x}_0 , \quad (90)$$

taking into account (36) and that $c(t|\mathbf{x}_0,0)$ is a deterministic function, (20), (38). Expression (90) becomes

$$\langle \mathbf{u}c\rangle = -\int_{\mathbf{x}_0} D_T(t)c(t|\mathbf{x}_0,0)\frac{\partial P_L(\mathbf{X},t|\mathbf{x}_0,0)}{\partial \mathbf{X}} d^3\mathbf{x}_0 = -D_T(t)\frac{\partial\langle c\rangle}{\partial \mathbf{X}} , \quad (91)$$

applying the Taylor expression (71) and (45), (36). Eq. (6) transforms to

$$\frac{\partial\langle c\rangle}{\partial t} = D_T(t)\frac{\partial^2\langle c\rangle}{\partial \mathbf{X}^2} + \langle W\rangle , \quad (92)$$

i.e. the expected reactive scalar $\langle c \rangle$ diffuses with the same turbulent diffusion coefficient $D_T(t)$ as an expected inert scalar.

Eqs (56), (64), (66), (76), (77), (79), (81), (86) are new. They are the principal result of the analysis conducted in the section 6.

7. The Turbulent Premixed Flame in BML Approximation

7.1 Equations for Mean Lagrangian, Eulerian and Mixed Lagrangian-Ealerian Quantities

As was already stated in the section 2 the Bray-Moss-Libby (BML) model for the turbulent premixed combustion [4] (e.g. [5], [6]), is based on the assumption that the flame front thickness is negligibly small in comparison with the length scales of the turbulent velocity field. In such a limit the progress variable takes only two values $c = 0$ in the reactants and $c = 1$ in the products, i.e. the probability of observation of intermediate values $0 < c < 1$ which is proportional to the flame front thickness is neglected. In other words, the infinitely thin flame front separates the reactants and products. The flame front propagates with the normal velocity of flame propagation S_L, supposed below to be constant, without loss of generality. The combustion signifies a jump of the progress variable c from $c = 0$ to $c = 1$. This jump takes place when the flame front hits a fluid particle in the reactants.

For the sake of simplicity we consider a statistically planar 1D flame propagating from left to right in a spatially homogeneous turbulence not affected by the combustion, i.e. passive reactive scalar. We assume that at initial moment $t = 0$ the products and reactants occupy the left half-space, and the right half-space, $x_0 > 0$, respectively. Therefore the initial distribution of the combustion progress variable is a step function

$$c_0(x_0) = 1 - H(x_0), \tag{93}$$

where $H(\varphi)$ is the Heaviside step function equal to zero for $\varphi < 0$ and unity for $\varphi > 0$.

The common Cartesian components of the vectors $\mathbf{X}$ and $\mathbf{x}_0$ are used below, i.e. $\mathbf{X} = (X,Y,Z)$, $\mathbf{x}_0 = (x_0, y_0, z_0)$. Eq. (24) reads

$$\frac{\partial c(t|\mathbf{x}_0,0)}{\partial t} = 0,\ x_0 < 0 \tag{94}$$

for the fluid particles in the products, $x_0 < 0$, where $c(t = 0|\mathbf{x}_0,0) = 1$, and

$$\frac{\partial c(t|\mathbf{x}_0,0)}{\partial t} = \delta(t - t_*(\mathbf{x}_0)), \tag{95}$$

for the fluid particles in the reactants, where $c(t = 0|\mathbf{x}_0,0) = 0$.

A notation $t_*(\mathbf{x}_0)$ stands for an instant when the flame front hits a particle in the reactants, $x_0 > 0$. The hitting time is a functional of the fluid particle trajectory, and depends also on the initial particle position $\mathbf{x}_0$, and the laminar flame velocity S_L. For simplicity of writing the dependence of the hitting time $t_*(\mathbf{x}_0)$ on the functional of the particle trajectory will be suppressed.

For the particles in the products

$$t_*(\mathbf{x}_0) = 0\,, x_0 < 0\,, \tag{96}$$

because they are already burned in correspondence with the initial condition (93).

Eqs (94), (95) can be put together

$$\frac{\partial c(t|\mathbf{x}_0,0)}{\partial t} = W(t,\mathbf{x}_0), \tag{97}$$

where

$$W(t,\mathbf{x}_0) = H(x_0)\delta(t - t_*(\mathbf{x}_0)) \tag{98}$$

denotes the chemical source. The chemical source $W(t, \mathbf{x}_0)$, as it is follows from Eq. (98), depends on the hitting time, and consequently it is a random function.

Eq. (97) is solved with the initial condition (93). The solution of the initial value problem (97), (93) reads

$$c(t|\mathbf{x}_0,0) = (1 - H(x_0)) + H(x_0)\int_0^t \delta(\tau - t_*(\mathbf{x}_0))d\tau = (1 - H(x_0)) + H(x_0)H(t - t_*(\mathbf{x}_0)) \quad (99)$$

Averaging Eqs (97), (98) yields

$$\frac{\partial \bar{c}(t|x_0,0)}{\partial t} = \overline{W(t,\mathbf{x}_0)} = H(x_0)\overline{\delta(t - t_*(\mathbf{x}_0))} = H(x_0)\Phi(t|x_0,0), \quad (100)$$

$$\overline{W(t,\mathbf{x}_0)} = H(x_0)\Phi(t|x_0,0), \quad (101)$$

where $\Phi(t|x_0,0)$ is the conditional Lagrangian probability density function of hitting time defined as follows

$$\Phi(t|x_0,0) = \overline{\delta(t - t_*(\mathbf{x}_0))}, t > 0, x_0 > 0. \quad (102)$$

It should be noted that $\Phi(t|x_0,0)$ depends only on x_0, by virtue of the assumption of homogeneity in the planes normal to X-axis.

Averaging Eq. (99), or, what is equivalent, integrating Eq. (100), yields

$$\bar{c}(t|x_0,0) = (1 - H(x_0)) + H(x_0)\int_0^t \Phi(\tau|x_0,0)d\tau. \quad (103)$$

$$\bar{c}(t|x_0,0) = \begin{cases} \int_0^t \Phi(\tau|x_0,0)d\tau, & x_0 > 0 \\ 1, & x_0 < 0 \end{cases}.$$

Lagrangian probability density function $P_L(c,t|\mathbf{x}_0,0)$ of the progress variable c takes only two values $c=0$ and $c=1$. Therefore

$$P_L(c,t|\mathbf{x}_0,0)=\bar{c}(t|x_0,0)\delta(c-1)+(1-\bar{c}(t|x_0,0))\delta(c-1), \tag{104}$$

taking into consideration the expressions (99), (103).

Let the initial coordinate of the particle in the reactants x_0, $x_0>0$, as large as desired, but of fixed value. Then Eq. (103) yields

$$\bar{c}(t=\infty|x_0,0)=\int_0^\infty \Phi(\tau|x_0,0)d\tau=1\,,x_0>0\,, \tag{105}$$

by virtue of the normalization condition of the probability density function $\Phi(\tau|x_0)$. Eq. (105) means simply, that the particle in the reactants with fixed initial position x_0 will be finally burned. It should be stressed that the assumption of fixed x_0 is important. Indeed, let us rewrite Eq. (105) in the form of the following limit at fixed x_0:

$$\lim_{T\to\infty}\bar{c}(T|x_0,0)=\lim_{T\to\infty}\int_0^T \Phi(\tau|x_0,0)d\tau=1\,,x_0>0 \text{ fixed.} \tag{106}$$

Let us analyse now what happens with (106) if the restriction on the fixed value of x_0 is disregarded. In other words we would like to know the limit behaviour of $\bar{c}(T|x_0,0)$ in the **double limit** $T\to\infty, x_0\to\infty$. This double limit depends on the order in what we carry out the limit, i.e. either $x_0\to\infty$ then $T\to\infty$, or $T\to\infty$ then $x_0\to\infty$. In the first case we obtain instead of (106)

$$\lim_{T\to\infty}\lim_{x_0\to\infty}\bar{c}(T|x_0,0)=\lim_{T\to\infty}\lim_{x_0\to\infty}\int_0^T \Phi(\tau|x_0,0)d\tau=0\,, \tag{107}$$

i.e. the fluid particle in the reactants will be never burned.

In the second limit

$$\lim_{x_0\to\infty}\lim_{T\to\infty}\bar{c}\left(T\middle|x_0,0\right)=\lim_{x_0\to\infty}\lim_{T\to\infty}\int_0^T\Phi(\tau|x_0,0)d\tau=1, \tag{108}$$

i.e. the fluid particle in the reactants will be eventually burned.

We turn now to the analysis of the Eulerian progress variable field $c(\mathbf{X},t)$. Inserting the instantaneous Lagrangian progress variable $c\left(t\middle|\mathbf{x}_0,0\right)$, given by Eq. (99), into the expression (42) yields

$$\begin{aligned} c(\mathbf{X},t) &= \int_{\mathbf{x}_0}(1-H(x_0))\delta(\mathbf{X}(t|\mathbf{x}_0,0)-\mathbf{X})d^3\mathbf{x}_0 + \\ &+\int_0^t\int_{\mathbf{x}_0}H(x_0)\delta\left(\tau-t_*\left(\mathbf{x}_0\right)\right)\delta(\mathbf{X}(t|\mathbf{x}_0,0)-\mathbf{X})d\tau d^3\mathbf{x}_0 \end{aligned}. \tag{109}$$

Averaging (109) results in

$$\langle c(X,t)\rangle=\langle c(\mathbf{X},t)\rangle=\int_{-\infty}^{0}\int_{-\infty}^{\infty}\int_{-\infty}^{\infty}P_L(\mathbf{X},t|\mathbf{x}_0,0)dx_0dy_0dz_0+\int_0^t\int_0^{\infty}\int_{-\infty}^{\infty}\int_{-\infty}^{\infty}P(\tau;\mathbf{X},t|\mathbf{x}_0,0)d\tau dx_0dy_0dz_0 \tag{110}$$

$$d^3\mathbf{x}_0=dx_0dy_0dz,$$

where $P(\tau;\mathbf{X},t|\mathbf{x}_0,0)$ denotes the conditional joint Lagrangian probability density function

$$P(\tau;\mathbf{X},t|\mathbf{x}_0,0)=\left\langle\delta\left(\tau-t_*\left(\mathbf{x}_0\right)\right)\delta(\mathbf{X}(t|\mathbf{x}_0,0)-\mathbf{X})\right\rangle,\ x_0>0,\ 0<\tau<t \tag{111}$$

of an event that a particle which starts at initial time $t=0$ from the position $\mathbf{x}_0$ in the reactants, hits the flame at time τ, it is burned and then it arrives at time t to the position $\mathbf{X}$. It is worth noting that in the definition (111) the time variables are in the following order $0<\tau<t$.

The probability density function $P(\tau;\mathbf{X},t|\mathbf{x}_0,0)$ obeys obviously to the following reduction properties

$$\int_{0}^{\infty} P(\tau;\mathbf{X},t|\mathbf{x}_0,0)d\tau = P_L(\mathbf{X},t|\mathbf{x}_0,0), \tag{112}$$

$$\int_{-\infty}^{\infty} P(\tau;\mathbf{X},t|\mathbf{x}_0,0)d^3\mathbf{X} = \Phi(\tau|x_0,0). \tag{113}$$

Expressions (112), (113) follow readily from the definition (111) of $P(\tau;\mathbf{X},t|\mathbf{x}_0,0)$. For example, to prove (112) we integrate (111) over all values of τ

$$\int_{0}^{\infty} \langle \delta(\tau - t_*(\mathbf{x}_0))\delta(\mathbf{X}(t|\mathbf{x}_0,0)-\mathbf{X})\rangle d\tau = \langle \delta(\mathbf{X}(t|\mathbf{x}_0,0)-\mathbf{X})\rangle = P_L(\mathbf{X},t|\mathbf{x}_0,0), \tag{114}$$

using first the property of the delta-function, and then the definition of $P_L(\mathbf{X},t|\mathbf{x}_0,0)$, (25).

Carring out integration in Eq. (110) in the transverse plane over whole values of y_0 and z_0 yields

$$\langle c\rangle(X,t) = \langle c(\mathbf{X},t)\rangle = \int_{-\infty}^{0} P_L(X,t|x_0,0)dx_0 + \int_{0}^{t}\int_{0}^{\infty} P(\tau;X,t|x_0,0)d\tau dx_0, \tag{115}$$

where we introduced the 1D probability density functions

$$P_L(X,t|x_0,0) = \int_{-\infty}^{\infty}\int_{-\infty}^{\infty} P_L(\mathbf{X},t|\mathbf{x}_0,0)dy_0 dz_0, \tag{116}$$

$$P(\tau;X,t|x_0,0) = \int_{-\infty}^{\infty}\int_{-\infty}^{\infty} P_L(\tau;\mathbf{X},t|\mathbf{x}_0,0)dy_0 dz_0. \tag{117}$$

Eqs (116), (117) follow from the assumption of homogeneity in the planes normal to x-axis.

Eq. (115) yields the expression for $\overline{\overline{c}}(t|X,x_0,0)$

$$\overline{\overline{c}}(t|X,x_0,0)P_L(X,t|x_0,0)=(1-H(x_0))P_L(X,t|x_0,0)+H(x_0)\int_0^t P(\tau;X,t|x_0,0)d\tau\,, \quad (118)$$

taking into account that $\langle c\rangle(X,t)$ and $\overline{\overline{c}}(t|X,x_0,0)$ are linked by the formula (45)

$$\langle c\rangle(X,t)=\int_{-\infty}^{\infty}\overline{\overline{c}}(t|X,x_0,0)P_L(X,t|x_0,0)dx_0\,.$$

Expression (118) for the mean mixed Eulerian-Lagrangian progress variable can be rewritten as follows

$$\overline{\overline{c}}(t|X,x_0,0)=(1-H(x_0))+H(x_0)\int_0^t P(\tau|X,t;x_0,0)d\tau\,, \quad (119)$$

$$\overline{\overline{c}}(t|X,x_0,0)=\begin{cases}\int_0^t P(\tau|X,t;x_0,0)d\tau, & x_0>0\\ 1, & x_0<0\end{cases}, \quad (120)$$

where

$$P(\tau|\mathbf{X},t;\mathbf{x}_0,0)=\frac{P(\tau;\mathbf{X},t|\mathbf{x}_0,0)}{P_L(\mathbf{X},t|\mathbf{x}_0,0)}=\frac{\langle\delta(\tau-t_*(\mathbf{x}_0))\delta(\mathbf{X}(t|\mathbf{x}_0,0)-\mathbf{X})\rangle}{\langle\delta(\mathbf{X}(t|\mathbf{x}_0,0)-\mathbf{X})\rangle}, x_0>0\,, \quad (121)$$

$$P(\tau;\mathbf{X},t|\mathbf{x}_0,0)=P(\tau|\mathbf{X},t;\mathbf{x}_0,0)P_L(\mathbf{X},t|\mathbf{x}_0,0)\,,\ \int_0^{\infty}P(\tau|\mathbf{X},t;\mathbf{x}_0,0)d\tau=1 \quad (122)$$

$P(\tau|\mathbf{X},t;\mathbf{x}_0,0)$ is a new conditional probability density function introduced using Bayes theorem. Eqs (122) follow from (111) and (112), respectively.

Eq. (120) for $\overline{\overline{c}}(t|X,x_0,0)$ is the counterpart of Eq. (105) for $\overline{c}(t|x_0,0)$.

The conditional probability density function $P(\tau;\mathbf{X},t|\mathbf{x}_0,0)$, according to (111) is the mean of the product of two random functionals of the velocity field $\mathbf{u}(\mathbf{X},t)$, $\delta(\tau-t_*(\mathbf{x}_0))$ and $\delta(\mathbf{X}(t|\mathbf{x}_0,0)-\mathbf{X})$. Therefore the quantity $P(\tau;\mathbf{X},t|\mathbf{x}_0,0)$ has some characteristic time and length scales that describe the decorrelation of two functional. Below only temporal decorrelation is discussed. $P(\tau;\mathbf{X},t|\mathbf{x}_0,0)$ depends on the two time moments, τ and t, $0<\tau<t$. It should be stressed that the correlation $\langle\delta(\tau-t_*(\mathbf{x}_0))\delta(\mathbf{X}(t|\mathbf{x}_0,0)-\mathbf{X})\rangle$ cannot be splitted (in other words, factorized) neither at $t\le T_L$ nor at $t>>T_L$. The claim for the first case, $t\le T_L$, $0<\tau<t$, is evident, because of the correlation of the velocity field $\mathbf{u}(\mathbf{X},t)$ in this time interval. In the second case, $t>>T_L$, the splitting can be justified only if to the condition $t>>T_L$ it is assumed, additionally, that the difference $t-\tau$ satisfies to the inequality $t-\tau>>T_L$. Therefore in the second case the splitting is not uniform with respect to x_0. Indeed, for each value of t, as large as you want, there is a particle in the reactants with the coordinate $\mathbf{x}_0$ such that the flame front hits this particle at an instant τ, such as $t-\tau\le T_L$. Therefore for this particle in the reactants there is a correlation between two random functionals of the velocity field $\mathbf{u}(\mathbf{X},t)$, $\delta(\tau-t_*(\mathbf{x}_0))$ and $\delta(\mathbf{X}(t|\mathbf{x}_0,0)-\mathbf{X})$. Recapitulating, it is shown that the following results are valid for the correlation $\langle\delta(\tau-t_*(\mathbf{x}_0))\delta(\mathbf{X}(t|\mathbf{x}_0,0)-\mathbf{X})\rangle$

$$\langle\delta(\tau-t_*(\mathbf{x}_0))\delta(\mathbf{X}(t|\mathbf{x}_0,0)-\mathbf{X})\rangle\neq\langle\delta(\tau-t_*(\mathbf{x}_0))\rangle\langle\delta(\mathbf{X}(t|\mathbf{x}_0,0)-\mathbf{X})\rangle,$$
$$t\le T_L, 0<\tau<t, \tag{123}$$

$$\langle\delta(\tau-t_*(\mathbf{x}_0))\delta(\mathbf{X}(t|\mathbf{x}_0,0)-\mathbf{X})\rangle\neq\langle\delta(\tau-t_*(\mathbf{x}_0))\rangle\langle\delta(\mathbf{X}(t|\mathbf{x}_0,0)-\mathbf{X})\rangle,$$
$$t>>T_L, t-\tau\le T_L, \tag{124}$$

$$\left\langle \delta(\tau - t_*(\mathbf{x}_0))\delta(\mathbf{X}(t|\mathbf{x}_0,0) - \mathbf{X}) \right\rangle \approx \left\langle \delta(\tau - t_*(\mathbf{x}_0)) \right\rangle \left\langle \delta(\mathbf{X}(t|\mathbf{x}_0,0) - \mathbf{X}) \right\rangle,$$

$$t >> T_L, t - \tau >> T_L. \qquad (125)$$

Let us consider the limit behaviour of $\bar{\bar{c}}(t|X, x_0, 0)$, (120) at $t \to \infty$, $x_0 > 0$. Similar to the analysis of the expected Lagrangian progress variable $\bar{c}(t|x_0,0)$, (107), (108) there are some subtleties. Namely, the double limit of $\bar{\bar{c}}(T|X, x_0, 0)$ at fixed X and $x_0 \to \infty$ depends on the order, i.e. either $x_0 \to \infty$ then $T \to \infty$, or $T \to \infty$ then $x_0 \to \infty$. In the first case the particle will be forever in the reactants, i.e. never burned, and (120) yields

$$\lim_{T\to\infty} \lim_{x_0\to\infty} \bar{\bar{c}}(T|X, x_0, 0) = \lim_{T\to\infty} \lim_{x_0\to\infty} \int_0^T P(\tau|X, t; x_0, 0)d\tau = 0. \qquad (126)$$

In the second case the particle will be eventually burned, and (120) yields

$$\lim_{x_0\to\infty} \lim_{T\to\infty} \bar{\bar{c}}(T|X, x_0, 0) = \lim_{x_0\to\infty} \lim_{T\to\infty} \int_0^T P(\tau|X, t; x_0, 0)d\tau = 1. \qquad (127)$$

If there are no restrictions on the value of X, we assume that the following relation is valid at fixed x_0

$$\int_0^t P(\tau|X, t; x_0, 0)d\tau = f(X - u_T t; x_0, 0), t \to \infty, \qquad (128)$$

where $f(\zeta; x_0, 0)$, $\zeta = X - u_T t$ is some monotonously decreasing function, $f(-\infty) = 1$, $f(\infty) = 0$, u_T is turbulent burning velocity (defined below, Eq. (150).

Steady flame propagation is assumed in writing (128) in the limit $t \to \infty$. Expression (119) becomes

$$\bar{\bar{c}}(t|X,x_0,0) \to f(X-u_T t;x_0,0),\ t\to\infty, x_0>0. \tag{129}$$

Eqs (129), (45) yield for the expected Eulerian progress variable

$$\langle c\rangle(X,t) \to F(X-u_T t), \tag{130}$$

where

$$F(X-u_T t) = \int_{x_0} f(X-u_T t;x_0,0)P_L(X,t|x_0,0)dx_0\ , t\to\infty. \tag{131}$$

Let us consider now the expected chemical source $\langle W(\mathbf{X},t)\rangle$ in the Eulerian frame. Averaging (44) yields

$$\langle W(\mathbf{X},t)\rangle = \int_{\mathbf{x}_0} \langle W(t|\mathbf{X},\mathbf{x}_0,0)\delta(\mathbf{X}(t|\mathbf{x}_0,0)-\mathbf{X})\rangle d^3\mathbf{x}_0\ . \tag{132}$$

Eq. (132) becomes

$$\langle W(\mathbf{X},t)\rangle = \int_{\mathbf{x}_0} H(x_0)\langle \delta(t-t_*(\mathbf{x}_0))\delta(\mathbf{X}(t|\mathbf{x}_0,0)-\mathbf{X})\rangle d^3\mathbf{x}_0\ , \tag{133}$$

using (98).

Eq. (133) transforms to

$$\langle W\rangle(X,t) = \langle W(\mathbf{X},t)\rangle = \int_{\mathbf{x}_0} H(x_0)P(t;\mathbf{X},t|\mathbf{x}_0,0)d^3\mathbf{x}_0 = \int_{\mathbf{x}_0} H(x_0)P(t|\mathbf{X},t;\mathbf{x}_0,0)P_L(\mathbf{X},t|\mathbf{x}_0,0)d^3\mathbf{x}_0\ ,$$

$$\langle W\rangle(X,t) = \langle W(\mathbf{X},t)\rangle = \int_0^\infty P(t;X,t|x_0,0)dx_0 = \int_0^\infty P(t|X,t;x_0,0)P_L(X,t|x_0,0)dx_0\ , \tag{134}$$

where $P(t;X,t|x_0,0)$ is given by (111), (117), the second rewriting in both lines in (134) uses (121).

The knowledge of $\langle W\rangle(X,t)$ permits to find the turbulent transport term in Eq. (6)

$$\frac{\partial\langle c\rangle}{\partial t}+\nabla\cdot\langle \mathbf{u}c\rangle=\langle W\rangle .$$

Indeed, differentiating Eq. (115) with respect to time yields

$$\frac{\partial\langle c\rangle}{\partial t}=\int_{-\infty}^{0}\frac{\partial P_L(X,t|x_0,0)}{\partial t}dx_0+\int_0^{\infty}P(t;X,t|x_0,0)dx_0+\int_0^t\int_0^{\infty}\frac{\partial P(\tau;X,t|x_0,0)}{\partial t}d\tau dx_0 . \quad (135)$$

Eq. (135) transforms to

$$\frac{\partial\langle c\rangle}{\partial t}=D_T(t)\frac{\partial^2}{\partial X^2}\int_{-\infty}^{0}P_L(X,t|x_0,0)dx_0+\int_0^t\int_0^{\infty}\frac{\partial P(\tau;X,t|x_0,0)}{\partial t}d\tau dx_0+\langle W\rangle(X,t), \quad (136)$$

$$\langle W\rangle(X,t)=\int_0^{\infty}P(t;X,t|x_0,0)dx_0=\int_0^{\infty}P(t|X,t;x_0,0)P_L(X,t|x_0,0)dx_0 ,$$

taking into account Eqs (73) and (134).

Eq. (115) yields for the integral in the second term of the RHS of Eq. (136)

$$\int_{-\infty}^{0}P_L(X,t|x_0,0)dx_0=\langle c\rangle(X,t)-\int_0^t\int_0^{\infty}P(\tau;X,t|x_0,0)d\tau dx_0 . \quad (137)$$

Plugging (137) into the third term in the RHS of Eq. (136) results in

$$\frac{\partial\langle c\rangle}{\partial t}=D_T(t)\frac{\partial^2\langle c\rangle}{\partial X^2}+\int_0^t\int_0^{\infty}\left(\frac{\partial P(\tau;X,t|x_0,0)}{\partial t}-D_T(t)\frac{\partial^2 P(\tau;X,t|x_0,0)}{\partial X^2}\right)d\tau dx_0+\langle W\rangle(X,t), \quad (138)$$

$$\langle W\rangle(X,t)=\int_0^{\infty}P(t;X,t|x_0,0)dx_0=\int_0^{\infty}P(t|X,t;x_0,0)P_L(X,t|x_0,0)dx_0 .$$

Comparison of Eqs (6) and (138) yields for the divergence of the turbulent flux

$$\nabla\cdot\langle\mathbf{u}c\rangle = -D_T(t)\frac{\partial^2\langle c\rangle}{\partial X^2} - \int_0^t\int_0^\infty\left(\frac{\partial P(\tau;X,t|x_0,0)}{\partial t} - D_T(t)\frac{\partial^2 P(\tau;X,t|x_0,0)}{\partial X^2}\right)d\tau dx_0 . \quad (139)$$

The integral term in Eq (139) is not equal to zero identically, because its integrand

$$\frac{\partial P(\tau;X,t|x_0,0)}{\partial t} - D_T(t)\frac{\partial^2 P(\tau;X,t|x_0,0)}{\partial X^2} \neq 0\,, x_0 > 0\,. \quad (140)$$

Inequality (140) is the consequence of the fact that the correlation $\langle\delta(\tau - t_*(\mathbf{x}_0))\delta(\mathbf{X}(t|\mathbf{x}_0,0)-\mathbf{X})\rangle$, $x_0 > 0$, can not be splitted uniformly with respect to x_0, Eqs (123)-(125). To elucidate (140), let us consider for determinacy the case $t >> T_L$, and assume for a moment on the contrary to Eq. (124), that the correlation $P(\tau;\mathbf{X},t|\mathbf{x}_0,0) = \langle\delta(\tau - t_*(\mathbf{x}_0))\delta(\mathbf{X}(t|\mathbf{x}_0,0)-\mathbf{X})\rangle$ can be splitted, i.e. the approximate expression

$$\langle\delta(\tau - t_*(\mathbf{x}_0))\delta(\mathbf{X}(t|\mathbf{x}_0,0)-\mathbf{X})\rangle \approx \langle\delta(\tau - t_*(\mathbf{x}_0))\rangle\langle\delta(\mathbf{X}(t|\mathbf{x}_0,0)-\mathbf{X})\rangle,$$
$$t >> T_L, \quad (141)$$

is valid, regardless of the value of the difference $t-\tau$ with respect to T_L. Then Eqs (111), (121) become

$$P(\tau;\mathbf{X},t|\mathbf{x}_0,0) = \langle\delta(\tau - t_*(\mathbf{x}_0))\delta(\mathbf{X}(t|\mathbf{x}_0,0)-\mathbf{X})\rangle \approx \langle\delta(\tau - t_*(\mathbf{x}_0))\rangle\langle\delta(\mathbf{X}(t|\mathbf{x}_0,0)-\mathbf{X})\rangle,$$

$$P(\tau;\mathbf{X},t|\mathbf{x}_0,0) \approx \Phi(\tau|x_0,0)P_L(\mathbf{X},t|\mathbf{x}_0,0)\,, \quad (142)$$

$$P(\tau|\mathbf{X},t;\mathbf{x}_0,0) = \frac{P(\tau;\mathbf{X},t|\mathbf{x}_0,0)}{P_L(\mathbf{X},t|\mathbf{x}_0,0)} \approx \Phi(\tau|x_0,0)\,, x_0 > 0\,. \quad (143)$$

Plugging (142) into the integrand in (138) results in

$$\left(\frac{\partial}{\partial t}-D_T(t)\frac{\partial^2}{\partial \mathbf{X}^2}\right)P(\tau;\mathbf{X},t|\mathbf{x}_0,0)\approx\Phi(\tau|x_0,0)\left(\frac{\partial}{\partial t}-D_T(t)\frac{\partial^2}{\partial \mathbf{X}^2}\right)P_L\approx 0,$$

$$P_L=P_L(\mathbf{X},t|\mathbf{x}_0,0), \qquad x_0>0 \tag{144}$$

by virtue of (73).

Then Eqs (138), (139) transform to

$$\frac{\partial\langle c\rangle}{\partial t}\approx D_T(\infty)\frac{\partial^2\langle c\rangle}{\partial X^2}+\langle W\rangle(X,t), \tag{145}$$

$$\nabla\cdot\langle\mathbf{u}c\rangle=-D_T(\infty)\frac{\partial^2\langle c\rangle}{\partial X^2}, \tag{146}$$

taking into consideration that the turbulent diffusion coefficient $D_T(t)$ (72) reaches its asymptotic value (74) at $t>>T_L$, i.e.

$$D_T(t)\approx D_T(\infty)=\langle u^2\rangle T_L. \tag{147}$$

We conclude that, the application of the splitting hypothesis (141) at $t>>T_L$ results in the Eqs (145), (146), i.e. the expected progress variable $\langle c\rangle$ would diffuse with the same asymptotic turbulent diffusion coefficient as an expected inert scalar. But this result is incorrect, because the splitting hypothesis (141) is not valid at $t>>T_L$ uniformly for all values of x_0, as was explained above in the discussion of the splitting hypothesis after the expressions (123)-(125). In short, for each value of t, as large as desired, a value of $\mathbf{x}_0$ exist that the flame hits a particle at an instant τ, such as $t-\tau\le T_L$. Therefore there is a correlation between two random functionals of the velocity field $\mathbf{u}(\mathbf{X},t)$ $\delta(\tau-t_*(\mathbf{x}_0))$ and $\delta(\mathbf{X}(t|\mathbf{x}_0,0)-\mathbf{X})$. The probability density function $P(\tau;\mathbf{X},t|\mathbf{x}_0,0)$ can not be factorized, as assume

(141), (142), and (140) is verified. As a consequence the integral term in (138) and (139) is not equal to zero.

We conclude from Eqs (139) and (140) that the expected combustion progress variable $\langle c \rangle$ does not diffuse as an expected inert scalar, i.e. with the turbulent diffusion coefficient $D_T(t)$ defined by (72). We consider in the Annex C with the aid of Eqs (138, (139) the case of the flame extinction which happens at some instant t_{ext}. It will be demonstrated that the transition of Eqs (138), (139) to (145), (146) after the flame extinction occurs during a relaxation time which is of the order of T_L.

Let us return to the discussion of the mean source $\langle W(\mathbf{X},t) \rangle$. Integrating Eq. (134) over the whole X yields

$$\int_{-\infty}^{\infty} \langle W \rangle (X,t) dX = \int_{-\infty}^{\infty} \int_{0}^{\infty} P(t;X,t|x_0,0) dx_0 dX = \int_{0}^{\infty} \Phi(t|x_0,0) dx_0 \,, \quad (148)$$

by virtue of (102), (116).

The turbulent burning velocity $u_T(t)$ is equal, by definition, to the spatial integral of the mean source

$$u_T(t) = \int_{-\infty}^{\infty} \langle W \rangle (X,t) dX \,. \quad (149)$$

Therefore from Eqs (148), (149) follows that

$$u_T(t) = \int_{0}^{\infty} \Phi(t|x_0,0) dx_0 \,. \quad (150)$$

Let us compare Eqs (134) and (150) with the corresponding results obtained in the framework of Eulerian combustion progress variable $c(\mathbf{x},t)$ equation Eq. (5) (e.g. [5], [6]).

$$\langle W(\mathbf{X},t) \rangle = S_L \langle |\nabla c(\mathbf{X},t)| \rangle, |\nabla c(\mathbf{X},t)| = \delta(G(\mathbf{x},t) - G_0) |\nabla G(\mathbf{x},t)| \,, \quad (151)$$

$$u_T(t) = S_L \int_{-\infty}^{\infty} \langle |\nabla c(\mathbf{X},t)| \rangle dX, \tag{152}$$

where S_L is the normal velocity of flame propagation, $c(\mathbf{X},t) = H(G(\mathbf{X},t))$ is the progress variable. The scalar $G(\mathbf{X},t) > 0$ verifies a field equation, [9], written below, Eq. (155). It is assumed that $G(\mathbf{X},t) > 0$ and $G(\mathbf{X},t) < 0$ in the products and the reactants ($G-$equation is considered in the paragraph 7.2), respectively. $G(\mathbf{X}_f, t) = 0$ is the equation for the flame surface. Comparing Eqs (134) and (150) yields

$$\langle W \rangle (X,t) = \int_{0}^{\infty} P(t; X, t | x_0, 0) dx_0 = S_L \langle |\nabla c(\mathbf{X},t)| \rangle, \tag{153}$$

$$u_T(t) = \int_{0}^{\infty} \Phi(t | x_0, 0) dx_0 = S_L \int_{\infty}^{\infty} \langle |\nabla c(\mathbf{X},t)| \rangle dX. \tag{154}$$

We recapitulate the principal new results obtained in the paragraph 7.2: 1) the expression (103) for the expected Lagrangian progress variable $\bar{c}(t | x_0, 0)$, 2) the expressions (110), (115) for the the expected Eulerian progress variable field $\langle c(\mathbf{X},t) \rangle$, 3) the expression (118) for the expected mixed Lagrangian-Eulerian progress variable $\bar{\bar{c}}(t | \mathbf{X}, \mathbf{x}_0, 0)$, 4) the analysis of the subtleties the double limit $t \to \infty$, $x_0 \to \infty$ presented by the expressions (107) and (108) for $\bar{c}(t | \mathbf{x}_0, 0)$, (126) and (127) for $\bar{\bar{c}}(t | \mathbf{X}, \mathbf{x}_0, 0)$. It is established that the result of such a limit depends on the order, i.e. either $x_0 \to \infty$ then $t \to \infty$, or $t \to \infty$ then $x_0 \to \infty$. 5) the expression (134) for the expected Eulerian chemical source $\langle W(\mathbf{X},t) \rangle$, 6) Eq. (138) and the expression (139) for the turbulent transport term, 7) the expression (150) for the turbulent flame velocity $u_T(t)$.

7.2. Derivation of the Equation for the Hitting Time Probability Function from G-Equation

Let us consider an isoscalar surface G_0 of a scalar field $G(\mathbf{X},t)$ which moves normal to itself at a speed S_L against a flow velocity $\mathbf{u}(\mathbf{X},t)$. In what follows S_L is assumed constant. The dynamics of the surface is modelled by the following IBVP [9]

$$\frac{\partial G}{\partial t}+\mathbf{u}\cdot\nabla G=S_L|\nabla G|,\ |\nabla G|=\sqrt{\frac{\partial G}{\partial X_i}\frac{\partial G}{\partial X_i}}, \tag{155}$$

$$G(\mathbf{X},t=0)=G_{in}(\mathbf{X}). \tag{156}$$

The particular choice of $G_{in}(\mathbf{X})$

$$G_{in}(\mathbf{X})=-X$$

for statistically 1D premixed flame, as was shown first time in [9], plays an important role in the theoretical analysis of IBVP

There is an implied summation over the repeated indices. Eq. (155) is a field equation, and its common name is G-equation. If applied to premixed combustion, the surface G_0 is associated with the flame front. It is assumed that $G(\mathbf{X},t)>G_0$ and $G(\mathbf{X},t)<G_0$ in the products and in the reactants, respectively. Initially, according to $G_{in}(\mathbf{X})=-X$, all level surfaces of $G(\mathbf{X},t=0)$ are planes normal to the X direction. For simplicity of writing it is assumed below that that $G_0=0$, and, consequently, the initial laminar flame position $X_f(t=0)=0$.

Once IBVP (155), (156) is solved, the field $G(\mathbf{X},t)$ defines the combustion progress variable $c(\mathbf{X},t)$ as

$$c(\mathbf{X},t)=H\big(G(\mathbf{X},t)\big),\ \nabla c(\mathbf{X},t)=\delta(G(\mathbf{X},t))\nabla G(\mathbf{X},t). \tag{157}$$

The equation for $c(\mathbf{X},t)$, which readily follows from (155), (157) reads

$$\frac{\partial c}{\partial t}+\mathbf{u}\cdot\nabla c=S_L\left|\nabla c\right|. \tag{158}$$

We are interested to solve the following problem. Let us consider a fluid particle in the reactants at initial time $t=0$ is at the spatial point $\mathbf{x}_0$, $x_0>0$ in correspondence with the initial conditions (156). The question is: at what time this fluid particle hits the flame surface (in other words the isosurface $G(\mathbf{X},t)=G_0=0$), the evolution of which is described by the IVBP (155), (156)

To answer this question let us consider the trajectory of a particle $\mathbf{X}(t|\mathbf{x}_0,0)$. The trajectory solves the following initial value problem (16)

$$\frac{\partial \mathbf{X}(t|\mathbf{x}_0,0)}{\partial t}=\mathbf{u}(\mathbf{X}(t|\mathbf{x}_0,0),t), \tag{159}$$

$$\mathbf{X}(t=0|\mathbf{x}_0,0)=\mathbf{x}_0,\, x_0>0\,. \tag{160}$$

Let us introduce the following function

$$F(t)=G(\mathbf{X}(t|\mathbf{x}_0,0),t). \tag{161}$$

Then, as it is readily seen, a root of the equation

$$F(t)=G(\mathbf{X}(t|\mathbf{x}_0,0),t)=G_0=0\,,x_0>0 \tag{162}$$

is the instant when the fluid particle in the reactants with the initial position $\mathbf{x}_0$ hits the flame front. Let us denote this instant $t_*(\mathbf{x}_0)$. The hitting time was already introduced in the section 7, Eqs (94)-(96). $t_*(\mathbf{x}_0)$ is a functional of the fluid particle trajectory, and depends also on the initial particle position $\mathbf{x}_0$, and the laminar flame velocity S_L. The fine-grained conditional probability density function of $t_*(\mathbf{x}_0)$ is $\delta(t - t_*(\mathbf{x}_0))$. Eq. (162) permits to link this fine-grained probability density function with $\delta(G(\mathbf{X}(t|\mathbf{x}_0,0),t))$. Indeed we transform $\delta(G(\mathbf{X}(t|\mathbf{x}_0,0),t))$

$$\delta(G(\mathbf{X}(t|\mathbf{x}_0,0),t)) = \delta(F(t)) = \left(\left. \frac{\partial F}{\partial t} \right|_{t=t_*} \right)^{-1} \delta(t - t_*(\mathbf{x}_0)), \tag{163}$$

using the properties of δ function. The derivative $\frac{\partial F}{\partial t}$ is equal to

$$\frac{\partial F}{\partial t} = \frac{\partial \mathbf{X}(t|\mathbf{x}_0,0)}{\partial t} \cdot \nabla_{\mathbf{X}} G + \frac{\partial G}{\partial t}, \tag{164}$$

where $\nabla_{\mathbf{X}} G$ stands for

$$\nabla_{\mathbf{X}} G = \nabla_{\mathbf{X}} G|_{\mathbf{X}=\mathbf{X}(t|\mathbf{x}_0,0)}, \ |\nabla_{\mathbf{X}} G| = |\nabla_{\mathbf{X}} G|\big|_{\mathbf{X}=\mathbf{X}(t|\mathbf{x}_0,0)} = \left. \sqrt{\frac{\partial G}{\partial X_i} \frac{\partial G}{\partial X_i}} \right|_{\mathbf{X}=\mathbf{X}(t|\mathbf{x}_0,0)}. \tag{165}$$

Inserting $\frac{\partial G}{\partial t}$ from (155) into (164) yields

$$\frac{\partial F}{\partial t} = \left(\frac{\partial \mathbf{X}(t|\mathbf{x}_0,0)}{\partial t} - \mathbf{u}(\mathbf{X}(t|\mathbf{x}_0,0),t) \right) \cdot \nabla_{\mathbf{X}} G + S_L |\nabla_{\mathbf{X}} G| = S_L |\nabla_{\mathbf{X}} G| \tag{166}$$

taking into account that the first term in the RHS of Eq. (166) is equal to zero due to (159).

Inserting the derivative $\frac{\partial F}{\partial t}$ given by (166) into (163) yields

$$\delta(t - t_*(\mathbf{x}_0)) = S_L |\nabla_{\mathbf{X}} G| \delta(G(\mathbf{X}(t|\mathbf{x}_0,0),t)). \tag{167}$$

Averaging (167) yields an alternative expression for the conditional Lagrangian probability density function $\Phi(t|\mathbf{x}_0,0)=\Phi(t|x_0,0)$ of hitting time

$$\Phi(t|x_0,0)=S_L\left\langle|\nabla_{\mathbf{X}}G|\delta\left(G\left(\mathbf{X}\left(t|\mathbf{x}_0,0\right),t\right)\right)\right\rangle,\ x_0>0 \tag{168}$$

which is equivalent to (101) obtained earlier.

Let us consider now the fine-grained conditional Lagrangian probability density function that a fluid particle in the reactants which starts at initial time $t=0$ from the position $\mathbf{x}_0$, hits the flame at time τ, and then arrives at time t at position $\mathbf{X}$, introduced by (111)

$$\delta\left(\tau-t_*\left(\mathbf{x}_0\right)\right)\delta(\mathbf{X}(t|\mathbf{x}_0,0)-\mathbf{X}),\ 0<\tau<t.$$

Eq. (111) becomes using (167)

$$\delta\left(\tau-t_*\left(\mathbf{x}_0\right)\right)\delta(\mathbf{X}(t|\mathbf{x}_0,0)-\mathbf{X})=S_L|\nabla_{\mathbf{X}}G|\delta\left(G\left(\mathbf{X}\left(t|\mathbf{x}_0,0\right),\tau\right)\right)\delta(\mathbf{X}(t|\mathbf{x}_0,0)-\mathbf{X}). \tag{169}$$

Averaging (169) yields an alternative expression for the conditional joint Lagrangian probability density function $P(\tau;\mathbf{X},t|\mathbf{x}_0,0)$, introduced above in this section 7 by (111).

$$P(\tau;\mathbf{X},t|\mathbf{x}_0,0)=S_L\left\langle|\nabla_{\mathbf{X}}G|\delta\left(G\left(\mathbf{X}\left(t|\mathbf{x}_0,0\right),\tau\right)\right)\delta(\mathbf{X}(t|\mathbf{x}_0,0)-\mathbf{X})\right\rangle,$$
$$0<\tau<t. \tag{170}$$

It is readily verified, using (163), that (170) satisfies the reduction properties (112), (113).

For illustration purposes, let us consider a simple case of the absence of turbulence, i.e. the velocity field $\mathbf{u}(\mathbf{X},t)=0$, and let $G_{in}(\mathbf{X})=-X$. Then the laminar flame front is a plane which propagates throughout the reactants with the velocity S_L from left to right. The front coordinate is equal to zero at the initial time $t = 0$, i.e. $X_f(t = 0) = 0$. The scalar field $G(X,t)$ and the trajectory of the particle $X(t|x_0,0)$ read

$$G(X,t) = -X + S_L t = -X + X_f(t), X_f(t) = S_L t\,, \tag{171}$$

$$X(t|x_0,0) = x_0 > 0\,. \tag{172}$$

The root of the equation $G(X(t|x_0,0),t) = G_0 = 0$, i.e.

$$G(X(t|x_0,0),t) = -X(t|x_0,0) + S_L t = -x_0 + X_f(t) = 0\,, \tag{173}$$

is a hitting time

$$t_*(x_0) = \frac{x_0}{S_L}, x_0 > 0\,. \tag{174}$$

Hence introduced above by Eqs (102), (111) and (121) the conditional Lagrangian probability density functions $\Phi(t|x_0,0)$, $P(\tau;X,t|x_0,0)$ and $P(\tau|X,t;x_0,0)$ reduce to δ-functions

$$\Phi(t|x_0,0) = S_L\langle|\nabla_X G|\delta(G(X(t|x_0,0),t))\rangle = S_L\delta(-x_0 + S_L t) = \delta\left(t - \frac{x_0}{S_L}\right), \tag{175}$$

$$P_L(X,t|x_0,0) = \delta(X - x_0),$$

$$P(\tau;X,t|x_0,0) = \delta\left(\tau - \frac{x_0}{S_L}\right)\delta(X - x_0)\,, P(\tau|X,t;x_0,0) = \delta\left(\tau - \frac{x_0}{S_L}\right), x_0 > 0\,. \tag{176}$$

Then the relations (103), (120), (115) yield

$$\bar{c}(t|X,x_0,0)=(1-H(x_0))+H(x_0)H\left(t-\frac{x_0}{S_L}\right)=H(S_Lt-x_0)=H(X_f(t)-x_0), \tag{177}$$

$$\bar{\bar{c}}(t|X,x_0,0)=(1-H(x_0))+H(x_0)H\left(t-\frac{x_0}{S_L}\right)=H(S_Lt-x_0)=H(X_f(t)-x_0), \tag{178}$$

$$\langle c\rangle(X,t)=(1-H(X))+H(X)H\left(t-\frac{X}{S_L}\right)=H(S_Lt-X)=H(X_f(t)-X), \quad 179)$$

respectively.

8. Approximate Expression and Equation for the Mean Progress Variable

8.1. Approximate Expression for the Mean Progress Variable

It should be stressed that the integration in Eq. (109), which is the consequence of (42), is performed over all values of $\mathbf{x}_0$, in other words the integration domain is fixed and does not depend on time. Eq. (42) can be transformed to the different but equivalent form with the integration over random time dependent volume $V_0(t)$, containing all products at the instant t

$$c(\mathbf{X},t)=\int_{V_0(t)}\delta(\mathbf{X}(t|\mathbf{x}_0,0)-\mathbf{X})d^3\mathbf{x}_0\,,\ \mathbf{x}_0\in V_0(t)\,. \tag{180}$$

The random volume $V_0(t)$ is the sum of the volume of the products at the initial time $t=0$ (the half-space $x_0<0$) and the part of the volume of reactants at the initial time the particles of which were hit by the flame front during the time interval $[0,t]$ and burned, i.e. the jumps from $c=0$ to $c=1$ took place. The temporal rate of change of the instantaneous volume $V_0(t)$ is obviously obeys the following equation

$$\frac{dV_0(t)}{dt} = S_L A(t), \tag{181}$$

where $A(t)$ is the area of the overall instantaneous flame front. $A(t)$ depends on the turbulent velocity field and S_L. It is a functional of the turbulent velocity field $\mathbf{u}(\mathbf{X},t)$.

It should be noted that the relation (180) looks simpler than (109), but it is deceptive conclusion. Indeed, such operations as taking time derivative and performing averaging is much simpler with (109) than with (180) because of the time dependence of the random function $V_0(t)$. Contrary to (180) the averaging (109) is directly applied to the integrand, Eqs (110), (115).

The averaging (180) results in

$$\langle c\rangle(\mathbf{X},t) = \left\langle \int\limits_{V_0(t)} \delta(\mathbf{X}(t|\mathbf{x}_0,0) - \mathbf{X})d^3\mathbf{x}_0 \right\rangle, \mathbf{x}_0 \in V_0(t). \tag{182}$$

The expected random integral (182) depends on the two statistically dependent random functionals of the velocity field $\mathbf{u}(\mathbf{X},t)$, $\delta(\mathbf{X}(t|\mathbf{x}_0,0) - \mathbf{X})$ and $V_0(t)$. There is no general approach to calculate this kind of expected quantities (182).

Because of this difficulty, to appreciate the goodness of (182), let us, as a first rough approximation, neglect by the statistical dependence of $\delta(\mathbf{X}(t|\mathbf{x}_0,0) - \mathbf{X})$ and $V_0(t)$. Even it is highly difficult to estimate the error introduced by this assumption, we would like to see what conclusion can be reached with it. According to the assumption of statistical independence of $\delta(\mathbf{X}(t|\mathbf{x}_0,0) - \mathbf{X})$ and $V_0(t)$, (182) transforms to the approximate equation

$$\langle c\rangle(\mathbf{X},t) \approx \int\limits_{\langle V_0(t)\rangle} \langle \delta(\mathbf{X}(t|\mathbf{x}_0,0) - \mathbf{X})\rangle d^3\mathbf{x}_0 = \int\limits_{\langle V_0(t)\rangle} P_L(\mathbf{X},t|\mathbf{x}_0,0)d^3\mathbf{x}_0. \tag{183}$$

Eq. (183) was postulated in [2] without justification. Analysis, conducted above, shows that Eq. (183) is not strict equation following only from BML model [4], but includes implicitly an additional assumption - neglecting the statistical dependence of $\delta(\mathbf{X}(t|\mathbf{x}_0,0) - \mathbf{X})$ and $V_0(t)$.

If the transition probability density function $P_L(\mathbf{X},t|\mathbf{x}_0,0)$ is assumed known, the problem of finding the mean Eulerian progress variable $\langle c\rangle(\mathbf{X},t)$, in the frame of (183), is reduced to the calculation of the expected volume $\langle V_0(t)\rangle$. One can guess that it is linked with a mean flame position.

Authors [2] derived from (183) an equation for mean progress variable $\langle c\rangle(\mathbf{X},t)$ invoking Taylor theory of turbulent diffusion [8], (73). We do not pursuit this derivation and limit ourselves in the next section by the consideration of 1D statistically planar turbulent flame.

8.2. Approximate Equation for the Expected Progress Variable

Let us consider a homogeneous isotropic stationary turbulent flow. Let the mean flame brush propagates from left to right in the positive X - direction. It is assumed that the right boundary of the expected volume $\langle V_0(t)\rangle$ is the mean flame coordinate $\langle X_f\rangle(t)$, [2]. Then Eq. (183) becomes

$$\langle c\rangle(X,t) \approx \int_{-\infty}^{\langle X_f\rangle} \langle \delta(\mathbf{X}(t|\mathbf{x}_0,0)-\mathbf{X})\rangle dx_0 = \int_{-\infty}^{\langle X_f\rangle} P_L(X,t|x_0,0)dx_0\,, \tag{184}$$

In what follows the exact equality sign will be written in (184).
Taking time derivative of (184) yields

$$\frac{\partial\langle c\rangle}{\partial t} = \int_{-\infty}^{\langle X_f\rangle} \frac{\partial P_L}{\partial t}dx_0 + \frac{d\langle X_f\rangle}{dt} P_L(X,t|x_0=\langle X_f\rangle,0)\,. \tag{185}$$

The second term in the RHS of Eq. (185) can be recast to the form

$$P_L(X,t|\langle X_f\rangle,0) = \int_{-\infty}^{\langle X_f\rangle} \frac{\partial P_L}{\partial x_0}dx_0 = -\int_{-\infty}^{\langle X_f\rangle} \frac{\partial P_L}{\partial X}dx_0 = -\frac{\partial}{\partial X}\int_{-\infty}^{\langle X_f\rangle} P_L(X,t|x_0,0)dx_0 = -\frac{\partial\langle c\rangle}{\partial X}, \tag{186}$$

taking into account (184), and the fact that for the homogeneous turbulence the transition probability density function $P_L(X,t|x_0,0)$ depends on $X - x_0$, i.e.

$$P_L(X,t|x_0,0) = P_L(X - x_0,t|0), \frac{\partial P_L}{\partial X} = -\frac{\partial P_L}{\partial x_0}. \tag{187}$$

The time derivative of $\langle X_f \rangle(t)$ is nothing but the turbulent burning velocity

$$\frac{d\langle X_f \rangle}{dt} = u_T(t). \tag{188}$$

Combining (185)-(188) yields

$$\frac{\partial\langle c \rangle}{\partial t} = \int_{-\infty}^{\langle X_f \rangle} \frac{\partial P_L}{\partial t} dx_0 - u_T(t)\frac{\partial\langle c \rangle}{\partial X}. \tag{189}$$

Let us transform the first term in the RHS of (189), applying the Taylor theory of turbulent diffusion [8]. According to this theory, $P_L(X,t|x_0,0)$ solves for 1D case 1D Eq. (73)

$$\frac{\partial P_L(X,t|x_0,0)}{\partial t} = D_T(t)\frac{\partial^2 P_L(X,t|x_0,0)}{\partial X^2}, \tag{190}$$

where $D_T(t)$ is the turbulent diffusion coefficient. Inserting the temporal derivative $\frac{\partial P_L}{\partial t}$ from (190) to (189) yields

$$\frac{\partial\langle c \rangle}{\partial t} = D_T(t)\int_{-\infty}^{\langle X_f \rangle} \frac{\partial^2 P_L(X,t|x_0,0)}{\partial X^2} dx_0 - u_T\frac{\partial\langle c \rangle}{\partial X} = D_T(t)\frac{\partial^2\langle c \rangle}{\partial X^2} - u_T\frac{\partial\langle c \rangle}{\partial X}, \tag{191}$$

The first term in the RHS of (191) was transformed with the aid of (184). We remind that (191) is written for the turbulent flame propagating from the left to right.

Another derivation of Eq. (191) can be done from the equations derived in the section 7 neglecting stochastic character of a hitting time $t_*(\mathbf{x}_0)$, assuming that $t_*(\mathbf{x}_0)$ is found from the equation

$$x_0 = \int_0^{t_*(x_0)} u_T(t)dt = \overline{X}_f(t_*(x_0)). \tag{192}$$

(192) signifies that $t_*(\mathbf{x}_0)$ is equal to a transit time of the mean flame to the fluid particle in the reactants with the mean longitudinal coordinate x_0. The derivation is presented in the Annex D.

Eq. (191) is nothing but the equation obtained theoretically (for the constant density turbulent velocity field) by the Prudnikov as early as in 1960 by a different method. Abridged presentation of this derivation can be found in [10]. Thus from historical viewpoint the credit to the derivation of (191) has to be ascribed to Prudnikov. Eq. (191) is a particular 1D case of the equation derived in [2] using a different method.

To conclude this section, we summarise the main results and do the critical comments.

1. Eq. (191) does not follow strictly from BML model. It is the consequence of the application of an approximate decorrelation hypotheses for the instantaneous volume of products $V_0(t)$ and the fine-grade probability density function $\delta(\mathbf{X}(t|\mathbf{x}_0,0)-\mathbf{X})$ to the strict (in the frame of BML model) expression (182). Hence the derivations of (191) in [2] and [10] do not prove that the expected progress variable $\langle c \rangle$, as one could conclude from the first term in the RHS of (191), is diffused with the same turbulent diffusion coefficient $D_T(t)$, as an inert scalar in the Taylor turbulent diffusion theory [8].

2. For the same reason the second term in the RHS of (191) cannot be considered as the expected chemical source, i.e. $\langle W \rangle \neq u_T \left| \frac{\partial \langle c \rangle}{\partial X} \right|$. Decorrelation hypothesis yields only the sum of two terms.

9. Discussion and the Relation to What Was Previously Published

Let us compare our results with the results of [1]. For convenience of the comparison the references to the Eqs from [1] will be written as (ZP.) with a corresponding number. The correspondence between the notations: $\langle c \rangle \leftrightarrow Y$, $\overline{\overline{c}} = \overline{Y}$, $t_0 = 0$, $\langle W \rangle = W$. First we discuss the Eq. (76). To compare (76) with the results derived in [1], we rewrite Eq. (76) in the different form. Namely, we eliminate the term $\overline{\overline{WP_L}}$ from the RHS of (76) with the aid of (73). Eq. (76) transforms to

$$\frac{\partial \overline{\overline{c}} P_L}{\partial t} = D_T \frac{\partial^2 \overline{\overline{c}} P_L}{\partial \mathbf{X}^2} + \left(P_L \frac{\partial \overline{\overline{c}}}{\partial t} - 2D_T \frac{\partial P_L}{\partial \mathbf{X}} \frac{\partial \overline{\overline{c}}}{\partial \mathbf{X}} - P_L D_T \frac{\partial^2 \overline{\overline{c}}}{\partial \mathbf{X}^2} \right). \tag{193}$$

Contrary to Eq. (76), the chemical source is hidden in (193). This particularity of writing (193) embarrasses the identification of the turbulent diffusion terms, as we will see soon. It should be noted that (193) can be obtained also directly from the equation for the transition probability density function (71), multiplying it by $\overline{\overline{c}}(t|\mathbf{X}, \mathbf{x}_0, 0)$ and performing the transformations [1]

$$\overline{\overline{c}} \frac{\partial P_L}{\partial t} = D_T \overline{\overline{c}} \frac{\partial^2 P_L}{\partial \mathbf{X}^2}, \tag{194}$$

$$\frac{\partial \overline{\overline{c}} P_L}{\partial t} - P_L \frac{\partial \overline{\overline{c}}}{\partial t} = D_T \frac{\partial}{\partial \mathbf{X}} \left(\overline{\overline{c}} \frac{\partial P_L}{\partial \mathbf{X}} \right) - 2D_T \frac{\partial P_L}{\partial \mathbf{X}} \frac{\partial \overline{\overline{c}}}{\partial \mathbf{X}},$$

$$\frac{\partial \overline{\overline{c}} P_L}{\partial t} - P_L \frac{\partial \overline{\overline{c}}}{\partial t} = D_T \frac{\partial}{\partial \mathbf{X}}\left(\frac{\partial \overline{\overline{c}} P_L}{\partial \mathbf{X}} - P_L \frac{\partial \overline{\overline{c}}}{\partial \mathbf{X}}\right) - D_T \frac{\partial P_L}{\partial \mathbf{X}} \frac{\partial \overline{\overline{c}}}{\partial \mathbf{X}},$$

$$\frac{\partial \overline{\overline{c}} P_L}{\partial t} - P_L \frac{\partial \overline{\overline{c}}}{\partial t} = D_T \frac{\partial^2 \overline{\overline{c}} P_L}{\partial \mathbf{X}^2} - 2D_T \frac{\partial P_L}{\partial \mathbf{X}} \frac{\partial \overline{\overline{c}}}{\partial \mathbf{X}} - P_L D_T \frac{\partial^2 \overline{\overline{c}}}{\partial \mathbf{X}^2}. \tag{195}$$

The first term in the RHS of (193), from the formal viewpoint, can be interpreted as the single transport term, which is just the turbulent diffusion term. Indeed, the integral of this term over the whole space in (193) is equal to zero. Hence the integral $\int_{\mathbf{X}} \overline{\overline{c}} P_L d\mathbf{X}$ is conserved by the action of this term. Thus the first term in the RHS of (193) only redistribute the product $\overline{\overline{c}} P_L$ in the space $\mathbf{X}$, similar to the analysis of Eq. (76) in the section 6. Contrary to the first term in the RHS of (193), the integral over the whole space $\mathbf{X}$ of the terms in the brackets is not equal to zero. Then it is tempting to consider the terms in the square brackets in RHS of (193), as a chemical source. Hence we could write

$$\overline{\overline{W}}_1 P_L = \left(P_L \frac{\partial \overline{\overline{c}}}{\partial t} - 2D_T \frac{\partial P_L}{\partial \mathbf{X}} \frac{\partial \overline{\overline{c}}}{\partial \mathbf{X}} - P_L D_T \frac{\partial^2 \overline{\overline{c}}}{\partial \mathbf{X}^2}\right). \tag{196}$$

We introduced here the new notation for this source $\overline{\overline{W}}_1$ to distinguish it from the real chemical source $\overline{\overline{W}}$ in (76), (77).

Let us note that the identification of the chemical source based only on the observation if the integral over whole space $\mathbf{X}$ of some group of terms in (193) equal or not to zero is not unique. Indeed, let us rewrite (193) in the different but an equivalent form

$$\frac{\partial \overline{\overline{c}} P_L}{\partial t} = \left\{D_T \frac{\partial^2 \overline{\overline{c}} P_L}{\partial \mathbf{X}^2} - D_T \frac{\partial}{\partial \mathbf{X}}\left(P_L \frac{\partial \overline{\overline{c}}}{\partial \mathbf{X}}\right)\right\} + \left(P_L \frac{\partial \overline{\overline{c}}}{\partial t} - D_T \frac{\partial P_L}{\partial \mathbf{X}} \frac{\partial \overline{\overline{c}}}{\partial \mathbf{X}}\right), \tag{197}$$

where (77) was used.

The integral of the terms in braces in the RHS of the Eq. (197) over the whole space $\mathbf{X}$ is equal to zero. Then, similarly to the above analysis, we could define a new chemical source as follows

$$\overline{\overline{W}}_2 P_L = \left(P_L \frac{\partial \overline{\overline{c}}}{\partial t} - D_T \frac{\partial P_L}{\partial \mathbf{X}} \frac{\partial \overline{\overline{c}}}{\partial \mathbf{X}} \right). \tag{198}$$

Obviously, the identifications of the source term presented by the relations (196) and (198) are quite different. Nevertheless the following integrals over the whole space $\mathbf{X}$ are equal

$$\int_{\mathbf{X}} \overline{\overline{W}}_1 P_L d\mathbf{X} = \int_{\mathbf{X}} \overline{\overline{W}}_2 P_L d^3\mathbf{X} = \int_{\mathbf{X}} \left(P_L \frac{\partial \overline{\overline{c}}}{\partial t} - D_T \frac{\partial P_L}{\partial \mathbf{X}} \frac{\partial \overline{\overline{c}}}{\partial \mathbf{X}} \right) d^3\mathbf{X}, \tag{199}$$

by virtue of the identity (78). In addition to just described difficulty of the identification of the chemical source in the Eq. (193), another problem raises, if the case of the premixed combustion is considered. Indeed, both (196) and (198) do not guarantee identically that $\overline{\overline{W}}_1$ and $\overline{\overline{W}}_2$ are positively defined, which is necessary condition for the case of the premixed combustion. Indeed the source is greater that zero, $W > 0$, according to (5), (98).

The authors [1] consider in the section B the general case of chemical source term on the base of Lagrangian approach and Taylor theory of turbulent diffusion [8]. The analysis is based on the use of the couple of the equations (ZP.8) and (ZP.18). These equations are equivalent to our Eqs (71) and (44), respectively. Then they used the transformation (194), (195) of the Eq. (ZP.8) multiplied by $\overline{Y}$ ($= \overline{\overline{c}}$) and integrating over $\mathbf{x}_0$ they obtained Eq. (ZP.19), which, using our notations, reads

$$\frac{\partial \langle c \rangle}{\partial t} = D_T \frac{\partial^2 \langle c \rangle}{\partial \mathbf{X}^2} + \int_{\mathbf{x}_0} \left(P_L \frac{\partial \overline{\overline{c}}}{\partial t} - 2D_T \frac{\partial P_L}{\partial \mathbf{X}} \frac{\partial \overline{\overline{c}}}{\partial \mathbf{X}} - P_L D_T \frac{\partial^2 \overline{\overline{c}}}{\partial \mathbf{X}^2} \right) d^3\mathbf{x}_0 . \tag{200}$$

Eq. (200) corresponds to Eq. (ZP.19) from [1]. The authors of [1] identify the integral term in RHS of the Eq. (200) as the mean source term, i.e.

$$\langle W\rangle = \langle W_1\rangle = \int_{\mathbf{x}_0}\left(P_L \frac{\partial \overline{\overline{c}}}{\partial t} - 2D_T \frac{\partial P_L}{\partial \mathbf{X}}\frac{\partial \overline{\overline{c}}}{\partial \mathbf{X}} - P_L D_T \frac{\partial^2 \overline{\overline{c}}}{\partial \mathbf{X}^2}\right) d^3\mathbf{x}_0 . \tag{201}$$

Expression (201) corresponds to the expression (ZP.20) [1]. Expression (201) is incorrect, because its consequence contradicts to the exact expression for the turbulent flux $\langle \mathbf{u}c\rangle$ (86). To demonstrate this claim, let us integrate (76) over $\mathbf{x}_0$. This results in

$$\begin{aligned} \frac{\partial\langle c\rangle}{\partial t} &= \int_{\mathbf{x}_0}\left(D_T \frac{\partial^2 \overline{\overline{c}} P_L}{\partial \mathbf{X}^2} - D_T \frac{\partial}{\partial \mathbf{X}}\left(P_L \frac{\partial \overline{\overline{c}}}{\partial \mathbf{X}}\right) - \frac{\partial \overline{\overline{\mathbf{v}'c'}} P_L}{\partial \mathbf{X}}\right) d^3\mathbf{x}_0 + \langle W\rangle \\ &= D_T \frac{\partial^2 \langle c\rangle}{\partial \mathbf{X}^2} + \int_{\mathbf{x}_0}\left(- D_T \frac{\partial}{\partial \mathbf{X}}\left(P_L \frac{\partial \overline{\overline{c}}}{\partial \mathbf{X}}\right) - \frac{\partial \overline{\overline{\mathbf{v}'c'}} P_L}{\partial \mathbf{X}}\right) d^3\mathbf{x}_0 + \langle W\rangle \end{aligned} , \tag{202}$$

by virtue of (44)-(46). We remind that, in correspondence with (87) the integral term in the RHS of the first rewriting of (202), and the sum of two first terms in RHS of the second rewriting of (202) with the sign minus is the divergence $\nabla \cdot \langle \mathbf{u}c\rangle$, (87). The turbulent flux $\langle \mathbf{u}c\rangle$ is given by the expression (86).

It should be stressed that the Eqs (200) and (202) are identical. The principal difference between two writings consists of: 1) the mean source $\langle W\rangle$ is hidden in (200), 2) the mean source $\langle W\rangle$ is explicit in (202).

Equating the RHS of two equations (200), (202) yields the following identity

$$\int_{\mathbf{x}_0}\left(P_L \frac{\partial \overline{\overline{c}}}{\partial t} - 2D_T \frac{\partial P_L}{\partial \mathbf{X}}\frac{\partial \overline{\overline{c}}}{\partial \mathbf{X}} - P_L D_T \frac{\partial^2 \overline{\overline{c}}}{\partial \mathbf{X}^2}\right) d^3\mathbf{x}_0 = \int_{\mathbf{x}_0}\left(- D_T \frac{\partial}{\partial \mathbf{X}}\left(P_L \frac{\partial \overline{\overline{c}}}{\partial \mathbf{X}}\right) - \frac{\partial \overline{\overline{\mathbf{v}'c'}} P_L}{\partial \mathbf{X}}\right) d^3\mathbf{x}_0 + \langle W\rangle \tag{203}$$

If we assume now, as is done in [1] that the mean source term is given by Eq. (201), then Eq. (203) reduces to the following constraint

$$\int_{\mathbf{x}_0}\left(- D_T \frac{\partial}{\partial \mathbf{X}}\left(P_L \frac{\partial \overline{\overline{c}}}{\partial \mathbf{X}}\right) - \frac{\partial \overline{\overline{\mathbf{v}'c'}} P_L}{\partial \mathbf{X}}\right) d^3\mathbf{x}_0 = \frac{\partial}{\partial \mathbf{X}}\left\{ \int_{\mathbf{x}_0}\left(- D_T \frac{\partial \overline{\overline{c}}}{\partial \mathbf{X}} - \overline{\overline{\mathbf{v}'c'}}\right) P_L d^3\mathbf{x}_0 \right\} = 0 , \tag{204}$$

which has to be verified identically. Therefore the following constraint has to be satisfied identically at arbitrary spatial point $\mathbf{X}$

$$\int_{\mathbf{x}_0}\left(D_T\frac{\partial\overline{\overline{c}}}{\partial\mathbf{X}}+\overline{\overline{\mathbf{v}'c'}}\right)P_L(\mathbf{X},t|\mathbf{x}_0,0)d^3\mathbf{x}_0=const \tag{205}$$

The $const$ in (205) is equal to zero

$$const=0.$$

This statement can be proven by the method similar to used after (82). The integral (205) over the whole space $\mathbf{x}_0$ can be considered as a limit of the integral over bounded space region V_R, e.g. a sphere with a radius R large, as desired. Let us take the point $\mathbf{X}$ in the domain exterior to the sphere V_R. Let the distance $|\mathbf{X}-\mathbf{x}_0|$ between $\mathbf{X}$ and all points $\mathbf{x}_0$ inside of the sphere large as desired. Then the integrand of (205) tends to zero, taking into account that $P_L(\mathbf{X},t|\mathbf{x}_0,0)\to 0$ in the limit $|\mathbf{X}-\mathbf{x}_0|\to\infty$.

Let us to recapitulate the obtained result: it is proven that the definition of the mean source used in [1], (ZP.20) requires fulfilling identically the constraint (205). In other words, the assumption (201) (ZP.20) is simply hidden use of (204), (205).

It is worth noting that the RHS of Eq. (205) enters into Eq. (86), the second term in the RHS of (86), for the turbulent flux $\langle\mathbf{u}c\rangle$ derived in the section 6. Therefore the assumption (201), (ZP.20) corresponds simply to disregarding the second term in the RHS of Eq (86). Consequently, Eqs (86) and (87) become

$$\langle\mathbf{u}c\rangle=-D_T\frac{\partial\langle c\rangle}{\partial\mathbf{X}}, \tag{206}$$

$$\frac{\partial\langle\mathbf{u}c\rangle}{\partial\mathbf{X}}=-D_T\frac{\partial^2\langle c\rangle}{\partial\mathbf{X}^2}, \tag{207}$$

respectively, by virtue of (205). The expressions (206), (207) are nothing but the gradient hypothesis for the turbulent flux $\langle \mathbf{u}c \rangle$! Eq. (206) signifies that the diffusion of the expected reactive scalar $\langle c \rangle$ proceeds with the same turbulent diffusion coefficient $D_T(t)$ as of the inert scalar.

It should be stressed that the constraint (205) can not be fulfilled in the general case. There are three cases which identically satisfy (205). The first case: the chemical source does not depend explicitly on the space coordinates, and, consequently, $c(t|\mathbf{x}_0,0)$ is a deterministic function that depends only from the elapsed time t and $\mathbf{x}_0$, (20). Then

$$c' = c(t|\mathbf{x}_0,0) - \overline{\overline{c}}(t|\mathbf{X},\mathbf{x}_0,0) = c(t|\mathbf{x}_0,0) - c(t|\mathbf{x}_0,0) = 0, \frac{\partial \overline{\overline{c}}}{\partial \mathbf{X}} = 0, \quad (208)$$

by virtue of (36), (38), (62), and it is readily seen that (205) is verified. The same conclusion follows from also (90).

The second case: the advection of the inert scalar, i.e. $W(c,\mathbf{X},t) = 0$, when (18) is valid. Third case: to apply the gradient hypothesis for the expected Lagrangian-Eulerian turbulent flux, i.e.
to assume that

$$\overline{\overline{\mathbf{v}'c'}} = -D_T \frac{\partial \overline{\overline{c}}}{\partial \mathbf{X}}. \quad (209)$$

Plugging (209) into the integrand of (205) results in

$$D_T P_L \frac{\partial \overline{\overline{c}}}{\partial \mathbf{X}} - D_T P_L \frac{\partial \overline{\overline{c}}}{\partial \mathbf{X}} = 0, \quad (210)$$

which is indeed equal identically to zero.

The combined using Taylor relation (71) and the gradient hypotheses (209) for the expected Lagrangian-Eulerian turbulent flux $\overline{\overline{\mathbf{v}'c'}}$ results in

$$\left(\overline{\overline{\mathbf{v}}}\,\overline{\overline{c}} + \overline{\overline{\mathbf{v}'c'}}\right)P_L = \overline{\overline{\mathbf{v}c}}P_L = -D_T \frac{\partial P_L}{\partial \mathbf{X}}\overline{\overline{c}} - D_T \frac{\partial \overline{\overline{c}}}{\partial \mathbf{X}} P_L = -D_T \frac{\partial \overline{\overline{c}} P_L}{\partial \mathbf{X}}. \qquad (211)$$

Integrating (211) over $\mathbf{x}_0$ yields (206)

$$\langle \mathbf{u}c \rangle = -D_T \frac{\partial \langle c \rangle}{\partial \mathbf{X}}.$$

It should be stressed that the gradient hypothesis (209) for the expected Lagrangian-Eulerian turbulent flux $\overline{\overline{\mathbf{v}'c'}}$ does not follow from the Taylor theory. It is an additional assumption to the Taylor theory [8]. In the general case, including premixed combustion in the BML limit, the integral term in (76) is not equal to zero, and the turbulent advection cannot be reduced only to diffusion term (first term in the RHS of Eq. (76)).

To conclude, the turbulent flux is presented by the relation (86)

$$\langle \mathbf{u}c \rangle = - D_T \frac{\partial \langle c \rangle}{\partial \mathbf{X}} + \int_{\mathbf{x}_0} \left(D_T \frac{\partial \overline{\overline{c}}}{\partial \mathbf{X}} + \overline{\overline{\mathbf{v}'c'}} \right) P_L(\mathbf{X},t|\mathbf{x}_0,0)\, d^3\mathbf{x}_0 ,$$

where the integral term in the RHS is not equal to zero in the general case. We can introduce formally the turbulent diffusion coefficient of the reactive scalar c by the relation

$$\langle \mathbf{u}c \rangle = - D_T \frac{\partial \langle c \rangle}{\partial \mathbf{X}} + \int_{\mathbf{x}_0} \left(D_T \frac{\partial \overline{\overline{c}}}{\partial \mathbf{X}} + \overline{\overline{\mathbf{v}'c'}} \right) P_L(\mathbf{X},t|\mathbf{x}_0,0)\, d^3\mathbf{x}_0 = -D_T^c \frac{\partial^2 \langle c \rangle}{\partial \mathbf{X}^2}. \qquad (212)$$

The turbulent diffusion coefficient of the reactive scalar $c(\mathbf{X},t)$ is not equal to Taylor turbulent diffusion coefficient $D_T(t)$

$$D_T^c \neq D_T,$$

contrary to the claim of [1].

Now we compare shortly the results obtained in the section 7 with the results [1], section C. Section C of [1] is devoted to the analysis of the premixed combustion in the BML approximation. Equation from [1] (one line above of (ZP.26)) in the limit $t \to \infty$

$$\frac{\partial}{\partial x}\int_{t_0}^{\infty}\Psi(t'|x,x_0,t_0)dt' = \frac{\partial^2}{\partial x^2}\int_{t_0}^{+\infty}\Psi(t'|x,x_0,t_0)dt' = 0$$

is incorrect, because of the subtleties the double limit $t \to \infty$, $x_0 \to \infty$. The result of such a limit depends on the order, i.e. either $x_0 \to \infty$ then $t \to \infty$, or $t \to \infty$ then $x_0 \to \infty$. This is illustrated in the section 7, Eqs (107) and (108) for the mean Lagrangian progress variable $\bar{c}(t|\mathbf{x}_0,0)$, (126) and (127) for the mean mixed Lagrangian-Eulerian progress variable $\bar{\bar{c}}(t|\mathbf{X},\mathbf{x}_0,0)$.

CONCLUSION

The systematic approach to cope with the Lagrangian formulation to treating the turbulent reacting flows is developed The case a passive reactive scalar is considered, i.e. it is assumed that the turbulence is not affected by the combustion. The relations connecting: 1) the instantaneous and expected Lagrangian and mixed Lagrangian-Eulerian quantities, 2) the instantaneous Eulerian and Lagrangian characteristics, and expected Eulerian and mixed Lagrangian-Eulerian quantities are derived. The equations for the expected mixed Lagrangian-Eulerian quantities are derived. The Taylor theory of turbulent diffusion is applied to transform the derived equation for the expected mixed Lagrangian-Eulerian reactive scalar. The derived equations are employed to the consideration of the turbulent premixed flame. The discussion addresses an asymptotic case of an infinitely thin flame front (flamelet) that separates the reactants and products in a turbulent flow and self-propagates at a speed S_L with respect to the products. The flame front hits the fluid particles of the reactants and they turn into the particles of the products. An exact analytical expression for the density probability function of the flame front hitting the unburned gas particle is derived with the aid of the well-known G-equation (the level set method for tracking interfaces). An approximate equation connecting expected Eulerian progress variable and the Lagrangian transition probability density function is derived. Then an approximate equation for the mean progress variable is derived. This equation coincides with well-known in the theory of the premixed combustion Prudnikov equation. The principal results of the derivation are compared to what was previously published.

ACKNOWLEDGMENTS

The work was supported by ONERA. The author is grateful to V. Zimont for helpful discussion.

ANNEX A. THE CASE OF STOCHASTIC $c_0(\mathbf{X})$

We consider two cases of the initial value problem (IVP): 1) the initial random field $c_0(\mathbf{X})$ is stochastically independent on the velocity field, 2) the initial random scalar field is given at instant t_1. The field $c(\mathbf{X}_1, t_1)$ is the solution of the first IVP, therefore it is stochastically dependent on the velocity field $\mathbf{u}(\mathbf{X}, t)$.

Let us consider the first IVP, assuming, without loss of generality, that chemical source is absent, i.e. $c(\mathbf{X}, t)$ is an inert scalar. Eq. (39) becomes

$$c(\mathbf{X},t) = \int_{\mathbf{x}_0} c_0(\mathbf{x}_0)\delta(\mathbf{x}_0(\mathbf{X},t) - \mathbf{x}_0)d^3\mathbf{x}_0 \,, \tag{A1}$$

$$c(t = 0|\mathbf{x}_0, 0) = c_0(\mathbf{x}_0),$$

by virtue of (18).

Ensemble averaging (A1) gives

$$\langle c(\mathbf{X},t)\rangle = \int_{\mathbf{x}_0} \langle c_0(\mathbf{x}_0)\delta(\mathbf{x}_0(\mathbf{X},t) - \mathbf{x}_0)\rangle d^3\mathbf{x}_0 = \int_{\mathbf{x}_0} \langle c_0(\mathbf{x}_0)\rangle P_L(\mathbf{X},t|\mathbf{x}_0,0)d^3\mathbf{x}_0 \,, \tag{A2}$$

$$P_L(\mathbf{X},t|\mathbf{x}_0,0) = \langle \delta(\mathbf{x}_0(\mathbf{X},t) - \mathbf{x}_0) \rangle .$$

The correlation

$$\langle c_0(\mathbf{x}_0)\delta(\mathbf{x}_0(\mathbf{X},t) - \mathbf{x}_0) \rangle = \langle c_0(\mathbf{x}_0) \rangle \langle \delta(\mathbf{x}_0(\mathbf{X},t) - \mathbf{x}_0) \rangle \quad \text{(A.3)}$$

was splitted in (A2) by virtue of the assumption of stochastic independence of the initial field $c_0(\mathbf{X})$ and the velocity field $\mathbf{u}(\mathbf{X}, t = 0)$.

Taking time derivative (A2) results in

$$\frac{\partial \langle c \rangle}{\partial t} = \int_{\mathbf{x}_0} \langle c_0(\mathbf{x}_0) \rangle \frac{\partial P_L(\mathbf{X}, t|\mathbf{x}_0, 0)}{\partial t} d^3\mathbf{x}_0 . \quad \text{(A4)}$$

Applying the Taylor theory (71-(73)

$$\frac{\partial P_L(\mathbf{X},t|\mathbf{x}_0,0)}{\partial t} = D_T(t) \frac{\partial^2 P_L(\mathbf{X},t|\mathbf{x}_0,0)}{\partial \mathbf{X}^2} ,$$

(A4) transforms to

$$\frac{\partial \langle c \rangle}{\partial t} = D_T(t) \frac{\partial^2 \langle c \rangle}{\partial \mathbf{X}^2} , \quad \text{(A.5)}$$

i.e. the mean scalar $\langle c \rangle$ diffuses with the turbulent diffusion coefficient $D_T(t)$, defined by Taylor theory, Eq. (73).

Let us consider the second IVP, formulating the following problem: a) the initial condition for the scalar c is taken at the instant $t_1 > 0$, b) this initial condition $c(\mathbf{X}_1, t_1)$, in its turn, is the solution of Eq. (17) taken at the instant t_1.

Then, similar to (A.1), we can write

$$c(\mathbf{X},t) = \int_{\mathbf{X}_1} c(\mathbf{X}_1,t_1)\delta(\mathbf{X}_1(\mathbf{X},t) - \mathbf{X}_1)d^3\mathbf{X}_1 \ , t_1 > 0, \tag{A6}$$

where

$$c(\mathbf{X}_1,t_1) = \int_{\mathbf{x}_0} c_0(\mathbf{x}_0)\delta(\mathbf{x}_0(\mathbf{X}_1,t_1) - \mathbf{x}_0)d^3\mathbf{x}_0 \ . \tag{A7}$$

Ensemble averaging (A6) yields

$$\langle c(\mathbf{X},t)\rangle = \int_{\mathbf{X}_1} \langle c(\mathbf{X}_1,t_1)\delta(\mathbf{X}_1(\mathbf{X},t) - \mathbf{X}_1)\rangle d^3\mathbf{X}_1 \ . \tag{A8}$$

The correlation $\langle c(\mathbf{X}_1,t_1)\delta(\mathbf{X}_1(\mathbf{X},t) - \mathbf{X}_1)\rangle$ in (A8) cannot be splitted, i.e.

$$\langle c(\mathbf{X}_1,t_1)\delta(\mathbf{X}_1(\mathbf{X},t) - \mathbf{X}_1)\rangle \neq \langle c(\mathbf{X}_1,t_1)\rangle\langle \delta(\mathbf{X}_1(\mathbf{X},t) - \mathbf{X}_1)\rangle = \langle c(\mathbf{X}_1,t_1)\rangle P_L(\mathbf{X},t|\mathbf{X}_1,t_1) \tag{A9}$$

because of the stochastic dependence of $c(\mathbf{X}_1,t_1)$ and $\delta(\mathbf{X}_1(\mathbf{X},t) - \mathbf{X}_1)$ on the velocity field $\mathbf{u}(\mathbf{X},t)$. Therefore it would be incorrect to write that

$$\langle c(\mathbf{X},t)\rangle = \int_{\mathbf{X}_1} \langle c(\mathbf{X}_1,t_1)\rangle\langle \delta(\mathbf{X}_1(\mathbf{X},t) - \mathbf{X}_1)\rangle d^3\mathbf{X}_1 = \int_{\mathbf{X}_1} \langle c(\mathbf{X}_1,t_1)\rangle P_L(\mathbf{X},t|\mathbf{X}_1,t_1)d^3\mathbf{X}_1 \ , \tag{A11}$$

and then to conclude from (A11) that the turbulent diffusion coefficient for the second IVP is equal to $D_T(t - t_1)$, and , in particular that it is equal to zero at instant t_1 .

The correct result is obtained, if (A7) is plugged into (A6). (A.6) becomes

$$c(\mathbf{X},t) = \int_{\mathbf{X}_1}\int_{\mathbf{x}_0} c_0(\mathbf{x}_0)\delta(\mathbf{X}_1(\mathbf{X},t_1) - \mathbf{X}_1)\delta(\mathbf{X}(t|\mathbf{x}_0,0) - \mathbf{X})d^3\mathbf{x}_0 d^3\mathbf{X}_1 \ . \tag{A12}$$

Eq. (A12) can be transformed to

$$c(\mathbf{X},t)=\int\limits_{\mathbf{X}_1}\int\limits_{\mathbf{x}_0} c_0(\mathbf{x}_0)\delta(\mathbf{X}(t|\mathbf{x}_0,0)-\mathbf{X})d^3\mathbf{x}_0 \,. \tag{A13}$$

taking into account the properties of the delta-function.

The relation (A13) is nothing but (A1). Therefore the turbulent diffusion coefficient for the second case is equal to $D_T(t)$. This conclusion is evident, because, the second IVP is, in fact the first one, divided in two steps.

Let us change slightly the formulation of the second IVP. Namely, let us introduce the following source into Eq. (17)

$$\frac{\partial c(t|\mathbf{x}_0,0)}{\partial t}=\delta(t-t_1)c_0(\mathbf{x}_0)\,,\, c(t=0|\mathbf{x}_0)=0\,, \tag{A14}$$

where the random scalar field $c_0(\mathbf{x}_0)$, as in Eq. (A1), is stochastically independent on the velocity field.

The solution of IVP (A14) reads

$$c(t|\mathbf{x}_0,0)=H(t-t_1)c_0(\mathbf{x}_0)\,. \tag{A15}$$

Then the relation (39) yields

$$c(\mathbf{X},t)=\int\limits_{\mathbf{X}_1}\int\limits_{\mathbf{x}_0} c_0(\mathbf{x}_0)\delta(\mathbf{X}(t|\mathbf{x}_0,0)-\mathbf{X})d^3\mathbf{x}_0=H(t-t_1)\int\limits_{\mathbf{X}_1}\int\limits_{\mathbf{x}_0} c_0(\mathbf{x}_0)\delta(\mathbf{X}(t|\mathbf{x}_0,0)-\mathbf{X})d^3\mathbf{x}_0 \tag{A16}$$

Averaging (A16) yields

$$\langle c(\mathbf{X},t)\rangle = H(t-t_1)\int\limits_{\mathbf{x}_0}\langle c_0(\mathbf{x}_0)\rangle P_L(\mathbf{X},t|\mathbf{x}_0,0)d^3\mathbf{x}_0\,, \tag{A17}$$

It is worth noting that Eq. (A17) is the particular case analysed in [11].

Let us derive Eq. (A17) from the general equations derived in the section 7. To this end, we observe, that Eq. (A14) describes an evolution of the scalar in a Lagrangian particle, which starts at instant $t=0$ from the spatial point $\mathbf{x}_0$ with initial scalar value $c(t=0|\mathbf{x}_0)=0$, and then at instant $t=t_1$ the scalar value at the particle jumps to $c_0(\mathbf{x}_0)$. Such a formulation of the problem corresponds formally to the premixed flame considered in the section 7. The expression for the conditional Lagrangian probability density function of hitting time $\Phi(t|\mathbf{x}_0,0)$, (102) becomes the delta-function

$$\Phi(t|\mathbf{x}_0,0) = \delta(t - t_1). \tag{A18}$$

Therefore the conditional probability density function $P(t;\mathbf{X},t|\mathbf{x}_0,0)$, Eq. (111) reads

$$P(t;\mathbf{X},t|\mathbf{x}_0,0) = \delta(t - t_1)\langle\delta(\mathbf{X}(t|\mathbf{x}_0,0) - \mathbf{X})\rangle = \delta(t - t_1)P_L(\mathbf{X},t|\mathbf{x}_0,0). \tag{A19}$$

Then Eq. (110) yields

$$\langle c(\mathbf{X},t)\rangle = \int_0^t \int_{\mathbf{x}_0} \langle c_0(\mathbf{x}_0)\rangle P(\tau;\mathbf{X},t|\mathbf{x}_0,0) d^3\mathbf{x}_0 d\tau. \tag{A20}$$

Eq. (A20), after integration over the time variable, becomes

$$\langle c(\mathbf{X},t)\rangle = H(t - t_1)\int_{\mathbf{x}_0} \langle c_0(\mathbf{x}_0)\rangle \langle\delta(\mathbf{X}(t|\mathbf{x}_0,0) - \mathbf{X})\rangle d^3\mathbf{x}_0 , \tag{A21}$$

by virtue of (A19). Therefore

$$\langle c(\mathbf{X},t)\rangle = H(t - t_1)\int_{\mathbf{x}_0} \langle c_0(\mathbf{x}_0)\rangle P_L(\mathbf{X},t|\mathbf{x}_0,0) d^3\mathbf{x}_0 , \tag{A22}$$

i.e. Eq. (A17). Eq. (A22) coincides with (A.2) at $t_1 > 0$. Therefore the mean scalar $\langle c(\mathbf{X},t)\rangle$ in the reformulated problem diffuse with the turbulent diffusion coefficient equal to $D_T(t)$ at $t > t_1$, but not with $D_T(t - t_1)$.

ANNEX B. THE SOURCE DOES NOT DEPEND ON THE REACTIVE SCALAR

If the source $W(c, \mathbf{X}, t)$ does not depend on c, $W(c, \mathbf{X}, t) = q(\mathbf{X}, t)$ then the IBVP (1), (2) reads

$$\frac{\partial c}{\partial t} + \mathbf{u} \cdot \frac{\partial c}{\partial \mathbf{X}} = q(\mathbf{X}, t), \tag{B1}$$

$$c(\mathbf{X}, t = 0) = c_0(\mathbf{X}), \tag{B2}$$

where $q(\mathbf{X}, t)$ is a deterministic function. It will be assumed, without loss of generality, that

$$c_0(\mathbf{X}) = 0 .$$

Then (17) transforms to

$$\frac{\partial c(t|\mathbf{x}_0, 0)}{\partial t} = q(\mathbf{X}(t|\mathbf{x}_0, 0), t) , c_0(\mathbf{x}_0) = 0 , t = 0 . \tag{B3}$$

$$c(t = 0|\mathbf{x}_0, 0) = c_0(\mathbf{x}_0) = 0 . \tag{B4}$$

The solution of (B3), (B4) is a functional of the particle trajectory $\mathbf{X}(t|\mathbf{x}_0, 0)$ and reads

$$c(t|\mathbf{x}_0, 0) = \int_0^t q(\mathbf{X}(\tau|\mathbf{x}_0), \tau) d\tau . \tag{B5}$$

Let us replace (B5) by the equivalent expression, presenting $q(\mathbf{X}(\tau|\mathbf{x}_0), \tau)$ in the form of the following integral

$$q(\mathbf{X}(\tau|\mathbf{x}_0), \tau) = \int_{\mathbf{X}_1} q(\mathbf{X}_1, \tau) \delta(\mathbf{X}(\tau|\mathbf{x}_0, 0) - \mathbf{X}_1) d^3\mathbf{X}_1 . \tag{B6}$$

(B6) is valid by virtue of the properties of the delta-function. (B5) becomes

$$c\left(t\middle|\mathbf{x}_0,0;\left[\mathbf{X}(\tau|\mathbf{x}_0)\right]\right)=\int_0^t\int_{\mathbf{X}_1} q\left(\mathbf{X}_1,t_1\right)\delta\left(\mathbf{X}(t_1|\mathbf{x}_0,0)-\mathbf{X}_1\right)d^3\mathbf{X}_1 dt_1\,, \quad \text{(B7)}$$

after changing the time variable τ by t_1.

Inserting (B5), (B7) into Eq. (39) yields

$$c(\mathbf{X},t)=\int_0^t\int_{\mathbf{x}_0} q\left(\mathbf{X}(\tau|\mathbf{x}_0,0),\tau\right)\delta(\mathbf{X}(t|\mathbf{x}_0,0)-\mathbf{X})d^3\mathbf{x}_0 d\tau\,, \quad \text{(B8)}$$

$$c(\mathbf{X},t)=\int_0^t\int_{\mathbf{x}_0}\int_{\mathbf{X}_1} q\left(\mathbf{X}_1,t_1\right)\delta(\mathbf{X}(t_1|\mathbf{x}_0,0)-\mathbf{X}_1)\delta(\mathbf{X}(t|\mathbf{x}_0,0)-\mathbf{X})d^3\mathbf{x}_0 d^3\mathbf{X}_1 dt_1\,, \quad \text{(B9)}$$

respectively. Using the fact that the Jacobian is equal to 1, Eq. (12), Eq. (B9) becomes

$$c(\mathbf{X},t)=\int_0^t\int_{\mathbf{x}_0}\int_{\mathbf{X}_1} q\left(\mathbf{X}_1,t_1\right)\delta(\mathbf{X}(t|\mathbf{x}_0,0)-\mathbf{X})\delta(\mathbf{x}_0(\mathbf{X}_1,t_1)-\mathbf{x}_0)d^3\mathbf{x}_0 d^3\mathbf{X}_1 dt_1\,. \quad \text{(B10)}$$

Eq. (B10) can be rewritten as

$$c(\mathbf{X},t)=\int_0^t\int_{\mathbf{X}_1} q\left(\mathbf{X}_1,t_1\right)\delta(\mathbf{X}(t|\mathbf{X}_1,t_1)-\mathbf{X})d^3\mathbf{X}_1 dt_1\,, \quad \text{(B11)}$$

by virtue of the properties of the delta function.

Averaging Eq. (B10) yields

$$\left\langle c(\mathbf{X},t)\right\rangle=\int_0^t\int_{\mathbf{x}_0}\int_{\mathbf{X}_1} q\left(\mathbf{X}_1,t_1\right)P_L(\mathbf{X},t;\mathbf{X}_1,t_1|\mathbf{x}_0,0)d^3\mathbf{x}_0 d^3\mathbf{X}_1 dt_1\,, \quad \text{(B12)}$$

where

$$P_L(\mathbf{X},t;\mathbf{X}_1,t_1|\mathbf{x}_0,0)=\left\langle\delta(\mathbf{X}(t|\mathbf{x}_0,0)-\mathbf{X})\delta(\mathbf{X}(t_1|\mathbf{x}_0,0)-\mathbf{X}_1)\right\rangle. \quad \text{(B13)}$$

Eq. (B12) transforms to

$$\langle c(\mathbf{X},t)\rangle=\int_0^t\int_{\mathbf{x}_0}\int_{\mathbf{X}_1}q(\mathbf{X}_1,t_1)P_L(\mathbf{X},t|\mathbf{X}_1,t_1;\mathbf{x}_0,0)P_L(\mathbf{X}_1,t_1|\mathbf{x}_0,0)d^3\mathbf{x}_0d^3\mathbf{X}_1dt_1, \quad \text{(B14)}$$

using Bayes theorem, and then to

$$\langle c(\mathbf{X},t)\rangle=\int_0^t\int_{\mathbf{x}_0}\int_{\mathbf{X}_1}q(\mathbf{X}_1,t_1)P_L(\mathbf{X},t|\mathbf{X}_1,t_1;\mathbf{x}_0,0)P_L(\mathbf{x}_0,0|\mathbf{X}_1,t_1)d^3\mathbf{x}_0d^3\mathbf{X}_1dt_1, \quad \text{(B15)}$$

taking into account that according to (12)

$$P_L(\mathbf{X}_1,t_1|\mathbf{x}_0,0)=P_L(\mathbf{x}_0,0|\mathbf{X}_1,t_1).$$

Carring out the integration over $\mathbf{x}_0$ in (B15) results in

$$\langle c(\mathbf{X},t)\rangle=\int_0^t\int_{\mathbf{X}_1}q(\mathbf{X}_1,t_1)P_L(\mathbf{X},t|\mathbf{X}_1,t_1)d^3\mathbf{X}_1dt_1. \quad \text{(B16)}$$

It is worth noting that Eq. (B16) follows readily averaging (B12). Taking time derivative (B16) yields

$$\frac{\partial\langle c\rangle}{\partial t}=\int_{\mathbf{X}_1}q(\mathbf{X}_1,t)P_L(\mathbf{X},t|\mathbf{X}_1,t)d^3\mathbf{X}_1+\int_0^t\int_{\mathbf{X}_1}q(\mathbf{X}_1,t_1)\frac{\partial P_L(\mathbf{X},t|\mathbf{X}_1,t_1)}{\partial t}d^3\mathbf{X}_1dt_1, \quad \text{(B17)}$$

$$\frac{\partial\langle c\rangle}{\partial t}=\int_{\mathbf{X}_1}q(\mathbf{X}_1,t)\delta(\mathbf{X}-\mathbf{X}_1)d^3\mathbf{X}_1+\int_0^t\int_{\mathbf{X}_1}q(\mathbf{X}_1,t_1)\frac{\partial P_L(\mathbf{X},t|\mathbf{X}_1,t_1)}{\partial t}d^3\mathbf{X}_1dt_1, \quad \text{(B18)}$$

taking into account that

$$P_L(\mathbf{X},t|\mathbf{X}_1,t)=\delta(\mathbf{X}-\mathbf{X}_1). \tag{B19}$$

Finally Eq. (B18) becomes

$$\frac{\partial\langle c\rangle}{\partial t}=q(\mathbf{X},t)+\int_0^t\int_{\mathbf{X}_1}q(\mathbf{X}_1,t_1)\frac{\partial P_L(\mathbf{X},t|\mathbf{X}_1,t_1)}{\partial t}d^3\mathbf{X}_1dt_1, \tag{B20}$$

by virtue of the properties of the delta-function.

Applying the Taylor theory [8], (71)-(73)

$$\frac{\partial P_L(\mathbf{X},t|\mathbf{X}_1,t_1)}{\partial t}=D_T(t-t_1)\frac{\partial^2 P_L(\mathbf{X},t|\mathbf{X}_1,t_1)}{\partial\mathbf{X}^2}=D_T(t-t_1)P_L(\mathbf{X},t|\mathbf{X}_1,t_1), \tag{B21}$$

$$D_T(t-t_1)=\int_{t_1}^t B_L(\tau)d\tau\,,\, B_L(\tau)=\frac{1}{3}\langle\mathbf{v}(t_1+\tau|\mathbf{x}_0,0)\cdot\mathbf{v}(t_1|\mathbf{x}_0,0)\rangle, \tag{B22}$$

Eq. (B20) becomes

$$\frac{\partial\langle c\rangle}{\partial t}=q(\mathbf{X},t)+\int_0^t\int_{\mathbf{X}_1}D_T(t-t_1)q(\mathbf{X}_1,t_1)\frac{\partial^2 P_L(\mathbf{X},t|\mathbf{X}_1,t_1)}{\partial\mathbf{X}^2}d^3\mathbf{X}_1dt_1. \tag{B23}$$

The integral term in (B23) with the sign minus is the divergence of the turbulent flux

$$\frac{\partial\langle\mathbf{u}c\rangle}{\partial\mathbf{X}}=-\int_0^t\int_{\mathbf{X}_1}D_T(t-t_1)q(\mathbf{X}_1,t_1)\frac{\partial^2 P_L(\mathbf{X},t|\mathbf{X}_1,t_1)}{\partial\mathbf{X}^2}d^3\mathbf{X}_1dt_1, \tag{B24}$$

as follows from the comparison with Eq. (6).

Eq. (B23) can be rewritten as

$$\frac{\partial\langle c\rangle}{\partial t}=-\frac{\partial\langle\mathbf{u}c\rangle}{\partial\mathbf{X}}+q(\mathbf{X},t), \tag{B25}$$

$$\langle \mathbf{u}c \rangle = -\frac{\partial}{\partial \mathbf{X}} \int_0^t \int_{\mathbf{X}_1} D_T(t-t_1) q(\mathbf{X}_1, t_1) P_L(\mathbf{X}, t | \mathbf{X}_1, t_1) d^3\mathbf{X}_1 dt_1 . \tag{B26}$$

It should be noted that the turbulent flux, (B26) cannot be presented as a diffusion term

$$\langle \mathbf{u}c \rangle = -\frac{\partial}{\partial \mathbf{X}} \int_0^t \int_{\mathbf{X}_1} D_T(t-t_1) q(\mathbf{X}_1, t_1) P_L(\mathbf{X}, t | \mathbf{X}_1, t_1) d^3\mathbf{X}_1 dt_1 \neq D_T(t) \frac{\partial \langle c \rangle}{\partial \mathbf{X}} \tag{B27}$$

even if $q(\mathbf{X}_1, t_1) = const$.

Let us consider a particular case

$$q(\mathbf{X}, t) = \sum_{k=1}^{k=N} q_k \delta(\mathbf{X} - \mathbf{X}_k) \delta(t - t_k). \tag{B28}$$

Eqs (B16), (B23) become

$$\langle c(\mathbf{X}, t) \rangle = \sum_{k=1}^{k=N} q_k P_L(\mathbf{X}, t | \mathbf{X}_k, t_k) , \tag{B29}$$

$$\frac{\partial \langle c \rangle}{\partial t} = q(\mathbf{X}, t) + \sum_{k=1}^{k=N} q_k \int_0^t \int_{\mathbf{X}_1} D_T(t-t_k) \delta(\mathbf{R} - \mathbf{X}_k) \delta(\tau - t_k) \frac{\partial^2 P_L(\mathbf{X}, t | \mathbf{R}, \tau)}{\partial \mathbf{X}^2} d^3\mathbf{R} d\tau \tag{B30}$$

If $t - t_k >> T_L$ for all $1 \le k \le N$, then (B30) transforms to

$$\frac{\partial \langle c \rangle}{\partial t} = q(\mathbf{X}, t) + D_T(\infty) \sum_{k=1}^{k=N} q_k \int_0^t \int_{\mathbf{X}_1} \delta(\mathbf{R} - \mathbf{X}_k) \delta(\tau - t_k) \frac{\partial^2 P_L(\mathbf{X}, t | \mathbf{R}, \tau)}{\partial \mathbf{X}^2} d^3\mathbf{R} d\tau , \tag{B31}$$

$$\frac{\partial \langle c \rangle}{\partial t} = q(\mathbf{X}, t) + D_T(\infty) \sum_{k=1}^{k=N} q_k \frac{\partial^2 P_L(\mathbf{X}, t | \mathbf{X}_k, t_k)}{\partial \mathbf{X}^2} , \tag{B32}$$

$$\frac{\partial\langle c\rangle}{\partial t} = q(\mathbf{X},t) + D_T(\infty)\frac{\partial^2\langle c\rangle}{\partial \mathbf{X}^2}. \tag{B33}$$

The expected progress variable $\langle c\rangle$, as follows from (B31), diffuses, at $t - t_k >> T_L$ for all $1 \leq k \leq N$, with the same asymptotic turbulent diffusion coefficient as an expected inert scalar. Asymptotic value of the turbulent diffusion coefficient is given by the relation (141)

$$D_T(\infty) = \langle u^2\rangle T_L.$$

ANNEX C. DIFFUSION OF THE PRODUCTS AFTER THE FLAME EXTINCTION

Let $t_{ext} > 0$ the moment of the flame extinction. Then the equations for the progress variable (94), (95) become

$$\frac{\partial c(t|\mathbf{x}_0,0)}{\partial t} = 0, \tag{C1}$$

for the fluid particles in the products (initially placed in the half-space $x_0 < 0$, where $c(t = 0|\mathbf{x}_0,0) = 1$), and

$$\frac{\partial c(t|\mathbf{x}_0,0)}{\partial t} = H(t_{ext} - t)\delta(t - t_*(\mathbf{x}_0)), \tag{C2}$$

for the fluid particles in the reactants (initially placed in the half-space $x_0 > 0$, where $c(t = 0|\mathbf{x}_0,0) = 0$).

$t_*(\mathbf{x}_0)$ stands for the time when the flame front hits the fluid particle in the reactants. Eqs (C1), (C2) can be put together

$$\frac{\partial c\left(t\middle|\mathbf{x}_0,0\right)}{\partial t}=W\left(t,\mathbf{x}_0\right), \tag{C3}$$

where

$$W\left(t,\mathbf{x}_0\right)=H(x_0)\delta\left(t-t_*\left(\mathbf{x}_0\right)\right)H(t_m-t). \tag{C4}$$

Eqs. (115) and (134) become

$$\left\langle c(\mathbf{X},t)\right\rangle=\begin{cases}\int\limits_{-\infty}^{0}P_L(X,t|x_0,0)dx_0+\int\limits_{0}^{t}\int\limits_{0}^{\infty}P(\tau;X,t|x_0,0)d\tau dx_0, & t<t_{ext}\\ \int\limits_{-\infty}^{0}P_L(X,t|x_0,0)dx_0+\int\limits_{0}^{t_{ext}}\int\limits_{0}^{\infty}P(\tau;X,t|x_0,0)d\tau dx_0, & t>t_{ext}\end{cases}, \tag{C5}$$

$$\left\langle W(\mathbf{X},t)\right\rangle=\begin{cases}\int\limits_{0}^{\infty}P(t;X,t|x_0,0)dx_0, & t<t_{ext}\\ 0, & t>t_{ext}\end{cases}, \tag{C6}$$

where $P(\tau;X,t|x_0)$ is defined by the relation (111).

Differentiating (C5) with respect to time yields

$$\frac{\partial\langle c\rangle}{\partial t}=\int\limits_{-\infty}^{0}\frac{\partial P_L(X,t|x_0,0)}{\partial t}dx_0+\int\limits_{0}^{t}\int\limits_{0}^{\infty}\frac{\partial P(\tau;X,t|x_0,0)}{\partial t}d\tau dx_0+\langle W\rangle(X,t)\,,\ t<t_{ext}, \tag{C7}$$

$$\langle W\rangle(X,t)=\int\limits_{0}^{\infty}P(t;X,t|x_0,0)dx_0=\int\limits_{0}^{\infty}P(t|X,t;x_0,0)P_L(X,t|x_0,0)dx_0\,,$$

$$\frac{\partial\langle c\rangle}{\partial t}=\int\limits_{-\infty}^{0}\frac{\partial P_L(X,t|x_0,0)}{\partial t}dx_0+\int\limits_{0}^{t_{t_{ext}}}\int\limits_{0}^{\infty}\frac{\partial P(\tau;X,t|x_0,0)}{\partial t}d\tau dx_0\,,\ t>t_{ext}, \tag{C8}$$

taking into account Eq. (C6).

Eqs (C7), (C8) become

$$\frac{\partial\langle c\rangle}{\partial t}=D_T(t)\frac{\partial^2}{\partial X^2}\int_{-\infty}^{0}P_L(X,t|x_0,0)dx_0+\int_0^t\int_0^\infty\frac{\partial P(\tau;X,t|x_0,0)}{\partial t}d\tau dx_0+\langle W\rangle(X,t)\,,t<t_{ext},\quad \text{(C9)}$$

$$\frac{\partial\langle c\rangle}{\partial t}=D_T(t)\frac{\partial^2}{\partial X^2}\int_{-\infty}^{0}P_L(X,t|x_0,0)dx_0+\int_0^{t_{ext}}\int_0^\infty\frac{\partial P(\tau;X,t|x_0,0)}{\partial t}d\tau dx_0\,,t>t_{ext},\quad \text{(C10)}$$

taking into account that $P_L(\mathbf{X},t|\mathbf{x}_0,0)$ obeys the Eq. (73)

$$\frac{\partial P_L(X,t|x_0,0)}{\partial t}=D_T(t)\frac{\partial^2 P_L(X,t|x_0,0)}{\partial X^2},$$

(C5) yields for the integral in the first term of the RHS of (C8)

$$\int_{-\infty}^{0}P_L(X,t|x_0,0)dx_0=\langle c(\mathbf{X},t)\rangle-\int_0^t\int_0^\infty P(\tau;X,t|x_0,0)d\tau dx_0\,,t<t_{ext},\quad \text{(C11)}$$

$$\int_{-\infty}^{0}P_L(X,t|x_0,0)dx_0=\langle c(\mathbf{X},t)\rangle-\int_0^{t_{ext}}\int_0^\infty P(\tau;X,t|x_0,0)d\tau dx_0\,.t>t_{ext}\quad \text{(C12)}$$

Plugging (C12) into (C10) yields

$$\frac{\partial\langle c\rangle}{\partial t}=D_T(t)\frac{\partial^2\langle c\rangle}{\partial X^2}+\int_0^{t_{ext}}\int_0^\infty\left(\frac{\partial P(\tau;X,t|x_0,0)}{\partial t}-D_T(t)\frac{\partial^2 P(\tau;X,t|x_0,0)}{\partial X^2}\right)d\tau dx_0\,,t>t_{ext}\,.\quad \text{(C13)}$$

The RHS of (C13) is the turbulent transport term in the equation for the expected scalar

$$\frac{\partial\langle c\rangle}{\partial t}+\frac{\partial\langle uc\rangle}{\partial X}=0\,,t>t_{ext}\,,\quad \text{(C14)}$$

$$\frac{\partial\langle uc\rangle}{\partial X} = -D_T(t)\frac{\partial^2\langle c\rangle}{\partial X^2} - \int_0^{t_{ext}}\int_0^{\infty}\left(\frac{\partial P(\tau;X,t|x_0,0)}{\partial t} - D_T(t)\frac{\partial^2 P(\tau;X,t|x_0,0)}{\partial X^2}\right)d\tau dx_0\,, t > t_{ext}, \quad \text{(C15)}$$

The integral term in the RHS of (C13), (C15) is not equal to zero identically, because

$$\frac{\partial P(\tau;X,t|x_0,0)}{\partial t} - D_T(t)\frac{\partial^2 P(\tau;X,t|x_0,0)}{\partial X^2} \neq 0\,, x_0 > 0\,. \quad \text{(C16)}$$

The discussion of the inequality (C16) is given in the section 7 after Eqs (138), (139) and is not repeated here.

Using results proven in the section 7, we conclude that after flame extinction the inequality (C16) is valid during the time interval $t - t_{ext} \leq T_L$. Therefore during this time interval the mean scalar $\langle c\rangle$ does not diffuse with the turbulent diffusion coefficient $D_T(t)$, defined by Taylor theory, Eq. (72). In other words, the solution of the following IBVP

$$\frac{\partial\langle c\rangle}{\partial t} = D_T(t)\frac{\partial^2\langle c\rangle}{\partial X^2}\,, t > t_{ext} \quad \text{(C17)}$$

$$\langle c\rangle(X,t_m) = \int_{-\infty}^{0} P_L(X,t|x_0,0)dx_0 + \int_0^{t_{ext}}\int_0^{\infty} P(\tau;X,t_m|x_0,0)d\tau dx_0\,, t = t_{ext} \quad \text{(C18)}$$

does not coincides with (C5) at $t > t_{ext}$, $t - t_{ext} \leq T_L$. "Initial" condition (C18) is obtained from (C5) setting $t = t_{ext}$.

The integral term in Eqs (C13), (C15) as it is shown in the section 7, can be set approximately to zero if $t - t_{ext} >> T_L$, (125).

Therefore asymptotically, at $t - t_{ext} >> T_L$, Eqs (C13), (C15) become

$$\frac{\partial\langle c\rangle}{\partial t} \approx D_T(\infty)\frac{\partial^2\langle c\rangle}{\partial X^2}\,, \quad \text{(C19)}$$

$$\frac{\partial \langle uc \rangle}{\partial X} \approx -D_T(\infty)\frac{\partial^2 \langle c \rangle}{\partial X^2}. \tag{C20}$$

where the asymptotic value of the turbulent diffusion coefficient is given by the relation (141)

$$D_T(\infty) = \langle u^2 \rangle T_L.$$

The mean progress variable $\langle c \rangle$, as follows from (C19), (C20), diffuses, at $t - t_{ext} >> T_L$, with the same asymptotic turbulent diffusion coefficient as an expected inert scalar. This result is expected from the physical reasoning. Indeed, after some relaxation time, which is of the order of T_L, the "initial" condition (C18) will be forgotten, and the mean scalar after the flame extinction will diffuse as an inert scalar.

ANNEX D. DERIVATION OF EQ. (191)

Eq. (183) can be obtained also neglecting stochastic character of a hitting time $t_*(x_0)$. Let us assume that the hitting time $t_*(x_0)$ is found from the following deterministic equation

$$x_0 = \int_0^{t_*(x_0)} u_T(t)dt = \overline{X}_f(t_*(x_0)), \tag{D1}$$

that means that $t_*(x_0)$ is equal to a transit time of the mean flame to the fluid particle in the reactants with the mean longitudinal coordinate x_0. The fine-grained probability density function of hitting time, as follows from (D1), is deterministic function

$$\delta(t - t_*(x_0)) = \delta(\overline{X}_f(t) - x_0)u_T(t) \tag{D2}$$

Therefore Eqs (101), (111), (120) become

$$\Phi(t|x_0,0) = \delta(\overline{X}_f(t) - x_0)u_T(t), \; x_0 > 0, \tag{D3}$$

$$P(\tau; X, t|x_0,0) = \delta(\overline{X}_f(\tau) - x_0)u_T(\tau)\langle\delta(X(t|x_0,0) - X)\rangle, \; x_0 > 0,$$

$$P(\tau; X, t|x_0,0) = \delta(\overline{X}_f(\tau) - x_0)u_T(\tau)P_L(X,t|x_0,0), \; x_0 > 0, \tag{D4}$$

$$P(\tau|X, t; x_0,0) = \Phi(\tau|x_0,0) = \delta(\overline{X}_f(\tau) - x_0)u_T(\tau). \tag{D5}$$

Eq. (97) with the source (98) becomes the following deterministic ODE

$$\frac{\partial c(t|x_0,0)}{\partial t} = W(t, x_0), \tag{D6}$$

$$W(t, x_0) = H(x_0)\delta(\overline{X}_f(t) - x_0)u_T(t). \tag{D7}$$

Eq. (D7) is solved with the initial condition (93). The solution of the initial value problem (D6), (D7), (93) reads

$$c(t|\mathbf{x}_0,0) = H(-x_0) + H(x_0)\int_0^t \delta(\overline{X}_f(\tau) - x_0)u_T(\tau)d\tau = H(-x_0) + H(x_0)H(\overline{X}_f(t) - x_0). \tag{D8}$$

The instantaneous progress variable $c(t|\mathbf{x}_0,0)$ is the deterministic function. Therefore Eqs (103), (119) read

$$\overline{c}(t|x_0,0) = (1 - H(x_0)) + H(x_0)H(x_0)H(\overline{X}_f(t) - x_0), \tag{D9}$$

$$\overline{\overline{c}}(t|X, x_0,0) = (1 - H(x_0)) + H(x_0)H(x_0)H(\overline{X}_f(t) - x_0), \tag{D10}$$

respectively, taking into consideration (D3), (D5).

Eq. (115) transforms to

$$\langle c\rangle(X,t) = \int_{-\infty}^{0} P_L(X,t|x_0,0)dx_0 + \int_0^t\int_0^\infty \delta(\overline{X}_f(\tau) - x_0)u_T(\tau)P_L(X,t|x_0,0)d\tau dx_0 \tag{D11}$$

$$\langle c\rangle(X,t)=\int_{-\infty}^{0}P_L(X,t|x_0,0)dx_0+\int_0^{\overline{X}_f(t)}\int_0^{\infty}\delta(\overline{X}_f(\tau)-x_0)P_L(X,t|x_0,0)d\overline{X}_f(\tau)dx_0 \quad \text{(D12)}$$

taking into account (D4). Integrating Eq. (D12) over $\overline{X}_f$ yields

$$\langle c\rangle(X,t)=\int_{-\infty}^{\overline{X}_f(t)}P_L(X,t|x_0,0)dx_0\,, \quad \text{(D13)}$$

Eq. (D13) coincides with Eq. (184), therefore the final result will be the same, Eq. (191). But the new derivation, in addition, permits to find the mean source in the Eulerian frame. Indeed, Eq. (134) becomes

$$\langle W\rangle(X,t)=\int_0^{\infty}P(t;X,t|x_0,0)dx_0=\int_0^{\infty}\delta(\overline{X}_f(t)-x_0)u_T(t)P_L(X,t|x_0,0)dx_0\,,$$

$$\langle W\rangle(X,t)=u_T(t)P_L(X,t|x_0=\overline{X}_f(t),0). \quad \text{(D14)}$$

by virtue of (D4).

Eq. (D14) transforms to

$$\langle W\rangle(X,t)=-u_T(t)\frac{\partial\langle c\rangle}{\partial X}\,, \quad \text{(D15)}$$

by virtue of the connection between the spatial derivative of the progress variable $\frac{\partial\langle c\rangle}{\partial X}$ and $P_L(X,t|\overline{X}_f(t),0)$

$$P_L(X,t|\overline{X}_f(t),0)=-\frac{\partial\langle c\rangle}{\partial X}. \quad \text{(D16)}$$

Indeed, differentiating (D13) with respect to X we get

$$\frac{\partial\langle c\rangle}{\partial X}=\frac{\partial}{\partial X}\int_{-\infty}^{\overline{X}_f(t)}P_L(X,t|x_0,0)dx_0=-\frac{\partial}{\partial x_0}\int_{-\infty}^{\overline{X}_f(t)}P_L(X,t|x_0,0)dx_0\,, \quad \text{(D17)}$$

using (187) to replace the derivative $\frac{\partial}{\partial X}$ in (D16) by the derivative $-\frac{\partial}{\partial x_0}$.

Finally (D16) becomes

$$\frac{\partial\langle c\rangle}{\partial X}=-\int_{-\infty}^{\overline{X}_f(t)}\frac{\partial}{\partial x_0}P_L(X,t|x_0,0)dx_0=-P_L(X,t|\overline{X}_f(t),0),$$

i.e. (D16).

The same critical comments, as at the end of the section (8) can be done to the derivation just presented. In particular, the expression for the mean source (D15) is an approximation. It is the consequence of the application of the deterministic function for the probability density function of hitting time (D3).

REFERENCES

[1] Zimont, V.L. and Pagnini, G. Lagrangian properties of turbulent diffusion with passive chemical reaction in the framework of the premixed combustion theory, *Phys. Fluids* 23, 035101 (2011).

[2] Pagnini, G. and Bonomi, E. *Lagrangian Formulation of Turbulent Premixed Combustion PRL* 107, 044503 (2011).

[3] Monin, A.S. and Yaglom, A.M. *Statistical Fluid Mechanics. Mechanics of Turbulence. vol 1.*, J. L. Lumley (Eds.) MIT Press (1971).

[4] Bray, K.N.C. and Moss, J.B. “A unified statistical model of the premixed turbulent flame,” *Acta Astronaut.* 4, 291, 1977.

[5] Peters, N. *Turbulent Combustion*, Cambridge University Press, Cambridge, 2000.

[6] Lipatnikov, A.N. *Fundamentals of Premixed Turbulent Combustion*, CRC Press, Boca Raton, Florida, 2012.

[7] Pope, S.B. *Turbulent Flows*, Cambridge University Press, Cambridge, 2000.

[8] Taylor, G.I. “Diffusion by continuous movements,” *Proc. London Math.Soc.* 2-20, 196, 1922.

[9] Kerstein, A.R., Ashurst, Wm.T., and Williams, F.A. Field equation for interface propagation in an unsteady homogeneous flow field, *Phys. Rev. A,* 1988, 37, 2728–2731.

[10] Prudnikov, A.G., in *Physical Basis of Processes in Combustion Chambers of Air-Breathing Engines*, edited by B.V. Raushenbakh et al. (Mashinostroenie, Moscow, (1964), page 255 (in Russian).

[11] Corrsin, S. Effect of passive chemical reaction on turbulent dispersion, *AIAA J.* 6, 1797, 1968.

In: Combustion
Editor: Joseph M. Grier

ISBN: 978-1-62948-967-4

Chapter 2

G-Equation in White Noise in a Time Turbulent Velocity Field: The Derivation of the Probability Density Function Equation

V. A. Sabelnikov*
ONERA - The French Aerospace Lab., Palaiseau, France

Abstract

A promising approach to simulating premixed turbulent combustion in gas turbine and spark ignition engines consists of tracking a flame surface in the turbulent velocity field in the framework of so called $G-$ equation, associating the evolution of the flame surface with the evolution of zero (or G_0) level set of a function $G(\mathbf{x},t)$. The goal of the present chapter is to derive the equation for the probability density function of G straightforwardly from the well-known G-equation for tracking an infinitely thin flame sheet (flamelet) in a white-noise in time statistically homogeneous isotropic constant density turbulent flow.

* E-Mail: vladimir.sabelnikov@onera.fr.

Keywords: Premixed turbulent combustion, $G-$equation, indicator function, probability density function, white-noise in time velocity field, combustion progress variable, random shear flow

1. INTRODUCTION

Level set methods are widely used in various fields of contemporary science [1, 2] and in premixed combustion, in particular, where this general approach is often called the $G-$equation [3]. If a ratio of the thickness of a flame to the lowest scales of velocity non-uniformities and flame wrinkling is sufficiently low, then, the flame may be considered to be an infinitely thin self-propagating sheet that separates reactants and products. Subsequently, the instantaneous position of the flame is linked to the evolution of zero (or G_0) level set of a function $G(\mathbf{x},t)$. However, since tracking an infinitely thin flame sheet in a turbulent flow is computationally expensive, this method has only been applied to very simple model problems (detailed presentation can be found, e.g. in [4]). To simulate premixed turbulent combustion in more realistic cases, the *G*-equation has to be averaged for tracking the mean flame brush in the mean velocity field.

The basic substantiation of the corresponding mean *G*-equation equation has not yet been elaborated well. The goal of the present paper is to derive the equation for the probability density function of G straightforwardly from the G-equation in a white-noise in time statistically homogeneous isotropic constant density turbulent flow. In section 2, the closed equation for the extended indicator function (or fine-grained probability density function) which includes both $G(\mathbf{x},t)$ and its spatial derivatives of the first and the second orders is derived. In section 3, the closed equation for the probability density function for both $G(\mathbf{x},t)$ and its spatial derivatives of the first and the second orders is derived averaging the equation for the extended indicator function in a white-noise in time statistically homogeneous isotropic constant density turbulent flow. In section 4, the closed equation for the probability density function for both $G(\mathbf{x},t)$ and its spatial derivatives of the first order is derived in a special case of a random shear flow, non homogeneous and non isotropic velocity field. In section 5, the principal results are summarized.

2. Derivation of the Equation for the Extended Indicator Function

Let us consider, without loss of generality, a statistically planar 1D premixed flame in an incompressible statistically isotropic homogeneous turbulent flow field, developing from the initially planar laminar flame normal to the x_1 direction. In the framework of the G - equation approach (the level set method for tracking interfaces) the instantaneous position of the flame surface is associated with the evolution of some level set of a scalar field $G(\mathbf{x},t)$ [3] (see also [4]). Let the initial position of the flame surface, i.e. at time $t=0$, is the level set G_0. It is assumed that the products and reactants occupy the domains $G(\mathbf{x},t=0)>G_0$, $G(\mathbf{x},t=0)<G_0$, respectively. The future evolution of this level set, and hence of the flame surface, is found from the solution of an initial boundary value problem (IBVP) of the following field equation

$$\frac{\partial G}{\partial t}+u_k\frac{\partial G}{\partial x_k}=u_F\left|\nabla G\right|=H_L\left(\nabla G\right),\ \left|\nabla G\right|=\sqrt{\frac{\partial G}{\partial x_i}\frac{\partial G}{\partial x_i}}\,, \qquad (1)$$

$$G(\mathbf{x},t=0)=G_{in}(x_1)\,. \qquad (2)$$

Here $\mathbf{u}(\mathbf{x},t)$ is a statistically isotropic homogeneous turbulent flow field with zero mean value $\left\langle \mathbf{u}\right\rangle =0$, u_F is the normal velocity of flame propagation with respect to the products, supposed below to be constant, without loss of generality [3]. According to (2) all level surfaces of $G(\mathbf{x},t=0)$ are planes normal to the x_1 direction, and the products and the reactants occupy at the initial instant $t=0$ the left half-space and the right half-space, respectively. Eq. (1) is a stochastic partial differential equation (SPDE) with multiplicative noise term $\mathbf{u}\cdot\nabla G$. In section 4 we consider a special case of a random shear flow, non homogeneous and non isotropic velocity field.

The particular choice of $G_{in}(x_1)$

$$G(\mathbf{x},t=0)=G_{in}(x_1)=-x_1 \qquad (3)$$

for statistically 1D premixed flame, as was shown first time in [3], plays an important role in the theoretical analysis of IBVP, because it essentially simplifies the expected equations.

For simplicity of writing it is assumed below that $G_0 = 0$ and, consequently, the initial plane laminar flame position $x_{1f}(t=0) = 0$. Thus

$$G(\mathbf{x}_{1f}, t=0) = G_{in}(x_{1f}) = G_{in}(0) = 0\,,\; G_{in}(x_1) > 0\,,\; x_1 < 0\,,$$
$$G(x_{1f}, t=0) = -x_{1f}(t=0) = G_0 = 0,\; G_{in}(x_1) < 0\,,\; x_1 > 0\,, \tag{4}$$

respectively for the initial conditions (2), (3).

The averaged flame, in accordance with (4), propagates from left to right. Let us designate the spatial derivatives of $G(\mathbf{x}, t)$ as follows

$$p_i = \frac{\partial G}{\partial x_i},\; p_{ij} = \frac{\partial^2 G}{\partial x_i \partial x_j} = \frac{\partial p_i}{\partial x_j},\; p_{ji} = p_{ij} = \frac{\partial^2 G}{\partial x_j \partial x_i} = \frac{\partial p_j}{\partial x_i},$$
$$p_{ijk} = \frac{\partial^3 G}{\partial x_i \partial x_j \partial x_k}. \tag{5}$$

Differentiating (1) with respect to x_i, and then with respect to x_i and x_j yields the PDEs for p_i and p_{ij}

$$\frac{\partial p_i}{\partial t} + u_k \frac{\partial p_i}{\partial x_k} + \frac{\partial u_k}{\partial x_i} p_k = \frac{\partial H_L}{\partial p_k} p_{ki}, \tag{6}$$

$$\frac{\partial p_{ij}}{\partial t} + u_k \frac{\partial p_{ij}}{\partial x_k} + \frac{\partial u_k}{\partial x_j} p_{ik} + \frac{\partial u_k}{\partial x_i} p_{kj} + \frac{\partial^2 u_k}{\partial x_i \partial x_j} p_k = \frac{\partial^2 H_L}{\partial p_k \partial p_l} p_{ki} p_{lj} + \frac{\partial H_L}{\partial p_k} p_{kij}. \tag{7}$$

where the following designations are introduced

$$H_L = u_F |\nabla G| = u_F p,\; p = (p_k p_k)^{1/2}, \tag{8}$$

$$\frac{\partial H_L}{\partial p_k} = u_F \frac{p_k}{p} = H_L \frac{p_k}{p^2}, \tag{9}$$

$$\frac{\partial^2 H_L}{\partial p_k \partial p_l} = u_F \left(\frac{\delta_{kl}}{p} - \frac{p_k p_l}{p^3} \right) = H_L \frac{1}{p^2} \left(\delta_{kl} - \frac{p_k p_l}{p^2} \right). \tag{10}$$

It is worth noting that H_L depends only on the p_k, i.e. on the gradient of $G(\mathbf{x},t)$. This fact will be used repeatedly below in deriving and transforming the equations.

SPDE's (6), (7) are not closed, because they include the third order derivatives $p_{ijk} = \dfrac{\partial^3 G}{\partial x_i \partial x_j \partial x_k}$. SPDE's are solved with the following initial conditions

$$p_i(\mathbf{x}, t=0) = \frac{\partial G_{in}(x_1)}{\partial x_1} \delta_{i1}, \ p_{ij}(\mathbf{x}, t=0) = \frac{\partial^2 G_{in}(x_1)}{\partial x_1 \partial x_1} \delta_{i1} \delta_{j1}, \tag{11}$$

$$p_i(\mathbf{x}, t=0) = -\delta_{i1}, \ p_{ij}(\mathbf{x}, t=0) = 0, \tag{12}$$

by virtue of the initial conditions (2) or (3), respectively, δ_{ij} is the Kronecker delta.

Let us introduce the extended indicator function (or the fine-grained probability density function) $\Phi(\mathbf{x}, t, \hat{G}, \hat{p}_i, \hat{p}_{ij})$, [5], [6]

$$\Phi(\mathbf{x}, t, \hat{G}, \hat{p}_i, \hat{p}_{ij}) = \delta(G(\mathbf{x}, t) - \hat{G}) \delta(p_i(\mathbf{x}, t) - \hat{p}_i) \delta(p_{ij}(\mathbf{x}, t) - \hat{p}_{ij}), \tag{13}$$

where $\hat{G}$, $\hat{p}_i$ and $\hat{p}_{ij}$ are the sample values of the random functions $G(\mathbf{x},t)$, $p_i(\mathbf{x},t)$ and $p_{ij}(\mathbf{x},t)$, respectively. It should be stressed that the sample values do not depend on $\mathbf{x}$ and t. The random functions $G(\mathbf{x},t)$, $p_i(\mathbf{x},t)$ and

$p_{ij}(\mathbf{x},t)$ obey to PDE's (1), (6) and (7), respectively. The closed Liouville equation for the extended indicator function Φ introduced by (13) is derived from (1), (6) and (7). This equation is obtained in a following way. Differentiating (13) with respect to time yields

$$\frac{\partial\Phi}{\partial t}=\frac{\partial\,\Phi}{\partial G}\frac{\partial G}{\partial t}+\frac{\partial\Phi}{\partial p_i}\frac{\partial p_i}{\partial t}+\frac{\partial\Phi}{\partial p_{ij}}\frac{\partial p_{ij}}{\partial t}=-\frac{\partial\,\Phi}{\partial\hat{G}}\frac{\partial G}{\partial t}-\frac{\partial\Phi}{\partial\hat{p}_i}\frac{\partial p_i}{\partial t}-\frac{\partial\Phi}{\partial\hat{p}_{ij}}\frac{\partial p_{ij}}{\partial t}, \tag{14}$$

$$\frac{\partial\Phi}{\partial t}+\frac{\partial\,\Phi}{\partial\hat{G}}\frac{\partial G}{\partial t}+\frac{\partial\Phi}{\partial\hat{p}_i}\frac{\partial p_i}{\partial t}+\frac{\partial\Phi}{\partial\hat{p}_{ij}}\frac{\partial p_{ij}}{\partial t}=0\,. \tag{15}$$

Inserting the temporal derivatives $\frac{\partial G}{\partial t}$, $\frac{\partial p_i}{\partial t}$ and $\frac{\partial p_{ij}}{\partial t}$ from PDE (1), (6), (7) into Eq. (15) results in

$$\frac{\partial\Phi}{\partial t}+\frac{\partial}{\partial\hat{G}}\left(-u_k\frac{\partial G}{\partial x_k}+H_L\right)\Phi+\frac{\partial}{\partial\hat{p}_i}\left(-u_k\frac{\partial p_i}{\partial x_k}-\frac{\partial u_k}{\partial x_i}p_k+\frac{\partial H_L}{\partial p_k}p_{ki}\right)\Phi+$$
$$+\frac{\partial}{\partial\hat{p}_{ij}}\left(-u_k\frac{\partial p_{ij}}{\partial x_k}-\frac{\partial u_k}{\partial x_j}p_{ik}-\frac{\partial u_k}{\partial x_i}p_{kj}-\frac{\partial^2 u_k}{\partial x_i\partial x_j}p_k+\frac{\partial^2 H_L}{\partial p_k\partial p_l}p_{ki}p_{lj}+\frac{\partial H_L}{\partial p_k}p_{kij}\right)\Phi=0 \tag{16}$$

The expression for the spatial derivative $\frac{\partial\Phi}{\partial x_k}$ of the indicator function Φ is needed to recast (16). It follows from the definition (13)

$$\frac{\partial\Phi}{\partial x_k}=\frac{\partial\,\Phi}{\partial G}\frac{\partial G}{\partial x_k}+\frac{\partial\Phi}{\partial p_i}\frac{\partial p_i}{\partial x_k}+\frac{\partial\Phi}{\partial p_{ij}}\frac{\partial p_{ij}}{\partial x_k}=-\frac{\partial\,\Phi}{\partial\hat{G}}\frac{\partial G}{\partial x_k}-\frac{\partial\Phi}{\partial\hat{p}_i}\frac{\partial p_i}{\partial x_k}-\frac{\partial\Phi}{\partial\hat{p}_{ij}}\frac{\partial p_{ij}}{\partial x_k}\,. \tag{17}$$

Collecting the convective terms in (16) yields

$$\frac{\partial \Phi}{\partial t}+u_k\frac{\partial \Phi}{\partial x_k}+\frac{\partial}{\partial \hat{G}}\left(H_L\Phi\right)+\frac{\partial}{\partial \hat{p}_i}\left(-\frac{\partial u_k}{\partial x_i}p_k+\frac{\partial H_L}{\partial p_k}p_{ki}\right)\Phi+$$
$$\frac{\partial}{\partial \hat{p}_{ij}}\left(-\frac{\partial u_k}{\partial x_i}p_{kj}-\frac{\partial u_k}{\partial x_j}p_{ik}-\frac{\partial^2 u_k}{\partial x_i\partial x_j}p_k+\frac{\partial^2 H_L}{\partial p_k\partial p_l}p_{ki}p_{lj}+\frac{\partial H_L}{\partial p_k}p_{kij}\right)\Phi=0 \quad (18)$$

by virtue of (17). Using sifting property of the delta-function (18) transforms to

$$\frac{\partial \Phi}{\partial t}+u_k\frac{\partial \Phi}{\partial x_k}+\frac{\partial}{\partial \hat{G}}\left(\hat{H}_L\Phi\right)+\frac{\partial}{\partial \hat{p}_i}\left(-\frac{\partial u_k}{\partial x_i}\hat{p}_k+\frac{\partial \hat{H}_L}{\partial \hat{p}_k}\hat{p}_{ki}\right)\Phi+$$
$$\frac{\partial}{\partial \hat{p}_{ij}}\left(-\frac{\partial u_k}{\partial x_i}\hat{p}_{kj}-\frac{\partial u_k}{\partial x_j}\hat{p}_{ik}-\frac{\partial^2 u_k}{\partial x_i\partial x_j}\hat{p}_k+\frac{\partial^2 \hat{H}_L}{\partial \hat{p}_k\partial \hat{p}_l}\hat{p}_{ki}\hat{p}_{lj}+\frac{\partial \hat{H}_L}{\partial \hat{p}_k}p_{kij}\right)\Phi=0 \quad (19)$$

The Liouville equation (19) is not closed yet, since it contains the third-order derivatives of Φ. However, the combination $\frac{\partial \Phi}{\partial t}-\frac{\partial \hat{H}_L}{\partial \hat{p}_k}\frac{\partial \Phi}{\partial x_k}$ will not include the third order derivatives (similar to the cases considered in [5], [6]). Indeed, the terms $\frac{\partial}{\partial \hat{p}_i}\left(\frac{\partial H_L}{\partial p_k}p_{ki}\Phi\right)$ and $\frac{\partial}{\partial \hat{p}_{ij}}\left(\frac{\partial H_L}{\partial p_k}p_{kij}\Phi\right)$ in (19) can be recast to

$$\frac{\partial}{\partial \hat{p}_i}\left(\frac{\partial H_L}{\partial p_k}p_{ki}\Phi\right)=\frac{\partial}{\partial \hat{p}_i}\left(\frac{\partial \hat{H}_L}{\partial \hat{p}_k}p_{ki}\Phi\right)=\frac{\partial^2 \hat{H}_L}{\partial \hat{p}_i\partial \hat{p}_k}\hat{p}_{ki}\Phi+\frac{\partial \hat{H}_L}{\partial \hat{p}_k}\frac{\partial \Phi}{\partial \hat{p}_i}\frac{\partial p_i}{\partial x_k}, \quad (20)$$

$$\frac{\partial}{\partial \hat{p}_{ij}}\left(\frac{\partial H_L}{\partial p_k}p_{kij}\Phi\right)=\frac{\partial \hat{H}_L}{\partial \hat{p}_k}\frac{\partial}{\partial \hat{p}_{ij}}\left(p_{kij}\Phi\right)=\frac{\partial \hat{H}_L}{\partial \hat{p}_k}\frac{\partial \Phi}{\partial \hat{p}_{ij}}\frac{\partial p_{ij}}{\partial x_k} \quad (21)$$

by virtue of the sifting property of the delta-function.
Hence the sum of (20) and (21) is given by the following equation

$$\frac{\partial}{\partial\hat{p}_i}\left(\frac{\partial H_L}{\partial p_k}p_{ki}\Phi\right)+\frac{\partial}{\partial\hat{p}_{ij}}\left(\frac{\partial H_L}{\partial p_k}p_{kij}\Phi\right)=\frac{\partial\hat{H}_L}{\partial\hat{p}_k}\left(\frac{\partial\Phi}{\partial\hat{p}_i}\frac{\partial p_i}{\partial x_k}+\frac{\partial\Phi}{\partial\hat{p}_{ij}}\frac{\partial p_{ij}}{\partial x_k}\right)+\frac{\partial^2\hat{H}_L}{\partial\hat{p}_i\partial\hat{p}_k}\hat{p}_{ki}\Phi \tag{22}$$

Using (17) the first term in the RHS of (22) becomes

$$\frac{\partial\hat{H}_L}{\partial\hat{p}_k}\left(\frac{\partial\Phi}{\partial\hat{p}_i}\frac{\partial p_i}{\partial x_k}+\frac{\partial\Phi}{\partial\hat{p}_{ij}}\frac{\partial p_{ij}}{\partial x_k}\right)=-\frac{\partial\hat{H}_L}{\partial\hat{p}_k}\frac{\partial\Phi}{\partial x_k}-\frac{\partial\hat{H}_L}{\partial\hat{p}_k}\frac{\partial\Phi}{\partial\hat{G}}\frac{\partial G}{\partial x_k}=-\frac{\partial\hat{H}_L}{\partial\hat{p}_k}\left(\frac{\partial\Phi}{\partial x_k}+\hat{p}_k\frac{\partial\Phi}{\partial\hat{G}}\right) \tag{23}$$

Eq. (23) can be rewritten as

$$\frac{\partial}{\partial\hat{p}_i}\left(\frac{\partial H_L}{\partial p_k}p_{ki}\Phi\right)+\frac{\partial}{\partial\hat{p}_{ij}}\left(\frac{\partial H_L}{\partial p_k}p_{kij}\Phi\right)=-\frac{\partial\hat{H}_L}{\partial\hat{p}_k}\frac{\partial\Phi}{\partial x_k}-\frac{\partial\hat{H}_L}{\partial\hat{p}_k}\hat{p}_k\frac{\partial\Phi}{\partial\hat{G}}+\frac{\partial^2\hat{H}_L}{\partial\hat{p}_i\partial\hat{p}_k}\hat{p}_{ki}\Phi. \tag{24}$$

Hence (19) becomes the closed equation for the extended indicator function $\Phi(\mathbf{x},t,\hat{G},\hat{p}_i,\hat{p}_{ij})$

$$\frac{\partial\Phi}{\partial t}+\left(u_k-\frac{\partial\hat{H}_L}{\partial\hat{p}_k}\right)\frac{\partial\Phi}{\partial x_k}+\left(\hat{H}_L-\hat{p}_k\frac{\partial\hat{H}_L}{\partial\hat{p}_k}\right)\frac{\partial\Phi}{\partial\hat{G}}-\frac{\partial u_k}{\partial x_i}\frac{\partial}{\partial\hat{p}_i}\left(\hat{p}_k\Phi\right)+$$
$$\frac{\partial}{\partial\hat{p}_{ij}}\left(-\frac{\partial u_k}{\partial x_i}\hat{p}_{kj}-\frac{\partial u_k}{\partial x_j}\hat{p}_{ik}-\frac{\partial^2 u_k}{\partial x_i\partial x_j}\hat{p}_k+\frac{\partial^2\hat{H}_L}{\partial\hat{p}_k\partial\hat{p}_l}\hat{p}_{ki}\hat{p}_{lj}\right)\Phi+\frac{\partial^2\hat{H}_L}{\partial\hat{p}_i\partial\hat{p}_k}\hat{p}_{ki}\Phi=0 \tag{25}$$

combining (22)-(24). It is readily verified from (8), (9) that

$$p_k\frac{\partial H_L}{\partial p_k}=H_L. \tag{26}$$

Therefore Liouville equation (25) transforms to

$$\frac{\partial\Phi}{\partial t}+\left(u_k-\frac{\partial\hat{H}_L}{\partial\hat{p}_k}\right)\frac{\partial\Phi}{\partial x_k}-\frac{\partial u_k}{\partial x_i}\frac{\partial}{\partial\hat{p}_i}(\hat{p}_k\Phi)+$$
$$\frac{\partial}{\partial\hat{p}_{ij}}\left(-\frac{\partial u_k}{\partial x_i}\hat{p}_{kj}-\frac{\partial u_k}{\partial x_j}\hat{p}_{ik}-\frac{\partial^2 u_k}{\partial x_i\partial x_j}\hat{p}_k+\frac{\partial^2\hat{H}_L}{\partial\hat{p}_k\partial\hat{p}_l}\hat{p}_{ki}\hat{p}_{lj}\right)\Phi+\frac{\partial^2\hat{H}_L}{\partial\hat{p}_i\partial\hat{p}_k}\hat{p}_{ki}\Phi=0 \tag{27}$$

Eq. (27) is a closed Liouville equation in the coordinate $\mathbf{x}$ and phase space $\hat{p}_i$, $\hat{p}_{ij}$. Eq. (27) is solved with the initial conditions

$$\Phi(\mathbf{x},t=0,\hat{G},\hat{p}_i,\hat{p}_{ij})=\delta(G_{in}(x_1)-\hat{G})\delta(\frac{\partial G_{in}(x_1)}{\partial x_1}\delta_{i1}-\hat{p}_i)\delta(\frac{\partial^2 G_{in}(x_1)}{\partial x_1\partial x_1}\delta_{i1}\delta_{j1}-\hat{p}_{ij}), \tag{28}$$

$$\Phi(\mathbf{x},t=0,\hat{G},\hat{p}_i,\hat{p}_{ij})=\delta(-x_1-\hat{G})\delta(-\delta_{1i}-\hat{p}_i)\delta(\hat{p}_{ij})=\delta(x_1+\hat{G})\delta(\delta_{1i}+\hat{p}_i)\delta(\hat{p}_{ij}). \tag{29}$$

corresponding to (2), (3), respectively.

The extended indicator function $\Phi(\mathbf{x},t,\hat{G},\hat{p}_i,\hat{p}_{ij})$ permits to find the different quantities, e.g. combustion progress variable $c(\mathbf{x},t)$, which is calculated as folows

$$c(\mathbf{x},t)=H(G(\mathbf{x},t))=\int_0^\infty\delta(G(\mathbf{x},t)-\hat{G})d\hat{G}, \tag{30}$$

where $H(\varphi)$ is the Heaviside step function equal to zero for $\varphi<0$ and unity for $\varphi>0$. Hence equation for $c(\mathbf{x},t)$, which readily follows from (1), reads

$$\frac{\partial c}{\partial t}+(\mathbf{u}+u_F\mathbf{n})\cdot\nabla c=0, \qquad \nabla c(\mathbf{x},t)=\delta(G(\mathbf{x},t))\nabla G(\mathbf{x},t), \tag{31}$$

where is the normal to the front surface pointed to the reactants, i. e.

$$\mathbf{n}(\mathbf{x},t) = -\frac{\nabla G(\mathbf{x},t)}{|\nabla G(\mathbf{x},t)|}\bigg|_{G(\mathbf{x},t)=0} \tag{32}$$

Eq. (31) can be transformed to

$$\frac{\partial c}{\partial t} + \mathbf{u}\cdot\nabla c = u_F |\nabla c|, \qquad |\nabla c(\mathbf{x},t)| = \delta(G(\mathbf{x},t))|\nabla G(\mathbf{x},t)|. \tag{33}$$

Let us obtain the equation for the progress variable $c(\mathbf{x},t) = H(G(\mathbf{x},t))$ from Liouville equations for the indicator function $\Phi(\mathbf{x},t,\hat{G},\hat{p}_i,\hat{p}_{ij})$ (19) (unclosed), and (27) (closed). It should noted that $c(\mathbf{x},t)$ defined by (30) can be found also, performing the integration of the extended indicator function $\Phi(\mathbf{x},t,\hat{G},\hat{p}_i,\hat{p}_{ij})$ over $\hat{G}$ in the half-space $\hat{G} \geq 0$ and over whole values of $\hat{p}_i$ and $\hat{p}_{ij}$. Indeed,

$$c(\mathbf{x},t) = \int_0^{\infty}\int_{-\infty}^{\infty}\int_{-\infty}^{\infty} \Phi(\mathbf{x},t,\hat{G},\hat{p}_i,\hat{p}_{ij})\, d\hat{G}\, d\hat{p}_i\, d\hat{p}_{ij} = \int_0^{\infty} \delta(G(\mathbf{x},t)-\hat{G})\, d\hat{G} = H(G(\mathbf{x},t)). \tag{34}$$

Integrating (19) over $\hat{G}$ in the half-space $\hat{G} \geq 0$ and over whole values of $\hat{p}_i$ and $\hat{p}_{ij}$ yields

$$\frac{\partial c}{\partial t} + u_k \frac{\partial c}{\partial x_k} - H_L \delta(G(\mathbf{x},t)) = 0,$$

$$\frac{\partial c}{\partial t} + u_k \frac{\partial c}{\partial x_k} = H_L \delta(G(\mathbf{x},t)) = u_F |\nabla G| \delta(G(\mathbf{x},t)) = u_F |\nabla c|,$$

i.e. Eq. (33).

The same result is obtained from (27). Indeed, integrating (27) over $\hat{G}$ in the half-space $\hat{G} \geq 0$ and over whole values of $\hat{p}_i$ and $\hat{p}_{ij}$ results in

$$\frac{\partial c}{\partial t} + u_k \frac{\partial c}{\partial x_k} - \frac{\partial}{\partial x_k} \int_0^{\infty} \int_{-\infty}^{\infty} \int_{-\infty}^{\infty} \frac{\partial \hat{H}_L}{\partial \hat{p}_k} \Phi d\hat{G} d\hat{p}_i d\hat{p}_{ij} + \int_0^{\infty} \int_{-\infty}^{\infty} \int_{-\infty}^{\infty} \frac{\partial^2 \hat{H}_L}{\partial \hat{p}_i \partial \hat{p}_k} \hat{p}_{ki} \Phi d\hat{G} d\hat{p}_i d\hat{p}_{ij} = 0 \tag{35}$$

Performing the integration of two last terms in the LHS of Eq. (35) it becomes

$$\frac{\partial c}{\partial t} + u_k \frac{\partial c}{\partial x_k} - \frac{\partial}{\partial x_k} \left(\frac{\partial H_L}{\partial p_k} H(G(\mathbf{x},t)) \right) + \frac{\partial^2 H_L}{\partial p_i \partial p_k} p_{ki} H(G(\mathbf{x},t)) = 0, \tag{36}$$

by virtue of the definition (13). Eq. (36) can be rewritten as follows

$$\frac{\partial c}{\partial t} + u_k \frac{\partial c}{\partial x_k} - \frac{\partial H_L}{\partial p_k} \delta(G(\mathbf{x},t)) \frac{\partial G}{\partial x_k} - \frac{\partial^2 H_L}{\partial p_i \partial p_k} \frac{\partial p_i}{\partial x_k} H(G(\mathbf{x},t)) + \frac{\partial^2 H_L}{\partial p_i \partial p_k} p_{ki} H(G(\mathbf{x},t)) = 0 \tag{37}$$

Taking into account the definitions (5), Eq. (37) becomes

$$\frac{\partial c}{\partial t} + u_k \frac{\partial c}{\partial x_k} - \frac{\partial H_L}{\partial p_k} p_k \delta(G(\mathbf{x},t)) - \frac{\partial^2 H_L}{\partial p_i \partial p_k} p_{ki} H(G(\mathbf{x},t)) + \frac{\partial^2 H_L}{\partial p_i \partial p_k} p_{ki} H(G(\mathbf{x},t)) = 0 \tag{38}$$

Finally Eq. (38) transforms to Eq. (33)

$$\frac{\partial c}{\partial t} + u_k \frac{\partial c}{\partial x_k} = H_L \delta(G(\mathbf{x},t)) = u_F |\nabla G| \delta(G(\mathbf{x},t)) = u_F |\nabla c|,$$

taking into account (26).

The G - equation can be obtained from (19), using the expression

$$G(\mathbf{x},t)=\int_{-\infty}^{\infty}\int_{-\infty}^{\infty}\int_{-\infty}^{\infty}\hat{G}\Phi(\mathbf{x},t,\hat{G},\hat{p}_i,\hat{p}_{ij})d\hat{G}d\hat{p}_i d\hat{p}_{ij}\,. \tag{39}$$

Multiplying (19) by $\hat{G}$ and integrating over whole values of $\hat{G}$, $\hat{p}_i$ and $\hat{p}_{ij}$ yields Eq. (1)

$$\frac{\partial G}{\partial t}+u_k\frac{\partial G}{\partial x_k}-\int_{-\infty}^{\infty}\int_{-\infty}^{\infty}\int_{-\infty}^{\infty}\hat{H}_L\Phi(\mathbf{x},t,\hat{G},\hat{p}_i,\hat{p}_{ij})d\hat{G}d\hat{p}_i d\hat{p}_{ij}=0\,, \tag{40}$$

$$\frac{\partial G}{\partial t}+u_k\frac{\partial G}{\partial x_k}=H_L=u_F\left|\nabla G\right| \tag{41}$$

taking into consideration (13).

The same result follows from (27). Indeed, multiplying (27) by $\hat{G}$ and then integrating over whole values of $\hat{G}$, $\hat{p}_i$ and $\hat{p}_{ij}$ yields

$$\frac{\partial G}{\partial t}+u_k\frac{\partial G}{\partial x_k}+\int_{-\infty}^{\infty}\int_{-\infty}^{\infty}\int_{-\infty}^{\infty}\hat{G}\left(-\frac{\partial\hat{H}_L}{\partial\hat{p}_k}\frac{\partial\Phi}{\partial x_k}+\frac{\partial^2\hat{H}_L}{\partial\hat{p}_i\partial\hat{p}_k}\hat{p}_{ki}\Phi\right)d\hat{G}d\hat{p}_i d\hat{p}_{ij}=0 \tag{42}$$

The last term in the LHS of (42) can be transformed, taking into account (24). The term in the LHS of (24) after multiplication by $\hat{G}$ and following integration over whole values of $\hat{G}$, $\hat{p}_i$ and $\hat{p}_{ij}$ is equal to zero

$$\int_{-\infty}^{\infty}\int_{-\infty}^{\infty}\int_{-\infty}^{\infty}\hat{G}\left(\frac{\partial}{\partial\hat{p}_i}\left(\frac{\partial H_L}{\partial\hat{p}_k}\hat{p}_{ki}\Phi\right)+\frac{\partial}{\partial\hat{p}_{ij}}\left(\frac{\partial H_L}{\partial\hat{p}_k}p_{kij}\Phi\right)\right)d\hat{G}d\hat{p}_i d\hat{p}_{ij}=0\,. \tag{43}$$

Therefore the integrated RHS of (24) multiplied by $\hat{G}$ becomes

$$\int_{-\infty}^{\infty}\int_{-\infty}^{\infty}\int_{-\infty}^{\infty}\hat{G}\left(-\frac{\partial\hat{H}_L}{\partial\hat{p}_k}\frac{\partial\Phi}{\partial x_k}-\frac{\partial\hat{H}_L}{\partial\hat{p}_k}\hat{p}_k\frac{\partial\Phi}{\partial\hat{G}}+\frac{\partial^2\hat{H}_L}{\partial\hat{p}_i\partial\hat{p}_k}\hat{p}_{ki}\Phi\right)d\hat{G}d\hat{p}_i d\hat{p}_{ij}=0\,. \tag{44}$$

Eq. (44) transforms to

$$\int_{-\infty}^{\infty}\int_{-\infty}^{\infty}\int_{-\infty}^{\infty}\hat{G}\left(-\frac{\partial \hat{H}_L}{\partial \hat{p}_k}\frac{\partial \Phi}{\partial x_k}+\frac{\partial^2 \hat{H}_L}{\partial \hat{p}_i \partial \hat{p}_k}\hat{p}_{ki}\Phi\right)d\hat{G}d\hat{p}_i d\hat{p}_{ij}=\int_{-\infty}^{\infty}\int_{-\infty}^{\infty}\int_{-\infty}^{\infty}\frac{\partial \hat{H}_L}{\partial \hat{p}_k}\hat{p}_k\hat{G}\frac{\partial \Phi}{\partial \hat{G}}d\hat{G}d\hat{p}_i d\hat{p}_{ij},$$

(45)

$$\int_{-\infty}^{\infty}\int_{-\infty}^{\infty}\int_{-\infty}^{\infty}\hat{G}\left(-\frac{\partial \hat{H}_L}{\partial \hat{p}_k}\frac{\partial \Phi}{\partial x_k}+\frac{\partial^2 \hat{H}_L}{\partial \hat{p}_i \partial \hat{p}_k}\hat{p}_{ki}\Phi\right)d\hat{G}d\hat{p}_i d\hat{p}_{ij}=\int_{-\infty}^{\infty}\int_{-\infty}^{\infty}\int_{-\infty}^{\infty}\hat{H}_L\hat{G}\frac{\partial \Phi}{\partial \hat{G}}d\hat{G}d\hat{p}_i d\hat{p}_{ij},$$

(46)

by virtue of (26). Integrating by parts the RHS in Eq. (46) results in

$$\int_{-\infty}^{\infty}\int_{-\infty}^{\infty}\int_{-\infty}^{\infty}\hat{G}\left(-\frac{\partial \hat{H}_L}{\partial \hat{p}_k}\frac{\partial \Phi}{\partial x_k}+\frac{\partial^2 \hat{H}_L}{\partial \hat{p}_i \partial \hat{p}_k}\hat{p}_{ki}\Phi\right)d\hat{G}d\hat{p}_i d\hat{p}_{ij}=-\int_{-\infty}^{\infty}\int_{-\infty}^{\infty}\int_{-\infty}^{\infty}\hat{H}_L\Phi d\hat{G}d\hat{p}_i d\hat{p}_{ij}=-H_L.$$

(47)

taking into account the definition (13), and the fact that $\hat{H}_L$ does not depends on $\hat{G}$, Eq. (8).

The last term in the LHS of (42), as follows from (47), is equal to $-H_L$, and Eq. (42) transforms to Eq. (1).

In the Annex A the unclosed equation for the reduced indicator function $\Phi_G(\mathbf{x},t,\hat{G})=\delta(G(\mathbf{x},t)-\hat{G})$ is derived.

3. Derivation of the Equation for the Probability Density Function in White-noise Velocity Field

Expected extended indicator function

$$\langle\Phi\rangle=\left\langle\delta(G(\mathbf{x},t)-\hat{G})\delta(p_i(\mathbf{x},t)-\hat{p}_i)\delta(p_{ij}(\mathbf{x},t)-\hat{p}_{ij})\right\rangle=P_{Gp_ip_{ij}}(\mathbf{x},t,\hat{G},\hat{p}_i,\hat{p}_{ij})$$

(48)

is the one-point joint probability density function of $G(\mathbf{x},t)$, $p_i(\mathbf{x},t)$ and $p_{ij}(\mathbf{x},t)$. As we can conclude readily, averaging (27) creates the problem of the averaging the products of the velocity field $\mathbf{u}(\mathbf{x},t)$ (and its space derivatives) and the indicator function Φ. The indicator function is the functional of the turbulent velocity field $\mathbf{u}(\mathbf{x},t)$ (through the functions $G(\mathbf{x},t)$, $p_i(\mathbf{x},t)$ and $p_{ij}(\mathbf{x},t)$ which obey the Eqs (1), (6), (7)). This problem, in general case of velocity field, is not solved. Averaging can be done if the velocity field is approximated by the delta-correlated in time Gaussian velocity field (white-in-time velocity field). The velocity correlation tensor in the framework of this approximation, reads [5], [6]

$$\left\langle u_i(\mathbf{x},t)u_j(\mathbf{x}',t')\right\rangle = B_{ij}(\mathbf{r},t,t') = 2\delta(t-t')B_{ij}^{eff}(\mathbf{r},t), \qquad \mathbf{r}=\mathbf{x}-\mathbf{x}', \tag{49}$$

$$B_{ij}^{eff}(\mathbf{r},t) = \frac{1}{2}\int_{-\infty}^{\infty} B_{ij}(\mathbf{r},t,t')dt' = \int_{0}^{\infty} B_{ij}(\mathbf{r},t,t')dt'. \tag{50}$$

The derivation is presented shortly in the Annex B. The technique of averaging SPDEs can be found, e.g. in [5], [6]. The resulting equation is the diffusion equation in the physical space $\mathbf{x}$ and in the sample variable space $\hat{p}_i$ and $\hat{p}_{ij}$. This equation reads ($P_{Gp_ip_{ij}} = P$ for simplicity of writing)

$$\begin{aligned}
&\left(\frac{\partial}{\partial t} - \frac{\partial}{\partial x_k}\frac{\partial \hat{H}_L}{\partial \hat{p}_k} + \frac{\partial}{\partial \hat{p}_{ij}}\frac{\partial^2 \hat{H}_L}{\partial \hat{p}_k \partial \hat{p}_l}\hat{p}_{ki}\hat{p}_{lj}\right)P = \left(U_{jk}\frac{\partial^2}{\partial x_j \partial x_k} + U_{jkml}\frac{\partial^2}{\partial x_k \partial \hat{p}_{ml}}\hat{p}_j + \right.\\
&+U_{jkim}\frac{\partial^2}{\partial \hat{p}_i \partial \hat{p}_m}\hat{p}_k\hat{p}_j + U_{jkim}\frac{\partial^2}{\partial \hat{p}_i \partial \hat{p}_{ml}}\hat{p}_k\hat{p}_{jl} + U_{jkil}\frac{\partial^2}{\partial \hat{p}_i \partial \hat{p}_{ml}}\hat{p}_k\hat{p}_{mj} +\\
&+U_{jkim}\frac{\partial^2}{\partial \hat{p}_m \partial \hat{p}_{in}}\hat{p}_j\hat{p}_{kn} + U_{jkim}\frac{\partial^2}{\partial \hat{p}_{in} \partial \hat{p}_{ml}}\hat{p}_{kn}\hat{p}_{jl} + U_{jkil}\frac{\partial^2}{\partial \hat{p}_{in} \partial \hat{p}_{ml}}\hat{p}_{kn}\hat{p}_{mj}\\
&+U_{jknm}\frac{\partial^2}{\partial \hat{p}_m \partial \hat{p}_{in}}\hat{p}_j\hat{p}_{ik} + U_{jknm}\frac{\partial^2}{\partial \hat{p}_{in} \partial \hat{p}_{ml}}\hat{p}_{ik}\hat{p}_{jl} + U_{jknl}\frac{\partial^2}{\partial \hat{p}_{in} \partial \hat{p}_{ml}}\hat{p}_{ik}\hat{p}_{mj}\\
&\left.+U_{jkin}\frac{\partial^2}{\partial x_j \partial \hat{p}_{in}}\hat{p}_k + U_{jkinml}\frac{\partial^2}{\partial \hat{p}_{in} \partial \hat{p}_{ml}}\hat{p}_k\hat{p}_j P - \frac{\partial^2 \hat{H}_L}{\partial \hat{p}_i \partial \hat{p}_k}\hat{p}_{ki}\right)P
\end{aligned} \tag{51}$$

where for the homogeneous isotropic turbulent velocity field

$$U_{ij} = B_{ij}^{eff}(0,t) = \int_0^\infty B_{ij}(0,t,t')dt' = \int_0^\infty \langle u_i(\mathbf{x},t)u_j(\mathbf{x},t')\rangle dt' = D_0\delta_{ij}, \quad (52)$$

$$D_0 = \frac{1}{3}B_{ii}^{eff}(0,t) = \frac{1}{3}\int_0^\infty \langle u_i(\mathbf{x},t)u_i(\mathbf{x},t')\rangle dt' = \frac{2}{3}\int_0^\infty \widetilde{E}(k,t)dk, \quad (53)$$

$$\widetilde{E}(k,t) = \int_0^\infty E(k,t,t')dt', \quad (54)$$

$$U_{klij} = \int_0^\infty \left\langle \frac{\partial u_k(\mathbf{x},t)}{\partial x_i}\frac{\partial u_l(\mathbf{x},t')}{\partial x_j}\right\rangle dt' = -\int_0^\infty \left.\frac{\partial^2 B_{kl}(\mathbf{r},t,t')}{\partial r_i \partial r_j}\right|_{r=0} dt' = -\left.\frac{\partial^2 B_{ij}^{eff}(\mathbf{r},t))}{\partial r_i \partial r_j}\right|_{r=0} \quad (55)$$

$$U_{klij} = D_2\left(\frac{1}{3}\delta_{kl}\delta_{ij} - I_{klij}\right), \quad (56)$$

$$I_{klij} = \frac{1}{15}\left(\delta_{kl}\delta_{ij} + \delta_{ki}\delta_{lj} + \delta_{kj}\delta_{li}\right), \quad (57)$$

$$D_2 = \int_0^\infty k^2\widetilde{E}(k,t)dk, \quad (58)$$

$$U_{klijmn} = \int_0^\infty \left\langle \frac{\partial^2 u_k(\mathbf{x},t)}{\partial x_i \partial x_j}\frac{\partial^2 u_l(\mathbf{x},t')}{\partial x_m \partial x_n}\right\rangle dt' = \int_0^\infty \left.\frac{\partial^4 B_{kl}(\mathbf{r},t,t')}{\partial r_i \partial r_j \partial r_m \partial r_n}\right|_{r=0} dt' = \left.\frac{\partial^4 B_{ij}^{eff}(\mathbf{r},t))}{\partial r_i \partial r_j \partial r_m \partial r_n}\right|_{r=0}, \quad (59)$$

$$U_{klijmn} = D_4\left(I_{ijmn}\delta_{kl} - I_{klijmn}\right), \quad (60)$$

$$I_{klijmn} = \frac{1}{105}(\delta_{kl}\delta_{ij}\delta_{mn} + \delta_{kl}\delta_{im}\delta_{jn} + \delta_{kl}\delta_{in}\delta_{jm} + \delta_{ki}\delta_{lj}\delta_{mn} + \delta_{ki}\delta_{lm}\delta_{jn} + \delta_{ki}\delta_{ln}\delta_{jm} + \delta_{kj}\delta_{li}\delta_{mn} + \delta_{kj}\delta_{lm}\delta_{in} + \delta_{kj}\delta_{ln}\delta_{im} + \delta_{km}\delta_{li}\delta_{jn} + \delta_{km}\delta_{lj}\delta_{in} + \delta_{km}\delta_{ln}\delta_{ij} + \delta_{kn}\delta_{li}\delta_{jm} + \delta_{kn}\delta_{lj}\delta_{im} + \delta_{kn}\delta_{lm}\delta_{ij}) \quad (61)$$

$$D_4 = \int_0^\infty k^4 \widetilde{E}(k,t)dk. \tag{62}$$

Using Eqs (52)-(62) the terms in the RHS of Eq. (51) become

$$U_{jk} \frac{\partial^2 P}{\partial x_j \partial x_k} = D_0 \frac{\partial^2 P}{\partial x_k \partial x_k}, \tag{63}$$

$$U_{jkml} \frac{\partial^2}{\partial x_k \partial \hat{p}_{ml}} \left(\hat{p}_j P\right) = \frac{D_2}{15} \frac{\partial^2}{\partial x_k \partial \hat{p}_{ml}} \left(4 \hat{p}_k \delta_{ml} - \hat{p}_m \delta_{kl} - \hat{p}_l \delta_{km}\right) P, \tag{64}$$

$$U_{jkim} \frac{\partial^2}{\partial \hat{p}_i \partial \hat{p}_m} \left(\hat{p}_k \hat{p}_j P\right) = \frac{D_2}{15} \frac{\partial^2}{\partial \hat{p}_i \partial \hat{p}_m} \left(4 \hat{p}_k \hat{p}_k \delta_{im} - 2 \hat{p}_m \hat{p}_i\right) P, \tag{65}$$

$$U_{jkim} \frac{\partial^2}{\partial \hat{p}_i \partial \hat{p}_{ml}} \left(\hat{p}_k \hat{p}_{jl} P\right) = \frac{D_2}{15} \frac{\partial^2}{\partial \hat{p}_i \partial \hat{p}_{ml}} \left(4 \hat{p}_k \hat{p}_{kl} \delta_{im} - \hat{p}_m \hat{p}_{il} - \hat{p}_i \hat{p}_{ml}\right) P, \tag{66}$$

$$U_{jkil} \frac{\partial^2}{\partial \hat{p}_i \partial \hat{p}_{ml}} \left(\hat{p}_k \hat{p}_{mj} P\right) = \frac{D_2}{15} \frac{\partial^2}{\partial \hat{p}_i \partial \hat{p}_{ml}} \left(4 \hat{p}_k \hat{p}_{mk} \delta_{il} - \hat{p}_l \hat{p}_{mi} - \hat{p}_i \hat{p}_{ml}\right) P, \tag{67}$$

$$U_{jkim} \frac{\partial^2}{\partial \hat{p}_m \partial \hat{p}_{in}} \left(\hat{p}_j \hat{p}_{kn} P\right) = \frac{D_2}{15} \frac{\partial^2}{\partial \hat{p}_m \partial \hat{p}_{in}} \left(4 \hat{p}_j \hat{p}_{jn} \delta_{im} - \hat{p}_i \hat{p}_{mn} - \hat{p}_m \hat{p}_{in}\right) P, \tag{68}$$

$$U_{jkim} \frac{\partial^2}{\partial \hat{p}_{in} \partial \hat{p}_{ml}} \left(\hat{p}_{kn} \hat{p}_{jl} P\right) = \frac{D_2}{15} \frac{\partial^2}{\partial \hat{p}_{in} \partial \hat{p}_{ml}} \left(4 \hat{p}_{kn} \hat{p}_{kl} \delta_{im} - \hat{p}_{mn} \hat{p}_{il} - \hat{p}_{in} \hat{p}_{ml}\right) P \tag{69}$$

$$U_{jkil}\frac{\partial^2}{\partial\hat{p}_{in}\partial\hat{p}_{ml}}\left(\hat{p}_{kn}\hat{p}_{mj}P\right)=\frac{D_2}{15}\frac{\partial^2}{\partial\hat{p}_{in}\partial\hat{p}_{ml}}\left(4\hat{p}_{jn}\hat{p}_{mj}\delta_{il}-\hat{p}_{ln}\hat{p}_{mi}-\hat{p}_{in}\hat{p}_{ml}\right)P,\tag{70}$$

$$U_{jknm}\frac{\partial^2}{\partial\hat{p}_{m}\partial\hat{p}_{in}}\left(\hat{p}_{j}\hat{p}_{ik}P\right)=\frac{D_2}{15}\frac{\partial^2}{\partial\hat{p}_{m}\partial\hat{p}_{in}}\left(4\hat{p}_{j}\hat{p}_{ij}\delta_{nm}-\hat{p}_{n}\hat{p}_{mi}-\hat{p}_{m}\hat{p}_{in}\right)P,\tag{71}$$

$$U_{jknm}\frac{\partial^2}{\partial\hat{p}_{in}\partial\hat{p}_{ml}}\left(\hat{p}_{ik}\hat{p}_{jl}P\right)=\frac{D_2}{15}\frac{\partial^2}{\partial\hat{p}_{in}\partial\hat{p}_{ml}}\left(4\hat{p}_{ij}\hat{p}_{jl}\delta_{nm}-\hat{p}_{mi}\hat{p}_{ln}-\hat{p}_{ml}\hat{p}_{in}\right)P,\tag{72}$$

$$U_{jknl}\frac{\partial^2}{\partial\hat{p}_{in}\partial\hat{p}_{ml}}\left(\hat{p}_{ik}\hat{p}_{mj}P\right)=\frac{D_2}{15}\frac{\partial^2}{\partial\hat{p}_{in}\partial\hat{p}_{ml}}\left(4\hat{p}_{ij}\hat{p}_{mj}\delta_{nl}-\hat{p}_{il}\hat{p}_{mn}-\hat{p}_{in}\hat{p}_{ml}\right)P,\tag{73}$$

$$U_{jkin}\frac{\partial^2}{\partial x_j\partial\hat{p}_{in}}\left(\hat{p}_{k}P\right)=\frac{D_2}{15}\frac{\partial^2}{\partial x_j\partial\hat{p}_{in}}\left(4\hat{p}_{j}\delta_{in}-\hat{p}_{n}\delta_{ji}-\hat{p}_{i}\delta_{jn}\right)P,\tag{74}$$

$$U_{jkinml}\frac{\partial^2}{\partial\hat{p}_{in}\partial\hat{p}_{ml}}\left(\hat{p}_{k}\hat{p}_{j}P\right)=\frac{2}{35}D_4\frac{\partial^2}{\partial\hat{p}_{ij}\partial\hat{p}_{mn}}3\left(\delta_{ij}\delta_{mn}+\delta_{im}\delta_{jn}+\delta_{in}\delta_{jm}\right)\hat{p}_{k}\hat{p}_{k}P-$$
$$-\frac{2}{105}D_4\frac{\partial^2}{\partial\hat{p}_{ij}\partial\hat{p}_{mn}}(\hat{p}_i\hat{p}_j\delta_{mn}+\hat{p}_i\hat{p}_m\delta_{jn}+\hat{p}_i\hat{p}_n\delta_{jm}+\hat{p}_j\hat{p}_m\delta_{in}+\hat{p}_j\hat{p}_n\delta_{im}+\hat{p}_m\hat{p}_n\delta_{ij})P\tag{75}$$

The parameters D_0, D_2 and D_4 are the diffusion coefficients in the space, and in the sample spaces $\hat{p}_i$, $\hat{p}_{ij}$, respectively. The diffusion coefficients D_2 and D_4 are related with the second and forth derivatives of the correlation function $B_{ij}^{eff}(\mathbf{r},t)$ taken at $\mathbf{r}=0$, (B30) and (B32) for D_2, (B35) and (B37) for D_4.

Remark. It is worth noting that Eqs (51)-(75) are valid in the general case of initial condition

$$G(\mathbf{x}, t=0) = G_{in}(\mathbf{x}), \tag{76}$$

where $G_{in}(\mathbf{x})$ is arbitrary function.

Equation (51) is solved with the following initial condition

$$P_{Gp_ip_{ij}}(\mathbf{x}, t=0, \hat{G}, \hat{p}_i, \hat{p}_{ij}) = \delta(G_{in}(x_1) - \hat{G})\delta(\frac{\partial G_{in}(x_1)}{\partial x_1}\delta_{i1} - \hat{p}_i)\delta(\frac{\partial^2 G_{in}(x_1)}{\partial x_1 \partial x_1}\delta_{i1}\delta_{j1} - \hat{p}_{ij}), \tag{77}$$

$$P_{Gp_ip_{ij}}(\mathbf{x}, t=0, \hat{G}, \hat{p}_i, \hat{p}_{ij}) = \delta(x_1 + \hat{G})\delta(\delta_{1i} + \hat{p}_i)\delta(\hat{p}_{ij}). \tag{78}$$

taking into consideration (28), (29), respectively.

The indexes in the Eqs (63), (64), (65) (74) take the values $k = l = m = 1$, $j = i = n = 1$.

A solution of Eq. (51) for statistically planar (1D) flame with the initial condition (78) has the following form [7]

$$P_{Gp_ip_{ij}}(\mathbf{x}, t, \hat{G}, \hat{p}_i, \hat{p}_{ij}) = P_{gp_ip_{ij}}(\hat{g}, t, \hat{p}_i, \hat{p}_{ij}),\ \hat{g} = \hat{G} - \langle G \rangle, \tag{79}$$

where

$$\langle G \rangle = -x_1 + X(t),\ X(t) = \int_0^t u_T(\tau) d\tau, \tag{80}$$

$$u_T(t) = \langle H_L \rangle = u_F \langle |\nabla G| \rangle, \tag{81}$$

$$G(\mathbf{x}, t) = \langle G \rangle + g(\mathbf{x}, t) = G(\mathbf{x}, t) + x_1 - X(t) + g(\mathbf{x}, t). \tag{82}$$

(82) is similar to Reynolds decomposition of a scalar in homogeneous turbulence with uniform mean scalar gradient, $g(\mathbf{x}, t)$ is the statistically

homogeneous fluctuating part of the field $G(\mathbf{x},t)$ in the decomposition (82), $\langle g(\mathbf{x},t)\rangle = 0$. $u_T(t)$ is the mean turbulent burning velocity.

Probability density function $P_{gp_ip_{ij}}(\hat{g},t,\hat{p}_i,\hat{p}_{ij})$ does not depend on x_1, and solves the following equation

$$\left(\frac{\partial}{\partial t}-u_T(t)\frac{\partial}{\partial \hat{g}}+\frac{\partial}{\partial \hat{g}}\frac{\partial \hat{H}_L}{\partial \hat{p}_1}+\frac{\partial}{\partial \hat{p}_{ij}}\frac{\partial^2 \hat{H}_L}{\partial \hat{p}_k \partial \hat{p}_l}\hat{p}_{ki}\hat{p}_{lj}\right)P_{gp_ip_{ij}}=\left(D_0\frac{\partial^2}{\partial \hat{g}^2}+U_{j111}\frac{\partial^2}{\partial \hat{g}\partial \hat{p}_{11}}\hat{p}_j+\right.$$
$$+U_{jkim}\frac{\partial^2}{\partial \hat{p}_i \partial \hat{p}_m}\hat{p}_k\hat{p}_j+U_{jkim}\frac{\partial^2}{\partial \hat{p}_i \partial \hat{p}_{ml}}\hat{p}_k\hat{p}_{jl}+U_{jkil}\frac{\partial^2}{\partial \hat{p}_i \partial \hat{p}_{ml}}\hat{p}_k\hat{p}_{mj}+$$
$$+U_{jkim}\frac{\partial^2}{\partial \hat{p}_m \partial \hat{p}_{in}}\hat{p}_j\hat{p}_{kn}+U_{jkim}\frac{\partial^2}{\partial \hat{p}_{in} \partial \hat{p}_{ml}}\hat{p}_{kn}\hat{p}_{jl}+U_{jkil}\frac{\partial^2}{\partial \hat{p}_{in} \partial \hat{p}_{ml}}\hat{p}_{kn}\hat{p}_{mj}$$
$$+U_{jknm}\frac{\partial^2}{\partial \hat{p}_m \partial \hat{p}_{in}}\hat{p}_j\hat{p}_{ik}+U_{jknm}\frac{\partial^2}{\partial \hat{p}_{in} \partial \hat{p}_{ml}}\hat{p}_{ik}\hat{p}_{jl}+U_{jknl}\frac{\partial^2}{\partial \hat{p}_{in} \partial \hat{p}_{ml}}\hat{p}_{ik}\hat{p}_{mj}$$
$$\left.+U_{1k11}\frac{\partial^2}{\partial \hat{g}\partial \hat{p}_{11}}\hat{p}_k+U_{jkinml}\frac{\partial^2}{\partial \hat{p}_{in} \partial \hat{p}_{ml}}\hat{p}_k\hat{p}_j-\frac{\partial^2 \hat{H}_L}{\partial \hat{p}_i \partial \hat{p}_k}\hat{p}_{ki}\right)P_{gp_ip_{ij}} \tag{83}$$

where $u_T(t)$ is given by Eq. (81) and the coefficients U_{j111}, U_{1k11} read

$$U_{j111}=\frac{2D_2}{15}\delta_{j1},\ U_{1k11}=\frac{2D_2}{15}\delta_{k1}. \tag{84}$$

Therefore

$$U_{j111}\frac{\partial^2}{\partial \hat{g}\partial \hat{p}_{11}}\hat{p}_j=U_{1k11}\frac{\partial^2}{\partial \hat{g}\partial \hat{p}_{11}}\hat{p}_k=\frac{2D_2}{15}\frac{\partial^2}{\partial \hat{g}\partial \hat{p}_{11}}\hat{p}_1. \tag{85}$$

The unclosed equation for the probability density function of the scalar G is derived in the Annex A. It is worth noting that this equation can be obtained directly from Eqs (51), (83) integrating over whole values of $\hat{p}_i$ and $\hat{p}_{ij}$.

The structure of Eq. (83) permits, using dimensional considerations, to write the general relation for the mean turbulent burning velocity $u_T(t)$, defined by the following equation

$$u_T(t) = u_F \int_{-\infty}^{+\infty} \langle |\nabla c(\mathbf{x},t)| \rangle dx_1 , \quad |\nabla c(\mathbf{x},t)| = \delta(G(\mathbf{x},t)) |\nabla G(\mathbf{x},t)| . \qquad (86)$$

For the initial conditions (3), (78) the relations (86) and (81) are equivalent, [4], [7]

Using dimensional analysis we can write

$$u_T(t) = u_F f\left(\frac{\sqrt{D_0 D_2}}{u_F}, \frac{\lambda_4}{\lambda_2}, tD_2\right). \qquad (87)$$

where f is function, depending on two dimensionless parameters and dimensionless time. That function can not be determined only from the dimensional analysis. The characteristic lengths λ_2 and λ_4 are defined by the expressions

$$D_2 = \frac{D_0}{\lambda_2^2}, \quad D_4 = \frac{D_0}{\lambda_4^4}. \qquad (88)$$

Writing (87) we took into consideration that the dimensions of D_0, D_2 and D_4 are the following

$$[D_0] = L^2 / T , \quad [D_2] = 1/T , \quad [D_4] = 1/L^2 T . \qquad (89)$$

The stationary regime (if exists) is described by the following equation

$$u_{Tst} = u_F f_{st}\left(\frac{\sqrt{D_0 D_2}}{u_F}, \frac{\lambda_4}{\lambda_2}\right). \qquad (90)$$

where

$$f_{st}\left(\frac{\sqrt{D_0D_2}}{u_F},\frac{\lambda_4}{\lambda_2}\right)=f\left(\frac{\sqrt{D_0D_2}}{u_F},\frac{\lambda_4}{\lambda_2},\infty\right).$$

4. Special Case: Random Shear flow

Let us apply the general equations obtained above to the 2D random shear flow. The velocity flow field reads

$$\mathbf{u}(x,y,t)=(u(x,y,t),v(x,y,t))=(\dot{W}(t)y,0),\qquad(91)$$
$$u(x,y,t)=\dot{W}(t)y,\ v(x,y,t)=0,$$

where $W(t)$ denotes the standard Wiener process $W(0)=0$, $\dot{W}(t)$ its temporal derivative, i.e. temporal white noise

$$\langle\dot{W}(t)\dot{W}(t')\rangle=\delta(t-t').\qquad(93)$$

Single nonzero velocity correlation is equal to

$$B(y,y',t,t')=\langle u(y,t)u(y',t')\rangle=\delta(t-t')yy'=2\delta(t-t')B_{ij}^{eff}(y,y',t),\qquad(93)$$

$$B^{eff}(y,y',t)=\frac{1}{2}\int_{-\infty}^{\infty}B(y,y',t,t')dt'=\frac{1}{2}yy'\int_{-\infty}^{\infty}\delta(t-t')dt'=\frac{1}{2}yy',\qquad(94)$$

$$B^{eff}(y,y,t)=\frac{1}{2}\int_{-\infty}^{\infty}B(y,y,t,t')dt'=\frac{1}{2}yy\int_{-\infty}^{\infty}\delta(t-t')dt'=\frac{1}{2}y^2.\qquad(95)$$

The equations (1), (6) become

$$\frac{\partial G}{\partial t}+\dot{W}(t)y\frac{\partial G}{\partial x}=u_Fp,\ H_L=u_Fp,\qquad p=\sqrt{p_x^2+p_y^2},\qquad(96)$$

$$\frac{\partial p_x}{\partial t} + \dot{W}(t) y \frac{\partial p_x}{\partial x} = u_F \frac{p_x}{p} p_{xx}, \tag{97}$$

$$\frac{\partial p_y}{\partial t} + \dot{W}(t) y \frac{\partial p_y}{\partial x} + \dot{W}(t) p_x = u_F \frac{p_y}{p} p_{yy}. \tag{98}$$

The initial conditions are

$$G = -x, \; p_x = -1, \; p_y = 0. \tag{99}$$

It is readily checked that the solution of the initial boundary value problem (IBVP) (96)-(99) is given by the following expression

$$G(x, y, t) = -x + W(t) y + u_F \int_0^t \sqrt{1 + W(\tau)^2}\, d\tau = -x + x_f(y, t), \tag{100}$$

$$x_f(y, t) = W(t) y + l(t), \; l(t) = u_F \int_0^t \sqrt{1 + W(\tau)^2}\, d\tau. \tag{101}$$

where $x_f(y,t)$ is the longitudinal flame front coordinate at given y. As it is seen from (101) in the problem under consideration the flame is single-valued.

From (100) follows

$$p_x(x, y, t) = -1, \; p_y(x, y, t) = W, \; p(x, y, t) = \sqrt{p_x p_x + p_y p_y} = \sqrt{1 + W^2}, \tag{102}$$

$$p_{xx}(x, y, t) = p_{yy}(x, y, t) = p_{xy}(x, y, t) = 0, \tag{103}$$

$$H_L = u_F p = u_F \sqrt{1 + W^2}. \tag{104}$$

The indicator function for this case becomes

$$\Phi(x,y,t,\hat{G},\hat{p}_x,\hat{p}_y) = \delta(G(x,y,t)-\hat{G})\delta(p_x(x,y,t)-\hat{p}_x)\delta(p_y(x,y,t)-\hat{p}_y) = \\ = \delta(G(x,y,t)-\hat{G})\delta(p_x+1)\delta(W-\hat{p}_y) \tag{105}$$

taking into account that, according to (103), $p_x(x,y,t)$ has a sharp value $p_x(x,y,t) = -1$. We consider below the reduced indicator function

$$\Phi'(x,y,t,\hat{G},\hat{p}_y) = \delta(G(x,y,t)-\hat{G})\delta(p_y(x,y,t)-\hat{p}_y) = \delta(G(x,y,t)-\hat{G})\delta(W-\hat{p}_y). \tag{106}$$

The Liouville equation for the reduced indicator function (106) can be obtained from (27)

$$\frac{\partial\Phi}{\partial t} + \left(u_k - \frac{\partial\hat{H}_L}{\partial\hat{p}_k}\right)\frac{\partial\Phi}{\partial x_k} - \frac{\partial u_k}{\partial x_i}\frac{\partial}{\partial\hat{p}_i}(\hat{p}_k\Phi) = 0\,, \tag{107}$$

taking into account the expression for the velocity field (91), and Eqs (101)-(103). Liouville equation and initial condition read

$$\frac{\partial\Phi'}{\partial t} + \dot{W}(t)y\frac{\partial\Phi'}{\partial x} - u_F\frac{\hat{p}_x}{\hat{p}}\frac{\partial\Phi'}{\partial x} - u_F\frac{\hat{p}_y}{\hat{p}}\frac{\partial\Phi'}{\partial y} + \dot{W}\frac{\partial\,\Phi'}{\partial\hat{p}_y} = 0, \tag{108}$$

$$\Phi'(x,y,t,\hat{G},\hat{p}_y) = \delta(x+\hat{G})\delta(\hat{p}_y)\,, \tag{109}$$

where (99) was used.

The sum of the third and forth terms in the LHS of (108) can be transformed using the expressions for $\dfrac{\partial\Phi'}{\partial x}$ and $\dfrac{\partial\Phi'}{\partial y}$ which follow from (106)

$$\frac{\partial\Phi'}{\partial x} = -\frac{\partial\,\Phi'}{\partial\hat{G}}\frac{\partial G}{\partial x} = -\frac{\partial\,\Phi'}{\partial\hat{G}}p_x = -\frac{\partial\,\Phi'}{\partial\hat{G}}\hat{p}_x, \tag{110}$$

$$\frac{\partial\Phi'}{\partial y} = -\frac{\partial\,\Phi'}{\partial\hat{G}}\frac{\partial G}{\partial y} = -\frac{\partial\,\Phi'}{\partial\hat{G}}\,p_y = -\frac{\partial\,\Phi'}{\partial\hat{G}}\,\hat{p}_y. \tag{111}$$

We obtain

$$u_F\frac{\hat{p}_x}{\hat{p}}\frac{\partial\Phi'}{\partial x} + u_F\frac{\hat{p}_y}{\hat{p}}\frac{\partial\Phi'}{\partial y} = u_F\frac{\hat{p}_x\hat{p}_x}{\hat{p}}\frac{\partial\,\Phi'}{\partial\hat{G}} + u_F\frac{\hat{p}_y\hat{p}_y}{\hat{p}}\frac{\partial\,\Phi'}{\partial\hat{G}} = u_F\sqrt{\hat{p}_x^2 + \hat{p}_y^2}\,\frac{\partial\,\Phi'}{\partial\hat{G}}. \tag{112}$$

Thus (108) becomes

$$\frac{\partial\Phi'}{\partial t} + \dot{W}(t)\,y\frac{\partial\Phi'}{\partial x} + u_F\sqrt{1+W^2}\,\frac{\partial\,\Phi'}{\partial\hat{G}} + \dot{W}\frac{\partial\,\Phi'}{\partial\hat{p}_y} = 0, \tag{113}$$

taking into account (102), $p_y(x, y, t) = W$.

The same result is obtained from (19) or using directly the relation

$$-\frac{\partial\hat{H}_L}{\partial\hat{p}_k}\frac{\partial\Phi}{\partial x_k} - \frac{\partial\hat{H}_L}{\partial\hat{p}_k}\,\hat{p}_k\,\frac{\partial\Phi}{\partial\hat{G}} = 0. \tag{114}$$

We rewrite (113) introducing for convenience purposes new notations φ and ω for the sample values of random functions G and $\hat{p}_y(x, y, t) = W$

$$\frac{\partial\Phi'}{\partial t} + \dot{W}(t)\,y\frac{\partial\Phi'}{\partial x} + u_F\sqrt{1+W^2}\,\frac{\partial\,\Phi'}{\partial\varphi} + \dot{W}\frac{\partial\,\Phi'}{\partial\omega} = 0. \tag{115}$$

Averaging (115) and (109) yields

$$\frac{\partial P_{WG}}{\partial t} = \frac{y^2}{2}\frac{\partial^2 P_{WG}}{\partial x^2} + \frac{1}{2}\frac{\partial^2 P_{WG}}{\partial\omega^2} - u_F\sqrt{1+\omega^2}\,\frac{\partial P_{WG}}{\partial\varphi}, \tag{116}$$

$$P_{WG}(x, y, t = 0, \omega, \varphi) = \delta(\omega)\delta(x+\varphi), \tag{117}$$

where

$$P_{WG}(x,y,t,\omega,\varphi)=\langle\Phi'(x,y,t,\varphi,\omega)\rangle \tag{118}$$

is the joint probability density function of W and G.

Integrating (116) over whole positive values of φ we obtain the mean progress variable equation

$$\frac{\partial\langle c\rangle}{\partial t}=\frac{y^2}{2}\frac{\partial^2\langle c\rangle}{\partial x^2}+u_F\int_0^{\infty}\int_{-\infty}^{\infty}\sqrt{1+\omega^2}\,P_{WG}\,d\omega d\varphi. \tag{119}$$

We conclude from (116) and (119) that the probability density function P_{WG} and $\langle c\rangle$ diffuse in the physical space with the diffusion coefficient equal to $\frac{y^2}{2}$. Probability density function P_{WG} diffuse in the sample space ω with the diffusion coefficient equal to $\frac{1}{2}$.

Let us rewrite (119) in another form, using the definition (30). Inserting the expression (100) into (30) yields

$$c(x,y,t)=H\left(Wy+u_F\int_0^t\sqrt{1+(W(\tau))^2}\,d\tau-x\right)=H\left(x_f(y,t)-x\right) \tag{120}$$

by virtue of (100). We conclude from (120) that from a formal viewpoint to find the mean value of $c(x,y,t)$ we need to calculate the integral over trajectories of Wiener process $W(t)$ in the time interval $0<\tau<t$.

Let us consider first the chemically neutral case, i.e. $u_F=0$. This case corresponds to turbulent diffusion of the burned gases placed initially in the left half-space. Then (120) becomes

$$\langle c(x,y,t)\rangle=\langle H(Wy-x)\rangle. \tag{121}$$

We can easily perform averaging (121) using only one-time Wiener process probability density function. Indeed, the derivative of the mean progress variable $\langle c(x,y,t)\rangle$ with respect to x reads

$$\frac{\partial\langle c(x,y,t)\rangle}{\partial x}=\langle\delta(Wy-x)\rangle=\int_{-\infty}^{\infty}\delta(\omega y-x)P_W(t,\omega)d\omega=-\frac{1}{\sqrt{2\pi}\sigma_c}\exp\left(-\frac{x^2}{2\sigma_c{}^2}\right), \tag{122}$$

$$\sigma_c{}^2=y^2t, \tag{123}$$

taking into account that

$$P_W(t,\omega)=-\frac{1}{\sqrt{2\pi}\sigma}\exp\left(-\frac{\omega^2}{2\sigma^2}\right), \qquad \sigma^2=t. \tag{124}$$

We conclude from (122), (123) that the equation for $\langle c(x,y,t)\rangle$ reads

$$\frac{\partial\langle c\rangle}{\partial t}=\frac{y^2}{2}\frac{\partial^2\langle c\rangle}{\partial x^2}. \tag{125}$$

Eq. (125) is the diffusion equation with the diffusion coefficient

$$D=\frac{y^2}{2}. \tag{126}$$

Let us return to the reactive case. To proceed we introduce the joint probability density function $P_{Wlx_f}(\omega,\eta,\xi)$ of W, l and x_f ,. We have, taking into accout (100)

$$P_{Wlx_f}(\omega,\eta,\xi)=P_{Wl}(\omega,\eta)\delta(\xi-y\omega-\eta), \tag{127}$$

where ω, η, ξ are the sample variables for the random functions W, l, x_f, respectively. Then the first expression in (120) can be rewritten in as follows

$$\langle c(x,y,t)\rangle = \int_{-\infty}^{\infty}\int_{-\infty}^{\infty}\int_{-\infty}^{\infty} H(y\omega+\eta-x)P_{Wlx_f}(\omega,\eta,\xi)d\omega d\eta d\xi = \\ = \int_{-\infty}^{\infty}\int_{-\infty}^{\infty}\int_{-\infty}^{\infty} H(y\omega+\eta-x)P_{Wl}(\omega,\eta)\delta(\xi-y\omega-\eta)d\omega d\eta d\xi \quad . \tag{128}$$

The second line of (128) becomes after integration over η

$$\langle c(x,y,t)\rangle = \int_{-\infty}^{\infty}\int_{-\infty}^{\infty} H(\xi-x)P_{Wl}(\omega,\xi-y\omega)d\omega d\xi d\omega d\xi = \int_{-\infty}^{\infty}\int_{x}^{\infty} P_{Wl}(\omega,\xi- \\ -y\omega)d\omega d\xi d\omega d\xi \tag{129}$$

by virtue of the sifting property of the delta-function. (129) can be presented in the form

$$\langle c(x,y,t)\rangle = \int_{-\infty}^{\infty} H(\xi-x)P_{x_f}(\xi)d\xi = \int_{x}^{\infty} P_{x_f}(\xi)d\xi \tag{130}$$

by virtue of the second expression in (120), where

$$P_{x_f}(\xi) = \int_{\omega} P_{Wl}(\omega,\xi-y\omega)d\omega \tag{131}$$

is the flame front position probability density function.

Let us derive now the equation for probability density function $P_{Wl}(\omega,\eta)$. The SDE for $W(t)$ and $l(t)$ read

$$\frac{dW}{dt} = \dot{W}\,, \tag{132}$$

$$\frac{dl}{dt} = u_F \sqrt{1+W^2} \, . \tag{133}$$

Hence the equation for probability density function $P_{Wl}(\omega,\eta)$ reads

$$\frac{\partial P_{Wl}}{\partial t} = \frac{1}{2}\frac{\partial^2 P_{Wl}}{\partial \omega^2} - u_F \frac{\partial}{\partial \eta}\left(\sqrt{(1+\omega^2)} P_{Wl}\right). \tag{134}$$

(134) is augmented by the following initial conditions

$$P_{Wl}(\omega,\eta) = \delta(\omega)\delta(\eta) \, , \; t = 0 \, , \tag{135}$$

because $W(0) = l(0) = 0$.

Let us differentiate (129) with respect to t

$$\frac{\partial \langle c \rangle}{\partial t} = \int_{-\infty}^{\infty}\int_{x}^{\infty} \frac{\partial}{\partial t} P_{Wl}(\omega, \xi - y\omega) d\omega d\xi \, . \tag{136}$$

Inserting the time derivative $\dfrac{\partial P_{Wl}}{\partial t}$ from (134) into (136) yields

$$\frac{\partial \langle c \rangle}{\partial t} = \int_{-\infty}^{\infty}\int_{x}^{\infty} \left[\frac{1}{2} \frac{\partial^2 P_{Wl}(\omega,\eta)}{\partial \omega^2} - u_F \frac{\partial}{\partial \eta}\left(\sqrt{(1+\omega^2)} P_{Wl}(\omega,\eta)\right) \right]_{\eta = \xi - y\omega} d\omega d\xi \tag{137}$$

After some simple algebra (137) becomes

$$\frac{\partial \langle c \rangle}{\partial t} = \frac{y^2}{2}\frac{\partial \langle c \rangle}{\partial x^2} + u_F \int_{-\infty}^{\infty}\int_{x}^{\infty} \sqrt{(1+\omega^2)} P_{Wl}(\omega, \xi - y\omega) d\omega d\xi \, . \tag{138}$$

The last term in (138) is nothing, but the expected chemical source $\langle W(\mathbf{x},t) \rangle = u_F \langle |\nabla c(\mathbf{x},t)| \rangle$ in (33). Indeed, we have

$$u_F\left|\nabla c(\mathbf{x},t)\right| = \delta(G(\mathbf{x},t))\left|\nabla G(\mathbf{x},t)\right|, \tag{139}$$

since, according to (31)

$$\left|\nabla c(\mathbf{x},t)\right| = \delta(G(\mathbf{x},t))\left|\nabla G(\mathbf{x},t)\right|. \tag{140}$$

We obtain from (100)

$$\nabla G(\mathbf{x},t) = \sqrt{1+W^2}\,,\ \delta(G(\mathbf{x},t)) = \delta(Wy + l(t) - x) = \delta(x_f(y,t) - x)\,, \tag{141}$$

$$u_F\left|\nabla c(\mathbf{x},t)\right| = u_F\sqrt{1+W^2}\,\delta(Wy + l(t) - x). \tag{142}$$

Expected (142) is the source in (138).

CONCLUSION

The closed equation for the extended indicator function (or fine-grained probability density function) which includes both $G(\mathbf{x},t)$ and its spatial derivatives of the first and the second orders is derived. Averaging the equation for the extended indicator function the closed equation for the probability density function for both $G(\mathbf{x},t)$ and its spatial derivatives of the first and the second orders is derived in a white-noise in time statistically homogeneous isotropic constant density turbulent flow. The closed equation for the probability density function for both $G(\mathbf{x},t)$ and its spatial derivatives of the first order is derived in a special case of a random shear flow, non homogeneous and non isotropic velocity field.

ACKNOWLEDGMENTS

The work was supported by ONERA. The author is grateful to V. I. Klyatskin for valuable discussion of Furutsu-Novikov formula.

ANNEX A. DERIVATION OF EQUATION FOR THE PROBABILITY DENSITY FUNCTION OF THE SCALAR G

Let us introduce a reduced indicator function

$$\Phi_G(\mathbf{x},t,\hat{G}) = \delta(G(\mathbf{x},t)-\hat{G}). \tag{A1}$$

Differentiating (A1) with respect to time

$$\frac{\partial\Phi_G}{\partial t} = \frac{\partial\,\Phi_G}{\partial G}\frac{\partial G}{\partial t} = -\frac{\partial\,\Phi_G}{\partial\hat{G}}\frac{\partial G}{\partial t}, \tag{A2}$$

and taking into account PDE's (1), (6), (7) yields the Liouville equation for the indicator function Φ_G

$$\frac{\partial\Phi_G}{\partial t} + \frac{\partial\,\Phi_G}{\partial\hat{G}}\frac{\partial G}{\partial t} = 0. \tag{A3}$$

$$\frac{\partial\Phi_G}{\partial t} + \frac{\partial}{\partial\hat{G}}\left[\left(-u_k\frac{\partial G}{\partial x_k} + H_L\right)\Phi_G\right] = 0. \tag{A4}$$

To recast (A4) we need the relation for the spatial derivative of the indicator function Φ, which follows from its definition (A1)

$$\frac{\partial\Phi_G}{\partial x_k} = \frac{\partial\,\Phi_G}{\partial G}\frac{\partial G}{\partial x_k} = -\frac{\partial\,\Phi_G}{\partial\hat{G}}\frac{\partial G}{\partial x_k}. \tag{A5}$$

Collecting the convective terms in (A.4) it becomes

$$\frac{\partial\Phi_G}{\partial t} + u_k\frac{\partial\Phi_G}{\partial x_k} + \frac{\partial}{\partial\hat{G}}\left(\hat{H}_L\Phi_G\right) = 0 \tag{A6}$$

by virtue of (A5). Eq. (A6) is not closed, since H_L contains the derivatives $\frac{\partial G}{\partial x_k}$, Eq. (8).

It is worth noting that Eq. (A6) can be obtained also from (27), taking into account the relation between the indicator functions Φ_G and Φ

$$\Phi_G(\mathbf{x},t,\hat{G}) = \int_{-\infty}^{\infty}\int_{-\infty}^{\infty} \Phi(\mathbf{x},t,\hat{G},\hat{p}_i,\hat{p}_{ij})d\hat{p}_i d\hat{p}_{ij} \,. \tag{A7}$$

Integrating (27) and (24) over $\hat{p}_i$ and $\hat{p}_{ij}$ yields

$$\frac{\partial \Phi_G}{\partial t} + u_k \frac{\partial \Phi_G}{\partial x_k} - \frac{\partial \hat{H}_L}{\partial \hat{p}_k}\frac{\partial \Phi_G}{\partial x_k} + \int_{-\infty}^{\infty}\int_{-\infty}^{\infty} \frac{\partial^2 \hat{H}_L}{\partial \hat{p}_i \partial \hat{p}_k} \hat{p}_{ki}\Phi(\mathbf{x},t,\hat{G},\hat{p}_i,\hat{p}_{ij})d\hat{p}_i d\hat{p}_{ij} = 0. \tag{A8)}$$

$$-\frac{\partial \hat{H}_L}{\partial \hat{p}_k}\frac{\partial \Phi_G}{\partial x_k} - \int_{-\infty}^{\infty}\int_{-\infty}^{\infty} \frac{\partial \hat{H}_L}{\partial \hat{p}_k}\hat{p}_k \frac{\partial}{\partial \hat{G}}\Phi d\hat{p}_i d\hat{p}_{ij} + \int_{-\infty}^{\infty}\int_{-\infty}^{\infty} \frac{\partial^2 \hat{H}_L}{\partial \hat{p}_i \partial \hat{p}_k} \hat{p}_{ki}\Phi d\hat{p}_i d\hat{p}_{ij} = 0, \tag{A9}$$

respectively. Expression (A9) transforms to

$$-\frac{\partial \hat{H}_L}{\partial \hat{p}_k}\frac{\partial \Phi_G}{\partial x_k} - \int_{-\infty}^{\infty}\int_{-\infty}^{\infty} \frac{\partial}{\partial \hat{G}}\hat{H}_L \Phi d\hat{p}_i d\hat{p}_{ij} + \int_{-\infty}^{\infty}\int_{-\infty}^{\infty} \frac{\partial^2 \hat{H}_L}{\partial \hat{p}_i \partial \hat{p}_k} \hat{p}_{ki}\Phi d\hat{p}_i d\hat{p}_{ij} = 0, \tag{A10}$$

by virtue of (26).

Excluding the last term in the LHS of (A8) with the aid of (A10) yields Eq. (A.6).

Equation for the progress variable, (30)

$$c(\mathbf{x},t) = \int_0^{\infty} \Phi_G(\mathbf{x},t,\hat{G})d\hat{G} = \int_0^{\infty} \delta(G(\mathbf{x},t) - \hat{G})d\hat{G} = H(G(\mathbf{x},t))$$

is obtained integrating Eq. (A6) over $\hat{G}$ in the range $0 \le \hat{G} \le \infty$

$$\frac{\partial c}{\partial t} + u_k \frac{\partial c}{\partial x_k} - H_L \delta(G(\mathbf{x},t)) = 0\,, \tag{A11}$$

$$\frac{\partial c}{\partial t} + u_k \frac{\partial c}{\partial x_k} = H_L \delta(G(\mathbf{x},t)) = u_F \left|\nabla G\right| \delta(G(\mathbf{x},t)) = u_F \left|\nabla c\right|. \tag{A12}$$

The probability density function $P_G(\mathbf{x},t,\hat{G})$ is a mean of the indicator function Φ_G, (A1)

$$\left\langle \Phi_G \right\rangle = \left\langle \delta(G(\mathbf{x},t) - \hat{G}) \right\rangle = P_G(\mathbf{x},t,\hat{G}). \tag{A13}$$

Averaging (A6) yields the equation for the probability density function $P_G(\mathbf{x},t,\hat{G})$

$$\frac{\partial P_G}{\partial t} = D_0 \frac{\partial^2 P_G}{\partial x_k \partial x_k} - \frac{\partial}{\partial \hat{G}} \left(\left\langle H_L \middle| G = \hat{G} \right\rangle P_G \right) = 0, \tag{A14}$$

where $\left\langle H_L \middle| G = \hat{G} \right\rangle$ is the conditional mean of $H_L = u_F \left|\nabla G\right| = u_F p$, i.e.

$$\left\langle H_L \middle| G = \hat{G} \right\rangle P_G(\mathbf{x},t,\hat{G}) = \int_{-\infty}^{\infty} \int_{-\infty}^{\infty} \hat{H}_L P_{Gp_i p_{ij}}(\mathbf{x},t,\hat{G},\hat{p}_i,\hat{p}_{ij}) d\hat{p}_i d\hat{p}_{ij}\,. \tag{A15}$$

Eq. (A14) can be obtained also from (51), taking into account the relation between two probability density functions $P_G(\mathbf{x},t,\hat{G})$ and $P_{Gp_i p_{ij}}(\mathbf{x},t,\hat{G},\hat{p}_i,\hat{p}_{ij})$

$$P_G(\mathbf{x},t,\hat{G}) = \int_{-\infty}^{\infty} \int_{-\infty}^{\infty} P_{Gp_i p_{ij}}(\mathbf{x},t,\hat{G},\hat{p}_i,\hat{p}_{ij}) d\hat{p}_i d\hat{p}_{ij} = \int_{-\infty}^{\infty} \int_{-\infty}^{\infty} P(\mathbf{x},t,\hat{G},\hat{p}_i,\hat{p}_{ij}) d\hat{p}_i d\hat{p}_{ij}\,. \tag{A16}$$

Integrating (51) over $\hat{p}_i$ and $\hat{p}_{ij}$ yields

$$\frac{\partial P_G}{\partial t} = D_0 \frac{\partial^2 P_G}{\partial x_k \partial x_k} + \frac{\partial}{\partial x_k} \int_{-\infty}^{\infty}\int_{-\infty}^{\infty} \frac{\partial \hat{H}_L}{\partial \hat{p}_k} P d\hat{p}_i d\hat{p}_{ij} - \int_{-\infty}^{\infty}\int_{-\infty}^{\infty} \frac{\partial^2 \hat{H}_L}{\partial \hat{p}_i \partial \hat{p}_k} \hat{p}_{ki} P d\hat{p}_i d\hat{p}_{ij}\,. \tag{A17}$$

The sum of the second and third terms in the RHS of Eq. (A17) can be transformed averaging (A10)

$$-\frac{\partial}{\partial x_k} \int_{-\infty}^{\infty}\int_{-\infty}^{\infty} \frac{\partial \hat{H}_L}{\partial \hat{p}_k} P d\hat{p}_i d\hat{p}_{ij} + \int_{-\infty}^{\infty}\int_{-\infty}^{\infty} \frac{\partial^2 \hat{H}_L}{\partial \hat{p}_i \partial \hat{p}_k} \hat{p}_{ki} P d\hat{p}_i d\hat{p}_{ij} - \int_{-\infty}^{\infty}\int_{-\infty}^{\infty} \frac{\partial}{\partial \hat{G}} \hat{H}_L P d\hat{p}_i d\hat{p}_{ij} = 0.$$

Therefore

$$\frac{\partial}{\partial x_k} \int_{-\infty}^{\infty}\int_{-\infty}^{\infty} \frac{\partial \hat{H}_L}{\partial \hat{p}_k} P d\hat{p}_i d\hat{p}_{ij} - \int_{-\infty}^{\infty}\int_{-\infty}^{\infty} \frac{\partial^2 \hat{H}_L}{\partial \hat{p}_i \partial \hat{p}_k} \hat{p}_{ki} P d\hat{p}_i d\hat{p}_{ij} = -\int_{-\infty}^{\infty}\int_{-\infty}^{\infty} \frac{\partial}{\partial \hat{G}} \hat{H}_L P d\hat{p}_i d\hat{p}_{ij}\,. \tag{A18}$$

Plugging (A18) into (A17) yields Eq. (A14)

$$\frac{\partial P_G}{\partial t} = D_0 \frac{\partial^2 P_G}{\partial x_k \partial x_k} - \frac{\partial}{\partial \hat{G}}\left(\left\langle H_L \middle| G = \hat{G}\right\rangle P_G\right) = 0, \tag{A19}$$

because

$$\frac{\partial}{\partial \hat{G}} \int_{-\infty}^{\infty}\int_{-\infty}^{\infty} \hat{H}_L P_{Gp_ip_{ij}}(\mathbf{x}, t, \hat{G}, \hat{p}_i, \hat{p}_{ij}) d\hat{p}_i d\hat{p}_{ij} = \frac{\partial}{\partial \hat{G}}\left(\left\langle H_L \middle| G = \hat{G}\right\rangle P_G\right).$$

Eqs (51) and (A14) can be simplified for the case of the initial condition (3), (78)

$$P_{Gp_ip_{ij}}(\mathbf{x}, t = 0, \hat{G}, \hat{p}_i, \hat{p}_{ij}) = \delta(x_1 + \hat{G})\delta(\delta_{1i} - \hat{p}_i)\delta(\hat{p}_{ij})\,. \tag{A20}$$

(A20) yields for the reduced probability density function $P_G(\mathbf{x}, t = 0, \hat{G})$

$$P_G(\mathbf{x},t=0,\hat{G}) = \delta(x_1+\hat{G}). \tag{A21}$$

In general case of initial condition (2) , the probability density functions $P_{Gp_ip_{ij}}(x_1,t,\hat{G},\hat{p}_i,\hat{p}_{ij})$ and $P_G(x_1,t,\hat{G})$ depend on x_1 and t, sample space variables $\hat{G},\hat{p}_i,\hat{p}_{ij}$ and $\hat{G}$, respectively. But for the case when $G(\mathbf{x},t)$ solves (1) with the initial condition (3), the field $g(\mathbf{x},t)$ in the decomposition (82) is statistically homogeneous. As a consequence, the $\hat{G}$ sample value and the coordinate x_1 enter into $P_{Gp_ip_{ij}}(x_1,t,\hat{G},\hat{p}_i,\hat{p}_{ij})$ and $P_G(x_1,t,\hat{G})$ through the combination $\hat{G}-\langle G\rangle$ [7], i.e.

$$P_{Gp_ip_{ij}}(\mathbf{x},t,\hat{G},\hat{p}_i,\hat{p}_{ij}) = P_{gp_ip_{ij}}(t,\hat{g},\hat{p}_i,\hat{p}_{ij}),\ \hat{g} = \hat{G}-\langle G\rangle, \tag{A22}$$

$$P_G(\mathbf{x},t,\hat{G}) = P_g(t,\hat{G}-\langle G\rangle), \tag{A23}$$

where

$$\langle G\rangle = -x_1 + X(t),\ X(t) = \int_0^t u_T(\tau)d\tau, \tag{A24}$$

$$u_T(t) = \langle H_L\rangle = u_F\langle|\nabla G|\rangle, \tag{A25}$$

$$g(\mathbf{x},t) = G(\mathbf{x},t) - \langle G\rangle = G(\mathbf{x},t) + x_1 - u_T(t),$$
$$G = \langle G\rangle + g(\mathbf{x},t) = -x_1 + u_T(t) + g(\mathbf{x},t), \tag{A26}$$

where $g(\mathbf{x},t)$ is a zero mean statistically homogeneous field.

We transform Eq. (A14) using the following expressions for the derivatives of P_G

$$\frac{\partial P_G}{\partial t}=\frac{\partial P_g}{\partial t}-\frac{\partial P_g}{\partial \hat{g}}\frac{\partial\langle G\rangle}{\partial t}=\frac{\partial P_g}{\partial t}-\frac{\partial P_g}{\partial \hat{g}}u_T(t),\ \frac{\partial\langle G\rangle}{\partial t}=u_T(t), \tag{A27}$$

$$\frac{\partial P_G}{\partial x_1}=\frac{\partial P_g}{\partial x_1}-\frac{\partial P_g}{\partial \hat{g}}\frac{\partial\langle G\rangle}{\partial x_1}=\frac{\partial P_g}{\partial \hat{g}},\ \frac{\partial\langle G\rangle}{\partial x_1}=-1, \tag{A28}$$

$$\frac{\partial^2 P_G}{\partial x_1{}^2}=\frac{\partial}{\partial x_1}\left(\frac{\partial P_g}{\partial \hat{g}}\right)-\frac{\partial}{\partial \hat{g}}\left(\frac{\partial P_g}{\partial \hat{g}}\right)\frac{\partial\langle G\rangle}{\partial x_1}=\frac{\partial^2 P_g}{\partial \hat{g}^2}, \tag{A29}$$

taking into account (A24) and the statistical homogeneity of the field $g(\mathbf{x},t)$, i.e.

$$\frac{\partial P_g}{\partial x_1}=0. \tag{A30}$$

Eq. (A14) transforms to

$$\frac{\partial P_g}{\partial t}-\frac{\partial P_g}{\partial \hat{g}}u_T(t)=D_0\frac{\partial^2 P_g}{\partial \hat{g}^2}-\frac{\partial}{\partial \hat{g}}\left(\langle H_L|\, g=\hat{g}\rangle P_g\right), \tag{A31}$$

$$\frac{\partial P_g}{\partial t}=D_0\frac{\partial^2 P_g}{\partial \hat{g}^2}-\frac{\partial}{\partial \hat{g}}\left(\langle H_L|g=\hat{g}\rangle-u_T(t)\right)P_g. \tag{A32}$$

where $u_T(t)$ is given by Eq. (A25).

Eq. (A32) transforms to

$$\frac{\partial P_g}{\partial t}=D_0\frac{\partial^2 P_g}{\partial \hat{g}^2}-\frac{\partial}{\partial \hat{g}}\left(\langle H_L|g=\hat{g}\rangle-\langle H_L\rangle\right)P_g, \tag{A33}$$

by virtue of (81)

$$u_T(t)=\langle H_L\rangle=u_F\langle|\nabla G|\rangle.$$

It is worth noting that applying the hypothesis

$$\langle H_L \,|\, g = \hat{g} \rangle - \langle H_L \rangle = \frac{\langle gH_L \rangle}{\langle g^2 \rangle}\hat{g}, \tag{A34}$$

(A33) becomes

$$\frac{\partial P_g}{\partial t} = D_0 \frac{\partial^2 P_g}{\partial \hat{g}^2} - \frac{\partial}{\partial \hat{g}} \frac{\langle gH_L \rangle}{\langle g^2 \rangle} \hat{g} P_g . \tag{A35}$$

The solution of (A35) is the Gaussion function

$$P_g(t, \hat{g}) = \frac{1}{\sqrt{2\pi \langle g^2 \rangle}} \exp\left(-\frac{\hat{g}^2}{2\langle g^2 \rangle} \right). \tag{A36}$$

Multiplying (A33) by $\hat{g}^2$ and performing integration over whole $\hat{g}$ yields the equation for the second moment $\langle g^2 \rangle$

$$\frac{d\langle g^2 \rangle}{dt} = 2D_0 + 2\langle gH_L \rangle . \tag{A37}$$

Stationary solution of (A37) reads

$$\langle gH_L \rangle = -u_F \langle g|\nabla G| \rangle = -u_F \left\langle g\sqrt{\left(-1+\frac{\partial g}{\partial x_1}\right)^2 + \left(\frac{\partial g}{\partial x_2}\right)^2 + \left(\frac{\partial g}{\partial x_3}\right)^2} \right\rangle = D_0 . \tag{A38}$$

The equation for the moment of order n $\langle g^n \rangle$ reads

$$\frac{d\langle g^n\rangle}{dt} + n\langle g^{n-1}\rangle u_T(t) = n(n-1)\langle g^{n-2}\rangle D_0 + n\langle g^{n-1}H_L\rangle. \quad \text{(A39)}$$

ANNEX B. DERIVATION OF THE EQUATION FOR THE PROBABILITY DENSITY FUNCTION

We assume that the solenoidal velocity field is statistically homogeneous and isotropic. Hence the correlation tensor $B_{ij}(\mathbf{x}-\mathbf{x}',t,t') = \langle u_i(\mathbf{x},t)u_j(\mathbf{x}',t')\rangle$ can be presented as follows [8].

$$\langle u_i(\mathbf{x},t)u_j(\mathbf{x}',t')\rangle = B_{ij}(\mathbf{r},t,t') = \int F_{kl}(\mathbf{k},t,t')\exp(i\mathbf{kr})d^3\mathbf{k},$$
$$\mathbf{r} = \mathbf{x}-\mathbf{x}', \quad \text{(B1)}$$

where $F_{kl}(\mathbf{k},t,t')$ is the spectral tensor

$$F_{kl}(\mathbf{k},t,t') = \frac{1}{(2\pi)^3}\int B_{ij}(\mathbf{r},t,t')\exp(-i\mathbf{kr})d^3\mathbf{k}, \quad \text{(B2)}$$

$$F_{kl}(\mathbf{k},t,t') = \frac{E(k,t,t')}{4\pi k^2}\left(\delta_{kl} - \frac{k_k k_l}{k^2}\right) = \frac{E(k,t,t')}{4\pi k^2}(\delta_{kl} - n_k n_l), \quad \text{(B3)}$$

$$n_l = \frac{k_l}{k},\ n_l n_l = \frac{k_l k_l}{k^2} = 1. \quad \text{(B4)}$$

The velocity field is delta-correlated, therefore

$$B_{ij}(\mathbf{r},t,t') = 2\delta(t-t')B_{ij}^{eff}(\mathbf{r},t), \quad \text{(B5)}$$

$$B_{ij}^{eff}(\mathbf{r},t) = \frac{1}{2}\int_{-\infty}^{\infty} B_{ij}(\mathbf{r},t,t')dt' = \int_0^{\infty} B_{ij}(\mathbf{r},t,t')dt', \quad \text{(B6)}$$

$$B_{ij}^{eff}(\mathbf{r},t)=\int F_{ij}^{eff}(\mathbf{k},t)\exp(i\mathbf{kr})d^3\mathbf{k}\,, \tag{B7}$$

$$F_{ij}^{eff}(\mathbf{k},t)=\frac{1}{(2\pi)^3}\int B_{ij}^{eff}(\mathbf{r},t)\exp(-i\mathbf{kr})d^3\mathbf{k}\,, \tag{B8}$$

$$F_{ij}^{eff}(\mathbf{k},t)=\frac{\widetilde{E}(k,t)}{4\pi k^2}\left(\delta_{kl}-\frac{k_k k_l}{k^2}\right)=\frac{\widetilde{E}(k,t)}{4\pi k^2}\left(\delta_{kl}-n_k n_l\right), \tag{B9}$$

$$\widetilde{E}(k,t)=\int_0^\infty E(k,t,t')dt'\,, \tag{B10}$$

where $d^3\mathbf{k}$ is an elementary volume in the $\mathbf{k}$-space (wave number space), $d\Omega$ is a solid angle

$$d^3\mathbf{k}=k^2 d\Omega\,,\ d\Omega=\sin\theta d\theta d\varphi\,. \tag{B11}$$

Below the following tensors appear

$$I_{ij}=\frac{1}{4\pi}\int_\Omega n_i n_j d\Omega\,,\ I_{klij}=\frac{1}{4\pi}\int_\Omega n_i n_j n_k n_l d\Omega\,,$$

$$I_{klijmn}=\frac{1}{4\pi}\int_\Omega n_i n_j n_m n_n n_k n_l d\Omega\,. \tag{B12}$$

These tensors are constructed on the base of the isotropic tensor δ_{ij}

$$I_{ij}=\frac{1}{4\pi}\int_\Omega n_i n_j d\Omega=\alpha_2\delta_{ij}\,, \tag{B13}$$

$$I_{klij}=\frac{1}{4\pi}\int_\Omega n_i n_j n_k n_l d\Omega=\alpha_4\left(\delta_{kl}\delta_{ij}+\delta_{ki}\delta_{lj}+\delta_{kj}\delta_{li}\right), \tag{B14}$$

$$I_{klijmn} = \alpha_6(\delta_{kl}\delta_{ij}\delta_{mn} + \delta_{kl}\delta_{im}\delta_{jn} + \delta_{kl}\delta_{in}\delta_{jm} + \delta_{ki}\delta_{lj}\delta_{mn} + \delta_{ki}\delta_{lm}\delta_{jn} + \delta_{ki}\delta_{ln}\delta_{jm} +$$
$$+ \delta_{kj}\delta_{li}\delta_{mn} + \delta_{kj}\delta_{lm}\delta_{in} + \delta_{kj}\delta_{ln}\delta_{im} + \delta_{km}\delta_{li}\delta_{jn} + \delta_{km}\delta_{lj}\delta_{in} + \delta_{km}\delta_{ln}\delta_{ij} +$$
$$+ \delta_{kn}\delta_{li}\delta_{jm} + \delta_{kn}\delta_{lj}\delta_{im} + \delta_{kn}\delta_{lm}\delta_{ij}) \tag{B15}$$

Constant α_2 is found performing the contraction in (B13)

$$I_{ii} = \frac{1}{4\pi}\int_{\Omega} n_i n_i d\Omega = 1 = 3\alpha_2\,,\ \alpha_2 = \frac{1}{3} \tag{B16}$$

Constant α_4 is found performing the contraction in (B14)

$$I_{kkii} = \frac{1}{4\pi}\int_{\Omega} n_k n_k d\Omega = \frac{1}{4\pi}\int_{\Omega} d\Omega = 1 = 15\alpha_4\,,\ \alpha_4 = \frac{1}{15}. \tag{B17}$$

Similarly, performing the contraction in (B15), we obtain the constant α_6

$$I_{kkiimm} = \frac{1}{4\pi}\int_{\Omega} n_m n_m d\Omega = \frac{1}{4\pi}\int_{\Omega} d\Omega = 1 = \alpha_6(3\cdot 3\cdot 3 + 3\cdot 3 + 3\cdot 3 +$$
$$+ 3\cdot 3 + 3 + 3 + 3\cdot 3 + 3 + 3 + 3 + 3 + 3\cdot 3 + 3 + 3 + 3\cdot 3) = 105\alpha_6$$
$$\alpha_6 = \frac{1}{105}. \tag{B18}$$

The averaging (27) is done with the aid of the Furutsu-Novikov formula, e.g. [5], [6]

$$\langle u_i(\mathbf{x},t)R[\mathbf{u}]\rangle = \int d\mathbf{x}'\int dt'\langle u_j(\mathbf{x}',t')u_i(\mathbf{x},t)\rangle\left\langle \frac{\delta R[\mathbf{u}]}{\delta u_j(\mathbf{x}',t')}\right\rangle,\quad 0 \le t' \le t, \tag{B19}$$

where $R[\mathbf{u}]$ is a functional of the random velocity field $\mathbf{u}(\mathbf{x},t)$, $\dfrac{\delta R[\mathbf{u}]}{\delta u_j(\mathbf{x}',t')}$ is the functional derivative, and extended Furutsu-Novikov formula

$$\langle L(u_i(\mathbf{x},t))R[\mathbf{u}]\rangle = \int d\mathbf{x}' \int dt' \langle L(u_j(\mathbf{x}',t')u_i(\mathbf{x},t))\rangle \left\langle \frac{\delta R[\mathbf{u}]}{\delta u_j(\mathbf{x}',t')} \right\rangle,$$

$$0 \le t' \le t, \tag{B20}$$

$$u_i = u_i(\mathbf{x},t),\ u_j' = u_j(\mathbf{x}',t'),$$

where $L(u_i(\mathbf{x},t))$ is a linear differential operator

$$L(u_i(\mathbf{x},t)) = -u_k(\mathbf{x},t)\frac{\partial}{\partial x_k} + a_{ki}(\mathbf{x},t)\frac{\partial u_k(\mathbf{x},t)}{\partial x_i} + a_{kin}(\mathbf{x},t)\frac{\partial^2 u_k(\mathbf{x},t)}{\partial x_i \partial x_n} + \ldots. \tag{B21}$$

If the velocity field $\mathbf{u}(\mathbf{x},t)$ is delta-correlated in time, then the functional derivative $\dfrac{\delta R[\mathbf{u}]}{\delta u_j(\mathbf{x}',t')}$ in (B19), (B20) is replaced by $\left.\dfrac{\delta R[\mathbf{u}]}{\delta u_j(\mathbf{x}',t')}\right|_{t'=t-0}$

$$\langle L(u_i(\mathbf{x},t))R[\mathbf{u}]\rangle = \int d\mathbf{x}' \langle L(u_j(\mathbf{x}',t')u_i(\mathbf{x},t))\rangle \left\langle \left.\frac{\delta R[\mathbf{u}]}{\delta u_j(\mathbf{x}',t')}\right|_{t'=t-0} \right\rangle. \tag{B22}$$

In our case the functional is the indicator function Φ governed by SPDE (27). Therefore the linear operator $L(u_i(\mathbf{x},t))$ reads

$$L(u_i(\mathbf{x},t)) = -u_k\frac{\partial}{\partial x_k} + \frac{\partial u_k}{\partial x_i}\frac{\partial}{\partial \hat{p}_i}(\hat{p}_k) + \frac{\partial}{\partial \hat{p}_{in}}\left(\frac{\partial u_k}{\partial x_i}\hat{p}_{kn} + \frac{\partial u_k}{\partial x_n}\hat{p}_{ik} + \frac{\partial^2 u_k}{\partial x_i \partial x_n}\hat{p}_k\right). \tag{B23}$$

The functional derivative in (B22) is found from (27). It reads

$$\left.\frac{\delta\Phi}{\delta u_j(\mathbf{x}',t')}\right|_{t'=t-0} =$$

$$=\left\{-\delta(\mathbf{r})\frac{\partial}{\partial x_j}+\frac{\partial\delta(\mathbf{r})}{\partial x_m}\left(\frac{\partial}{\partial p_m}p_j+\frac{\partial}{\partial p_{ml}}p_{jl}\right)+\frac{\partial\delta(\mathbf{r})}{\partial x_l}\frac{\partial}{\partial p_{ml}}p_{mj}+\frac{\partial^2\delta(\mathbf{r})}{\partial x_m\partial x_l}\frac{\partial}{\partial p_{ml}}p_j\right\}\Phi(\mathbf{x},t) \tag{B24}$$

$$\mathbf{r}=\mathbf{x}-\mathbf{x}',$$

using the following formula

$$\frac{\delta u_i(\mathbf{x},t)}{\delta u_j(\mathbf{x}',t')}=\delta_{ij}\delta(\mathbf{x}-\mathbf{x}')\delta(t-t')H(t-t'),\ 0\le t'\le t. \tag{B25}$$

Averaging (27) and taking into account (B22)-(B24) results in the equation for the probability density function

$$\frac{\partial P}{\partial t}-\frac{\partial}{\partial x_k}\left(\frac{\partial \hat{H}_L}{\partial \hat{p}_k}P\right)+\frac{\partial}{\partial \hat{p}_{ij}}\left(\frac{\partial^2 \hat{H}_L}{\partial \hat{p}_k\partial \hat{p}_l}\hat{p}_{ki}\hat{p}_{lj}P\right)+\frac{\partial^2 \hat{H}_L}{\partial \hat{p}_i\partial \hat{p}_k}\hat{p}_{ki}P=$$

$$=\int d\mathbf{x}'\left\langle\left\{-B_{jk}^{eff}\frac{\partial}{\partial x_k}+\frac{\partial B_{jk}^{eff}}{\partial x_i}\frac{\partial}{\partial \hat{p}_i}(\hat{p}_k)+\frac{\partial}{\partial \hat{p}_{ij}}\left(\frac{\partial B_{jk}^{eff}}{\partial x_i}\hat{p}_{kj}+\frac{\partial B_{jk}^{eff}}{\partial x_j}\hat{p}_{ik}+\frac{\partial^2 B_{jk}^{eff}}{\partial x_i\partial x_j}\hat{p}_k\right)\right\}\right\rangle$$

$$\left\{-\delta(\mathbf{r})\frac{\partial}{\partial x_j}+\frac{\partial\delta(\mathbf{r})}{\partial x_m}\left(\frac{\partial}{\partial p_m}p_j+\frac{\partial}{\partial p_{ml}}p_{jl}\right)+\frac{\partial\delta(\mathbf{r})}{\partial x_l}\frac{\partial}{\partial p_{ml}}p_{mj}+\frac{\partial^2\delta(\mathbf{r})}{\partial x_m\partial x_l}\frac{\partial}{\partial p_{ml}}p_j\right\}P \tag{B26}$$

$$P=\langle\Phi\rangle.$$

Finally, using the properties of the delta-function in (B26), equation for the probability density function equation becomes

$$\frac{\partial P}{\partial t}-\frac{\partial}{\partial x_k}\left(\frac{\partial \hat{H}_L}{\partial \hat{p}_k}P\right)+\frac{\partial}{\partial \hat{p}_{ij}}\left(\frac{\partial^2 \hat{H}_L}{\partial \hat{p}_k \partial \hat{p}_l}\hat{p}_{ki}\hat{p}_{lj}P\right)=U_{jk}\frac{\partial^2 P}{\partial x_j \partial x_k}+U_{jkml}\frac{\partial^2}{\partial x_k \partial \hat{p}_{ml}}\left(\hat{p}_j P\right)+$$
$$+U_{jkim}\frac{\partial^2}{\partial \hat{p}_i \partial \hat{p}_m}\left(\hat{p}_k \hat{p}_j P\right)+U_{jkim}\frac{\partial^2}{\partial \hat{p}_i \partial \hat{p}_{ml}}\left(\hat{p}_{jl}\hat{p}_k P\right)+U_{jkil}\frac{\partial^2}{\partial \hat{p}_i \partial \hat{p}_{ml}}\left(\hat{p}_{mj}\hat{p}_k P\right)+$$
$$+U_{jkim}\frac{\partial^2}{\partial \hat{p}_m \partial \hat{p}_{in}}\left(\hat{p}_{kn}\hat{p}_j P\right)+U_{jkim}\frac{\partial^2}{\partial \hat{p}_{in}\partial \hat{p}_{ml}}\left(\hat{p}_{kn}\hat{p}_{jl}P\right)+U_{jkil}\frac{\partial^2}{\partial \hat{p}_{in}\partial \hat{p}_{ml}}\left(\hat{p}_{kn}\hat{p}_{mj}P\right)$$
$$+U_{jknm}\frac{\partial^2}{\partial \hat{p}_m \partial \hat{p}_{in}}\left(\hat{p}_{ik}\hat{p}_j P\right)+U_{jknm}\frac{\partial^2}{\partial \hat{p}_{in}\partial \hat{p}_{ml}}\left(\hat{p}_{ik}\hat{p}_{jl}P\right)+U_{jknl}\frac{\partial^2}{\partial \hat{p}_{in}\partial \hat{p}_{ml}}\left(\hat{p}_{ik}\hat{p}_{mj}P\right)$$
$$+U_{jkin}\frac{\partial^2}{\partial x_j \partial \hat{p}_{in}}\left(\hat{p}_k P\right)+U_{jkinml}\frac{\partial^2}{\partial \hat{p}_{in}\partial \hat{p}_{ml}}\left(\hat{p}_k \hat{p}_j P\right)-\frac{\partial^2 \hat{H}_L}{\partial \hat{p}_i \partial \hat{p}_k}\hat{p}_{ki}P$$

(B27)

where

$$U_{ij}=B_{ij}^{eff}(0,t)=\int_0^\infty B_{ij}(0,t,t')dt'=\int_0^\infty \left\langle u_i(\mathbf{x},t)u_j(\mathbf{x},t')\right\rangle dt'=D_0\delta_{ij},$$

(B28)

$$D_0=\frac{1}{3}B_{ii}^{eff}(0,t)=\frac{1}{3}\int_0^\infty \left\langle u_i(\mathbf{x},t)u_i(\mathbf{x},t')\right\rangle dt'=\frac{2}{3}\int_0^\infty \widetilde{E}(k,t)dk, \quad \text{(B29)}$$

$$U_{klij}=\int_0^\infty \left\langle \frac{\partial u_k(\mathbf{x},t)}{\partial x_i}\frac{\partial u_l(\mathbf{x},t')}{\partial x_j}\right\rangle dt'=-\int_0^\infty \left.\frac{\partial^2 B_{kl}(\mathbf{r},t,t')}{\partial r_i \partial r_j}\right|_{r=0}dt'=-\left.\frac{\partial^2 B_{ij}^{eff}(\mathbf{r},t)}{\partial r_i \partial r_j}\right|_{r=0}$$

(B30)

$$U_{klij}=\int k_i k_j F_{kl}^{eff}(\mathbf{k},t)d^3\mathbf{k}=\int k_i k_j \frac{\widetilde{E}(k,t,t')}{4\pi k^2}\left(\delta_{kl}-n_k n_l\right)d^3\mathbf{k},$$

(B31)

$$U_{klij}=D_2\left(\frac{1}{3}\delta_{kl}I_{ij}-I_{ijkl}\right), \quad \text{(B32)}$$

$$D_2 = \int_0^\infty k^2 \widetilde{E}(k,t)dk, \tag{B33}$$

$$U_{klijmn} = \int_0^\infty \left\langle \frac{\partial^2 u_k(\mathbf{x},t)}{\partial x_i \partial x_j} \frac{\partial^2 u_l(\mathbf{x},t')}{\partial x_m \partial x_n} \right\rangle dt' = \int_0^\infty \left. \frac{\partial^4 B_{kl}(\mathbf{r},t,t')}{\partial r_i \partial r_j \partial r_m \partial r_n} \right|_{r=0} dt' = \left. \frac{\partial^4 B_{ij}^{eff}(\mathbf{r},t))}{\partial r_i \partial r_j \partial r_m \partial r_n} \right|_{r=0}, \tag{B34}$$

$$U_{klijmn} = \int k_i k_j k_m k_n F_{kl}^{eff}(\mathbf{k},t) d^3\mathbf{k} = \int k_i k_j k_m k_n \frac{\widetilde{E}(k,t,t')}{4\pi k^2} (\delta_{kl} - n_k n_l) d^3\mathbf{k}, \tag{B35}$$

$$U_{klijmn} = D_4 \left(I_{ijmn} \delta_{kl} - I_{klijmn} \right) \tag{B36}$$

$$D_4 = \int_0^\infty k^4 \widetilde{E}(k,t)dk. \tag{B37}$$

The tensors I_{ij}, I_{ijkl} and I_{klijmn} in (B32) and (B36) are given by the expressions (B13)-(B15).

REFERENCES

[1] Sethian, J.A., Level Set Methods and Fast Marching Methods: Evolving Interfaces in Computational Geometry, Fluid Mechanics, *Computer Vision, and Materials Science,* 2nd edn., Cambridge University Press, Cambridge, 1999.

[2] Osher, S.J. and Fedkiw, R.P., *Level Set Methods and Dynamic Implicit Surfaces,* Springer-Verlag, Berlin, 2002.

[3] Kerstein, A.R., Ashurst, W.T. and Williams, F.A. Field equation for interface propagation in an unsteady homogeneous flow field, Phys. *Rev. A,* 1988, 37, 2728–2731.

[4] Peters, N. Turbulent Combustion, Cambridge University Press, Cambridge, 2000.

[5] Klyatskin, V.I. *Dynamics of Stochastic Systems,* Elsevier, 2005

[6] Klyatskin, V.I. *Stochastic Equations through the Eye of the Physicist (Basic Concepts, Exact Results, and Asymptotic Approximations),* Elsevier, 2005.

[7] Sabelnikov, V.A. and Kerstein, A.R. *Derivation from the field equation of the mean consumption velocity for front propagation in homogeneous turbulent flow, International Journal of Spray and Combustion Dynamics,* 4, no 1, 61-76, (2012).

[8] Monin, A.S. and Yaglom, A.M. *Statistical Fluid Mechanics. Mechanics of Turbulence.* 1971 (Vol.I) and 1975 (Vol.II), MIT Press, ed. J. Lumley, 769 pp. and 890 pp.

In: Combustion
Editor: Joseph M. Grier
ISBN: 978-1-62948-967-4

Chapter 3

DEPOSITION OF THIN FUNCTIONAL COATINGS AT ATMOSPHERIC PRESSURE USING COMBUSTION CHEMICAL VAPOUR DEPOSITION

I. Zunke**, *P. Rüffer, T. Tölke, A. Heft, B. Grünler and A. Schimanski
Innovent e.V., Jena, Germany

ABSTRACT

The deposition of thin functional layers under atmospheric pressure conditions using flame-based techniques is a cost-effective solution for several applications in comparison with the well known vacuum techniques. The basic principles of the flame-based deposition techniques are reviewed and the technical realization is shown by means of selected examples of the current research and development of Innovent e.V.

In the case of combustion chemical vapour deposition (CCVD) a combustible gas-air mixture is ignited in a burner. The temperature of the flame depends on the gas-air ratio and strongly on the zone of the flame. In the simplest case oxydizing and reducing flame are differed. In dependence of both of these main flame-properties, the chemistry of precursor reactions and film-deposition is influenced. The film morphology can also be adjusted for example by the distance of the

* Corresponding author: I. Zunke, iz@innovent-jena.de.

substrate to the flame. Additionally a specialized remote CCVD system is developed in order to deposit hydrolyzing precursers.

The formation of silicone oxide layers with a wide range of properties is investigated in detail. The most famous applications like adhesive, optical effective, barrier and matrix layers for embedding particles are reviewed. Fundamental investigations were performed to deposit zinc oxide layers with the objective of transparent and conducting coatings. Further coating systems like photocatalytic active (TiO_2), electrochromic (wolfram oxide), silver, aluminium oxide and rare-earth layers are also introduced.

1. Introduction

The deposition of thin films for the modification and functionalisation of surfaces takes place largely through the application of vacuum processes, for example by means of typical physical vapour deposition (PVD) such as solid state atomisation (sputtering), thermal or electron beam evaporation and chemical gas phase deposition (CVD) such as low pressure CVD. By evacuating the reaction chamber, a defined and controllable process management with a high degree of purity is made possible at the expense of high costs for plant engineering and the on-going operation, in particular for large area coatings and high throughputs.

For these reasons, trials on depositing thin films under atmospheric pressure conditions are of great interest to the user. Thermal normal pressure CVD methods have long been known and are applied in a variety of ways. In the process, thin films are formed on hot surfaces by feeding gaseous or finely atomised fluid substances, so-called precursors, these being chemically converted on the thermally reactive surface. Due to the necessarily high substrate temperature, the material selection for this process is considerably limited.

One alternative is offered by plasma discharges under atmospheric pressure conditions. In this case the conversion of the precursors takes place in the thermal or non-thermal plasmas. Due to the low energy input, the latter in particular are not only of interest for thermally sensitive substrates. The application of jet plasmas in particular is currently experiencing an enormous growth in significance. However, the treatment of large areas with high throughput rates is still a problem with these systems.

A further method of depositing thin films under atmospheric pressure conditions is flame supported deposition. In contrast to the flaming of plastics

[1, 2] for the purpose of surface activation which has been established for many years (Kreidl process) and which proceeds without precursor additive, with flame supported deposition substances are fed which are either converted directly in the flame or in its immediate vicinity. As it is a chemical transformation process, this procedure isdesignated Combustion Chemical Vapour Deposition (CCVD).

The historical coating method, also known as flame pyrolysis, was originally developed in the 1980's for improving the adhesion of metal-plastic composites in dentistry (Silicoater® process [3]). The surfaces treated in this case amounted to a few mm^2. In the past years this process has undergone comprehensive further development with INNOVENT e.V. to become an applicable large-area coating technology.

Accompanying this is a comprehensive development both in burner technology and in overall process control and monitoring. Initially, to improve adhesion and to inhibit corrosion, more or less undefined silicon oxide layers (SiO_x) a few 10 nm's thick were deposited (PYROSIL® process [4, 5]). The present state of the art allows the deposition of large-area coatings of several metres in width by means of flaming and with it the adjustment of thickness and quality. The range of depositable materials has also developed strongly; besides SiO_x, in principle virtually all metal oxides [6, 7] and even phosphates [7] or precious metals such as gold and silver [8] can nowadays be deposited by means of CCVD.

The advantage of the flame supported coating process clearly lies in the significantly lower plant costs when compared to vacuum coating plants of comparable coating performance. Large-area coating with an adhesion promoter, e.g. for powder painting applications on float glass, only starts to make economic sense with this process. This also applies to simple optical functional coatings for anti-reflective coating of large areas.

Of course, these advantages also have to be seen against a range of not inconsiderable disadvantages which limit the process. In comparison to physical and chemical low pressure processes which necessarily have to run in a closed system, an open system exists in the atmospheric pressure process. Due to the nature of the flame and the chemical processes taking place therein, the selection of substances for deposition is more or less limited to strongly oxidised materials with the exception of some precious metals.

The thermal stressing of the substrate limits the deployment of the CCVD methods; however, with suitable parameters and plant modifications, even heat sensitive materials can be coated.

This paper is an introduction to the basic principles of CCVD and some approaches to coating solutions based on selected examples of applications already implemented or current development work.

2. The CCVD Process

Combustion is understood in general to be the oxidation of a fuel (methane, propane, etc.) under the formation of a flame. It is an exothermic reaction which gives off energy in the form of light and heat. In the process, the appearance of the flames, i.e. their size, shape and temperature is dependant on different factors such as the ratio of fuel to oxidant (e.g. propane to air) and the burner used by influencing the gas flow. The temperatures of propane-air flames range from 800 °C up to 1800 °C [9]. In principle, the temperatures in the oxidising range are higher than in the reductive range. A large part of the chemical reaction proceeds in the occurrent flames.

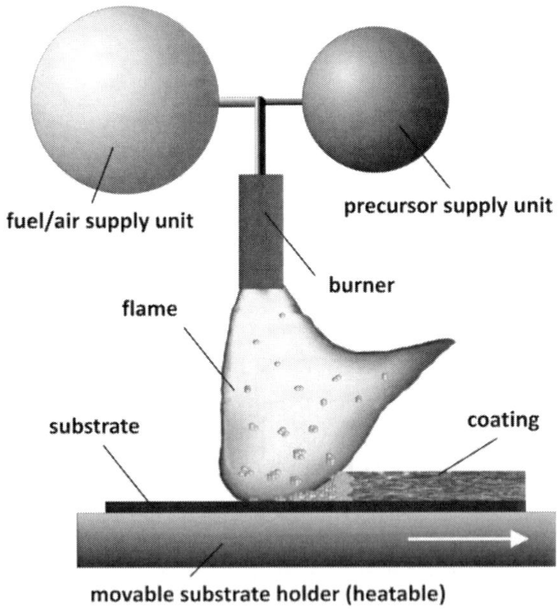

Figure 2.1. Schematic presentation of the CCVD deposition method.

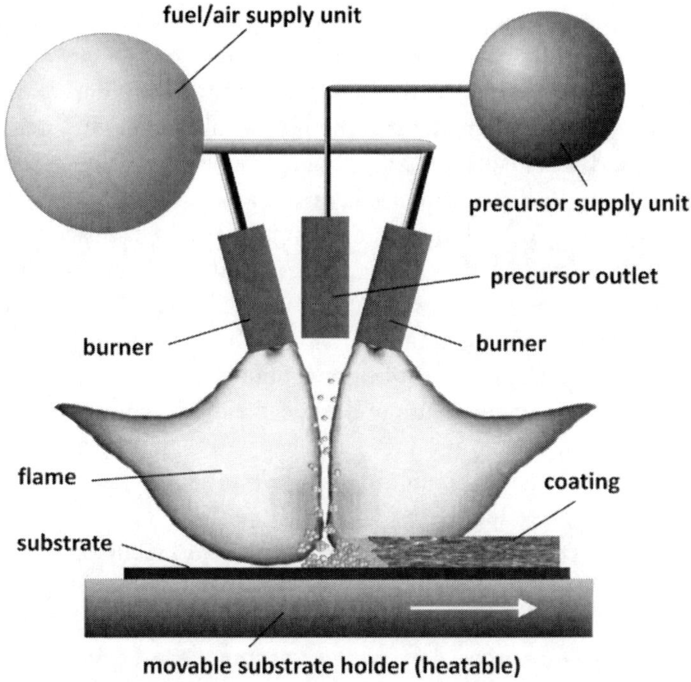

Figure 2.2. Schematic presentation of the remote CCVD (r-CCVD) deposition method.

In the case of purely thermal CVD processes, gaseous precursor substances or vapours are converted into film forming material only by the high temperatures of the substrate surface. This conversion can in a broader sense also take place in a correspondingly hot volume of gas. With CCVD processes the flame is not only the supplier of energy but also produces highly reactive species such as $OH^{\bullet}$ hydroxide radicals.

The principle of all flame supported film deposition processes is the chemical conversion of at least one precursor which is also why the designation CCVD is used. With flame pyrolysis the chemical compounds are broken down thermally by high temperatures ($T > 500$ °C) directly in the flame (Figure 2.1).

Hence the substrates to be coated do not have to possess high temperatures in comparison to low pressure and plasma enhanced CVD processes in order to enable film deposition on the surface. With the classic CCVD processes, the precursors are conducted directly through the flame and converted there while with the remote CCVD processes (r-CCVD), the precursors are chemically transformed in the immediate vicinity of the flame (remote zone). Depending on the chemical composition of the precursor substances, pyrolytic and/or hydrolytic reactions thus lead to their breakdown or, in the case of solid nanoparticles, to their surface activation. Precursors can be introduced into the flame in various ways depending in which physical or chemical state these are available. Liquid precursor substances with an adequate vapour pressure can be introduced into the combustion gas or the applied air. Solids can likewise be fed in the form of solutions or nanoparticles in a suspension. Flame pyrolysis is thus a versatile method for manufacturing thin films under atmospheric conditions. Several publications [6, 7, 10 - 14] show the diversification of application options.

The main difference between the conventional CCVD processes and the r-CCVD processes lies in the nature of the proceeding reaction. With the CCVD process, the precursors are introduced directly into the flame and pyrolytically broken down there. The resultant broken down products react with the highly reactive combustion products.

In the r-CCVD process, on the other hand, the precursor substances are indeed conducted into the direct vicinity of the flame but they do not end up en masse in its hot zone. Hence pyrolytic reactions play only a subordinate role here. In fact, with this process hydrolytic reactions can be exploited whereby the flame serves as a supplier of energy and water. By means of a suitable configuration of the geometric arrangement of burners and precursor nozzles in relation to each other and to the substrate, a defined reaction zone can be created (Figure 2.2) which is limited by the flames. Moreover, the carrier gas stream prevents the flames from coming directly in contact with the substrate which is advantageous in the case of sensitive materials.

2.1. Fundamentals of the Combustion

In the case of CCVD, a mixture of propane C_3H_8 and air is used as the combustion gas which is premixed in a gas generating unit to a specific, adjustable ratio to the volume flows and subsequently conducted to the burner element and burnt there. The set propane-air mixture has a decisive influence on the combustion and thereby also on the film formation process. This ratio is referred to as the air ratio λ_{air} [15] and can be described with

$$\lambda_{air} = m_{L,actual} / m_{L,min} \quad (2.1)$$

whereby $m_{L,actual}$ is the available air mass and $m_{L,min}$ is the stoichiometric air mass. At the same time, the value λ_{air} can assume less than (combustion with air deficiency), equal to (stoichiometric combustion) and greater than 1 (combustion with excess air).

Because the combustion of propane constitutes the basis of the CCVD process further developed by INNOVENT e.V., it is this process which will now be dealt with in detail. The precise description of the many individual activities taking place during combustion is very elaborate and is described as an example in [16 - 18]. Due to the large number of reactions, only a selection of elemental reactions of the precursor substances participating in the film formation will be gone into. The use of propane as a combustion gas in the flame pyrolysis has the advantage over natural gas in that no fluctuations occur in the composition of the combustion gas which could have an influence on the process.

In the case of a burner system with a laminar flow profile, the flame can in principle be subdivided into several zones [19]. Different temperature gradients result from this and reactive species can be found. Directly below the burner outlet is the reductive zone (Figure 2.3). This is the coldest zone of a flame. The hydrogen present is oxidised to water and the single free radicals are hydrogen radicals $H^{\bullet}$ which react with the hydrocarbons and oxygen present thereby preventing the formation of further radicals. This zone is irrelevant in the technical utilisation for a film deposition or activation as no oxidation of the substrate surface occurs.

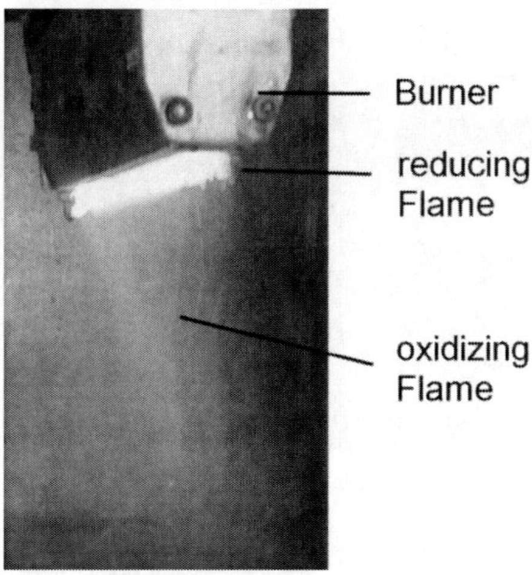

Figure 2.3. Depiction of the most important flame-regions of propane-air flames.

The zone which is primarily utilised is the oxidising zone which follows the reductive zone. The temperatures lie in a range of around 1800 °C [9] and are thus the highest in the flame. Free radicals are present in excess to the detriment of the reactant whereby this zone is heavily oxidising. The composition and the associated characteristics of the flame can be altered within certain limits. An almost transparent flame is produced when the oxygen component is strongly increased and thus a lean mixture is burnt whereby methyl radicals $CH^{\bullet}$ are increasingly excited. By contrast, the flame in an extreme case with a rich mixture becomes yellowish through the excitement of carbon dioxide radicals $C_2^{\bullet}$ and C particles. The zone following this is the largest in the case of laminar flames. The temperature remains constantly high over a wide range due to the exothermic oxidation of carbon monoxide CO. Here, $H^{\bullet}$, $OH^{\bullet}$ and $O^{\bullet}$ radicals can also be confirmed among the typical molecules of the oxidising zone [20].

In general, the oxidation of hydrocarbons consists of a complex reaction mechanism with many elemental reactions. This is already clear with the oxidation of the simplest aliphatic hydrocarbon, methane, which consists of approx. 400 elemental reactions. The global reaction of the propane oxidation is given with

$$C_3H_{8(g)} + 5O_{2(g)} \rightarrow 3CO_{2(g)} + 4H_2O_{(g)}. \quad (2.2)$$

The required stoichiometric ratio of propane to oxygen for the complete combustion of propane can then be calculated. Allowing for a volume component of oxygen in air, this ratio is 1:23.9. For a complete combustion of the propane, a propane/air ratio of at least 1:24 should be selected. However, combustion processes never proceeds fully [19]. With the combustion of hydrocarbons, both uncombusted species and carbon compounds such as CO and others are always present. When air is used for combustion, there are also typically nitrogen compounds NO_x [21 - 23] present.

Turns [16] describes by way of example the oxidation of aliphatic hydrocarbons of a higher order (C_nH_{2n+2}, $n > 2$) in eight steps. Initially, the hydrocarbon radicals form ethyl and methyl from the C-C bonding with the participation of a collision partner M

$$C_3H_8 + M \rightarrow C_2H_5^{\bullet} + CH_3^{\bullet} + M. \quad (2.3)$$

In a second step, these hydrocarbon radicals react with the unsaturated hydrocarbons ethylene and methylene forming hydrogen radicals

$$C_2H_5^{\bullet} + M \rightarrow C_2H_4 + H^{\bullet} + M \tag{2.4}$$

$$CH_3^{\bullet} + M \rightarrow CH_2 + H^{\bullet} + M. \tag{2.5}$$

With the emergence of hydrogen radicals, a range of new radicals such as $O^{\bullet}$ and $OH^{\bullet}$ are formed

$$H^{\bullet} + O_2 \rightarrow O^{\bullet} + OH^{\bullet}. \tag{2.6}$$

In the subsequent reaction of these radicals with new propane molecules, n-propyl or isopropyl radicals emerge

$$OH^{\bullet} + C_3H_8 \rightarrow C_3H_7^{\bullet} + H_2O \tag{2.7}$$

$$H^{\bullet} + C_3H_8 \rightarrow C_3H_7^{\bullet} + H_2 \tag{2.8}$$

$$O^{\bullet} + C_3H_8 \rightarrow C_3H_7^{\bullet} + OH^{\bullet}. \tag{2.9}$$

These radicals break down to

$$C_3H_7^{\bullet} + M \rightarrow C_3H_6 + H^{\bullet} + M \tag{2.10}$$

$$C_3H_7^{\bullet} + M \rightarrow C_2H_4 + CH_3^{\bullet} + M. \tag{2.11}$$

The unsaturated hydrocarbons from the previous steps react further to form formyl radicals and formaldehyde

$$C_3H_6 + O^{\bullet} \rightarrow C_2H_5^{\bullet} + CHO^{\bullet} \tag{2.12}$$

$$C_3H_6 + O^{\bullet} \rightarrow C_2H_4 + CH_2O. \tag{2.13}$$

The formation of carbon monoxide from products of earlier reactions

$$CH_2 + O^{\bullet} \rightarrow CO + H_2 \tag{2.14}$$

$$CH_2 + O_2 \rightarrow CO + H_2O \tag{2.15}$$

and its oxidation

$$CO + O_2 \rightarrow CO_2 + O^\bullet \tag{2.16}$$

$$O^\bullet + H_2O \rightarrow OH^\bullet + OH^\bullet \tag{2.17}$$

$$CO + OH^\bullet \rightarrow CO_2 + H^\bullet \tag{2.18}$$

are likewise an important component. For both the activation of surfaces and for the deposition of thin films and the directly associated decomposition and reaction of the precursors in the flame, the radicals formed, in particular $O^\bullet$, $OH^\bullet$ and $H^\bullet$, play a decisive role.

Glassman [24] and Warnatz [25] further describe the processes of combustion and its thermodynamic in detail.

2.2. Technical Implementation

The precursor is in general fed as a vapour or aerosol directly into the combustion gas/air mixture and thereby conducted to the burner element. There, it is chemically converted in the flame and conveyed to the substrate surface. To guarantee an even coating of the substrate – glass, ceramic, metals, plastics or even textiles are conceivable here – it is guided through the flame at a specific velocity. The resulting contact duration between the substrate surface and flame determines, along with the burner-to-substrate distance "bsd" (distance between the bottom edge of the burner and the substrate surface) and the precursor concentration, determines the film growth. This procedure is repeated cyclically until a specific film thickness is achieved and is termed burner passes "bp".

Table 2.1. Overview of process parameters associated with typical limits, which vary strongly depending on the layer system

Process parameters	Unit	Typical range
Propane flow rate	l/min	0.5 to 50
Air flow rate	l/min	15 to 1000
Precursor flow rate	ml/min	0.5 to 10
Burner-to-substrate distance (bsd)	mm	2 to 40
Substrate speed	mm/s	20 to 400
Burner passes (bp)	–	1 to >100

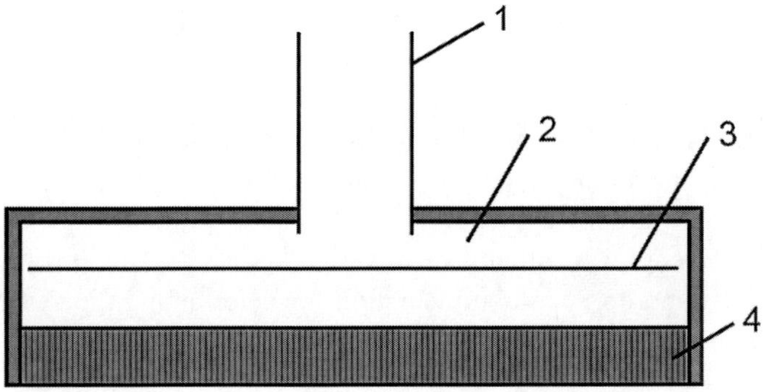

Figure 2.4. Schematic build-up of a burner with the main components: gas-air supply (1), pressure compensation chamber (2), baffle plate (3), effusion nozzles (4).

An overview of the variable process parameters is given in Table 2.1. These parameters are set separately from each other and interact to influence the resulting film properties. More detailed explanations follow in the next section 2.3.

Apart from the parameters identified, the selection of burner type is a decisive factor in the depositing of thin films. Figure 2.4 shows the schematic structure of a burner for large-area, planar substrates. Small hand-held equipment with tailored storage vessels for the gas-precursor mixture which, for example, are suitable in particular for the surface activation and pre-treatment for adhesive bonding of small three dimensional parts, can likewise be used.

The combustion gas/air mixture containing the precursor is conducted (1) directly into the burner and distributed homogenously over the breadth of the burner in a pressure equalisation chamber (2). In so doing, the built-in baffle plate (3) effects homogenisation over the width of the burner via pressure equalisation. Depending on design, this mixture arrives at the burner outlet (4) via round holes or slot-shaped outlet openings and is combusted at this point. For surface activation, typical commercially available systems (SURA Instruments, Webber Brennertechnik) are used which are not, however, conceived for the deposition of thin films. The capacity of such burner designs is in the range of 0.3 – 0.8 kW/cm burner breadth. Thermally demanding materials can only be coated conditionally by this means. Nevertheless, coating of plastics, for example, is feasible by means of shortening the contact time and introducing pauses between the passes which is, however, very time-consuming. A development in new burner types should obviate this problem. These evolved slot burners, then, offer the advantage of working with a capacity of only 50 – 100 W/cm burner breadth whereby the heat input is reduced. Due to the homogenous flame front, substrates with marginal deviations in the film thickness can be produced [26].

CCVD or r-CCVD systems basically consist of three fundamental component parts: a gas controller, a precursor dosing unit and a gas outlet system – the burner. Figure 2.5 shows both systems schematically. A conveying system for substrates, extraction and filtering equipment for exhaust gases and the appropriate instrumentation for determining the system-specific parameters complete the set-up.

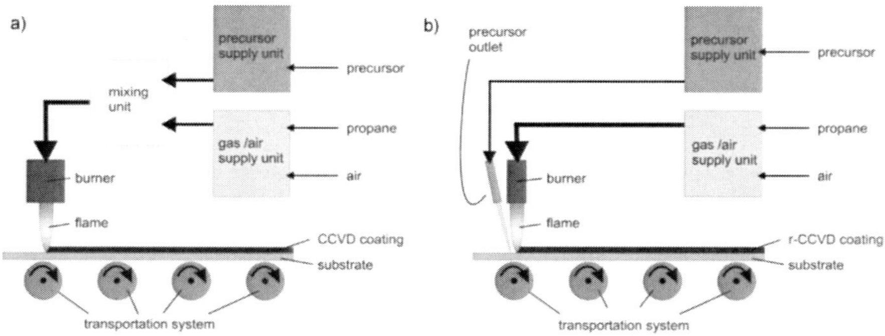

Figure 2.5. Installation of a CCVD system (a) and a r-CCVD system (b).

Figure 2.6. The flame pattern of an optimized coating dual-slot burner (30 cm breadth).

Figure 2.7. Typical r-CCVD coating system with two borehole burners and an slot-style exhaust mechanism for the precursor between the burners.

The ratio of the mixture of propane and air is regulated by means of a gas controller by which the respective flow is predetermined. Burners vary in execution which is determined according to substrate size and shape. The required power as well as the homogeneity of the coatings likewise plays an important role here. Ribbon burners, with breadths from a few centimetres up to over a metre are just as conceivable as slot burners. Ribbon burners were originally developed for the surface activation and heating of substrates and therefore work with a capacity of approximately 500 W/cm which makes them unsuitable for homogenous coatings. These burners, specifically developed for coatings, therefore mostly have fairly complex facilities for the homogenisation of the gas flow and adapted outlet openings which enable a homogenous flame fringe (Figure 2.6).

By means of specific modifications to the burner, the required total quantity of gas can be reduced considerably so that both the energy consumption (< 100 W/cm) as well as the thermal stressing of the substrate is reduced.

These assertions apply equally to the r-CCVD process with the difference that additional discharge equipment for the precursor is deployed which delivers the latter into the vicinity of the flame (Figure 2.7).

Generally the precursor feed takes place in the carrier gas stream, preferably air, but nitrogen, argon or oxygen are also conceivable. Depending on its form, the respective precursor substance must be injected into this carrier stream in a suitable way. Gaseous substances can be admixed by means of an appropriate controller. Liquid substances with an adequate vapour pressure can be transformed into the gas phase via bubble or vaporiser systems. In the case of these precursors, heating of the entire precursor feed line up to the burner or the discharge opening can be necessary in order to prevent recondensation. Solid precursor chemicals can also be used. These are available either dissolved in a fluid medium or as nanoparticles in a suspension. In order to achieve a homogenous distribution of these substances, the liquid carrier medium must be atomised in a carrier gas stream. This can take place, for instance, through appropriate nozzle arrangements [6]. The precursor gas in the case of the CCVD process is conducted directly into the combustion gas air flow and thence fed to the burner. On the other hand using the r-CCVD process, the precursor gas is fed via a separate precursor discharge aperture.

2.3. Coating Morphology in CCVD

With the injection of the precursor molecules, the conversion takes place in accordance with the injected chemicals per equation 2.2. Using the example of the extensively deployed SiO_x coating system, one arrives, using hexamethyldisiloxane (HMDSO), at

$$C_6H_{18}OSi_2 + 12O_2 \rightarrow 6CO_2 + 9H_2O + 2SiO_2 \quad (2.19)$$

for thin films. The flame gases thereby incorporate the material to be condensated in ample form. The phase change is induced in accordance with the classic nucleation theory (Volmer, Weber [27], Becker, Döring [28], Frenkel [29] and Zeldovich [30]) through the formation of critical clusters. Further growth is determined by Brownsche coagulation. The precursor concentration is the primary cause for the particle formation in that it has a direct influence on the number of critical seed crystals through oversaturation of the flame.

The manner of the film formation depends not only on the specific properties of the materials to be deposited, system-determined parameters also play a strong part in the development of specific structural features. The reaction time t_r, which elapses in the gas stream between the commencement of the chemical reactions and the appearance on the substrate, is of a decisive duration in this connection. This is determined by the burner-to-substrate distance, the exit area of the gas A_{burner} and the exit velocity of the gas stream air Q_{air}, propane Q_{gas} and precursor Q_{prec}

$$t_r = (bsd \cdot A_{burner}) \,/\, (Q_{air} + Q_{gas} + Q_{prec}). \tag{2.20}$$

It lies in the 4 ms range and as an initial approximation the burner-to-substrate distance and the exit velocity are inversely proportional to each other whereby each of these magnitudes also exerts its own influence on the result of the deposition process.

After the formation of the first seed crystals as a consequence of critical clusters, the further growth is determined through enlargement of the particle diameter and subsequent particle agglomeration. In general a large burner-to-substrate distance coupled with a longer reaction time in the discharging gas volume means that an increased agglomeration takes place which leads to the depositing of larger particles. Further details on particle growth, particularly in flames, can be found in [31 - 34]. In extreme cases, the formation of dust particles which are no longer cross-linked with each other takes place and these land on the substrate as a wipable layer or are borne away by the gas stream. By preventing the volume agglomeration, relatively smooth and dense films can be produce with a roughness of $R_a \leq 1$ nm. In the case of stronger agglomerate formation, porous and rough films with values of $R_a > 10$ nm develop. The morphology of the resultant films is shown in Figure 2.8.

In the case of flame pyrolysis, the liquid precursors are thus brought into the hot reaction zone whereby the solvent rapidly vaporises or breaks down and forms precursor gas. By this means the desired products can be formed through breakdown or other chemical reactions.

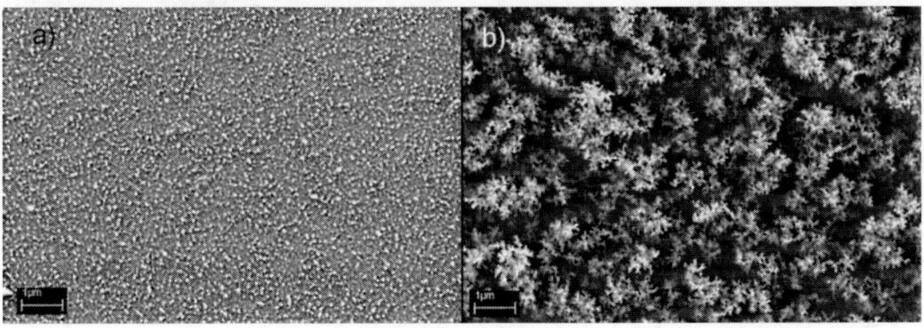

Figure 2.8. SEM images of flame pyrolysis deposited SiO_x layers with different structures: more dense and smooth (a) or very porous (b).

The choice of suitable precursors and process parameters is important in transforming the whole precursor to the gas phase, otherwise a spray pyrolysis would proceed rather than a true CVD process [6]. Thus seed crystals initially form in the flame and already partially agglomerate in the progression of the process.

A likewise decisive process parameter is the substrate temperature. This is determined by the effect of the flames subject to substrates-specific characteristics such as heat conductivity or degree of IR absorption and system-specific magnitudes such as substrate traverse speeds but also the distance to the burner (see section 2.1) and the discharge velocity of the gas stream.

The theoretically achievable film thickness can be roughly determined based on physical considerations. Based on the concentration of the precursor c_{prec} and its dosing rate q_{prec} and with the given traversing speed of the substrate V_{sub} and the width to be coated b_{sub}, the resultant film thickness d_f can be determined. In considering the density of the resulting film ρ_f, this yields

$$d_f = c \cdot (c_{prec} \cdot q_{prec}) \,/\, (V_{sub} \cdot b_{sub} \cdot \rho_f). \qquad (2.21)$$

This theoretically calculated film thickness will not, however, be achieved due to a few factors. For example, the density of the developed film can vary strongly and thus deviate from literary values. Depending on the dosing of the precursors, it is likewise conceivable that due to recondensation on the side walls or, in the case of aerosols simple wall contact, the number of convertible particles in the reaction space decreases. The conversion rates of the precursor in the flame also play a decisive role. These affect, on the whole, a decrease of the resultant film thickness and are embraced in the correction factor c. Not taken into account here are the particle growth and its agglomeration in the flame which likewise have a direct influence on the density and thickness of the film.

The multiplicity of the influencing parameters makes it clear that film deposition with the CCVD and r-CCVD processes is a very complicated procedure and suitable conditions have to be found for each species to be deposited, each substrate material and each substrate geometry in order to achieve the desired coating parameters.

3. Selected Examples

3.1. Transmission Enhancement (On Glass, Plastics)

The flame pyrolytic deposition of nanoscale SiO_x films represents a simple but economical coating alternative for enhancing the light transmission of transparent substrates by reducing the surface reflection. These types of solutions are at present being urgently sought in various technical fields. An increase in light transmission is not only of interest for the glazing of buildings but also for raising the effectiveness of plant for the exploitation of solar energy ranging from greenhouse glazing to photovoltaic and solar heat plants.

For a while now, attempts have been made to achieve such effects by applying so-called moth eye structures [35], by the use of Sol-Gel systems [36 - 38], by blooming the glass substrate by means of sputtered coating systems [39] or by etching. These technologies have indeed exhibited partially good results in terms of the gain in transmission but are nonetheless very investment and operating costs intensive. The Extra Clear P coating developed by INNOVENT e.V. is a system which can be produced economically and on large areas.

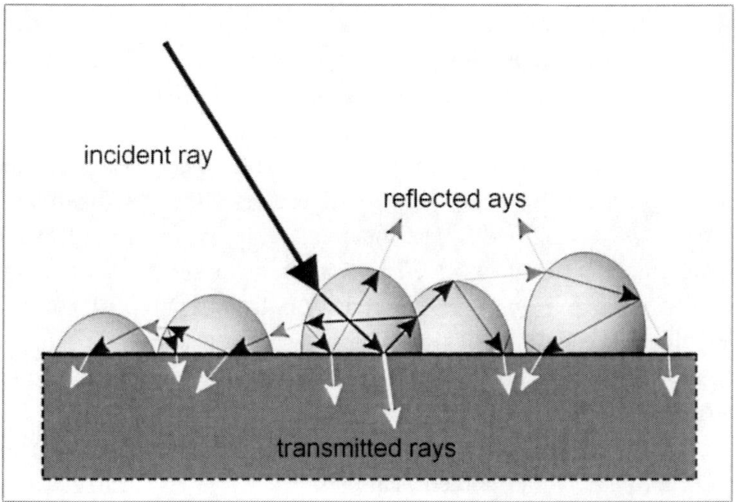

Figure 3.1. Principle of light refraction and reflection on the transmission increasing SiO_x coating.

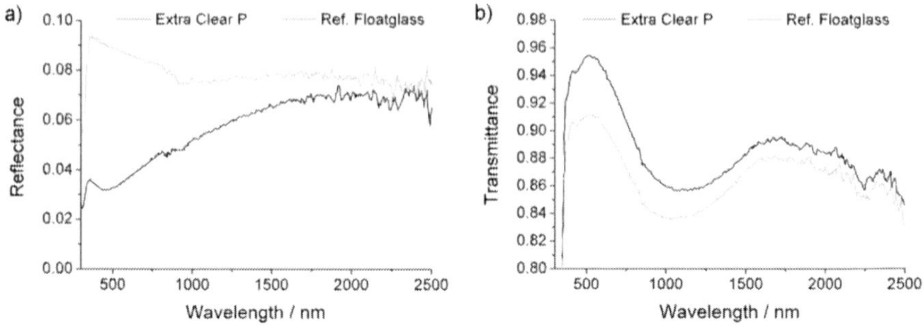

Figure 3.2. Reflectance (a) and transmittance (b) of untreated float glass and SiO_x coated Extra Clear P glass.

The effect of the transmission increase is based, as with the above mentioned processes, on a nanoscale roughening of the substrate surfaces, e.g. of float glass. Due to the deposition of stochastic distributed SiO_x particles controlled via the concentration ratio of the organosilicon precursors in the combustion gas mixture and the matched burner-to-substrate distances, a porous film results (see Figure 3.1) with an effective refraction index which is lower than that of the glass substrate. As a consequence of the multiple refraction and reflection on the rough surface structure, a feedback of the light takes place in the direction of the substrate. Because the density of the porous film increases towards the substrate, the effective refractive index, which is yielded as an averaging of the indices of air and SiO_x, converges in this direction on the value of the glass substrate (Figure 3.1).

Float glass samples treated on both sides which had been produced using this process in a plant for coating up to 1.2 meters wide flat substrates (called Extra Clear P) were certified by the independent test institute ift Rosenheim in accordance with EN 410:1998-04. The light transmission index was increased through the flame pyrolytic SiO_x deposition from $\tau_v = 0.91$ to $\tau_v = 0.95$, correspondingly the light reflection index was lowered from $\rho_v = 0.08$ to $\rho_v = 0.03$, whereby the uncoated float glass by the company Guardian was used as a comparison (Figure 3.2) [40].

To test the long term stability of the Extra Clear P coating, a verification of the durability of the coated samples was likewise carried out at ift Rosenheim in accordance with DIN EN 1096-2, Class A. In so doing, the samples were tested for their resistance to condensation on the surfaces, their resistance to sulphur dioxide SO_2, their resistance to abrasion and their stability against salt spray mist. Before and after the tests, the light transmission of the individual probes was determined at 550 nm and 900 nm.

In none of the specified endurance tests any alteration in the coating properties and thereby the transmission index did occur. Thus the coating Extra Clear P fulfils the requirements of DIN EN 1096-2, class A and is thereby suitable for the highest DIN demands on coated glass. This guaranteed and unlimited deployability of this type of coated glazing allows the application of the external side in the facade during weathering [41].

In addition to improvement of the transmission properties, the glass plates provided with flame pyrolytic SiO_x also exhibit a lower tendency to soiling due to the hydrophilic effect of the coating.

The two effects together demonstrate their value directly in an increase of solar yield. This was shown in an outdoor weathering test on the Canary Islands over a period of 7 months by a 6.5% increase in capacity of the treated modules in comparison to untreated modules [42].

A transmission increasing effect is in principle also achievable on transparent plastic materials. However, the samples have to be moved under the burner at a significantly higher speed in order not to damage the material itself. There are thus more burner passes necessary to achieve a similar effect to that of glass substrates.

An increase of the diffuse transmission of approximately 1.5% (in the 550 nm range) for polymethylmethacrylate PMMA can be achieved for instance coating both sides using 8 bp each at a substrate speed of 400 mm/s (Figure 3.3).

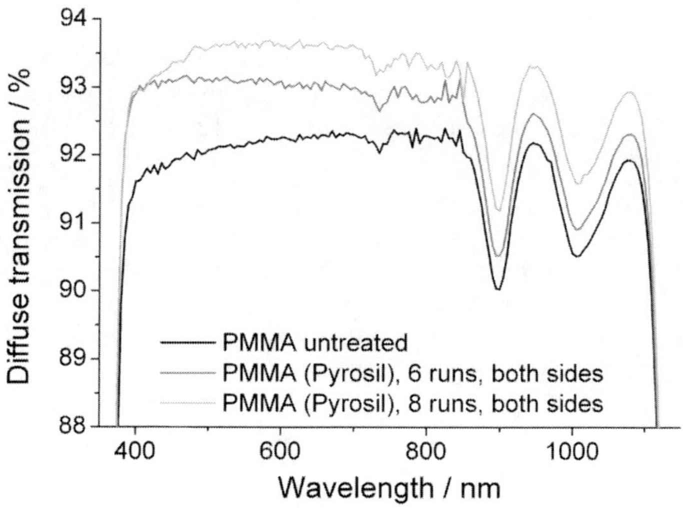

Figure 3.3. Transmission enhancement on SiO_x coated acrylic glass (PMMA).

This value could still be increased using more burner passes but this would have no advantage due to economic considerations. At this juncture, therefore, there still exists a considerable development potential to arrive at higher coating rates with less stressing of the substrates using adapted burners.

3.2. Activation of Plastics

A further area of application for the flaming process is the surface activation of materials, in particular plastic materials. These types of surface pre-treatments are deployed by industry in many cases and are necessary in order to configure the subsequent process steps, such as painting or adhesive bonding, more efficiently.

In addition to flaming, alternative pre-treatment processes are also deployed which are executed in both the low pressure area as well as under atmospheric pressure. In addition to flame treatment, the corona discharge process is the most frequently used for surface activation of polymer substrates [43]. A decisive step is constituted here with the oxidation of CH_3 methyl groups on the surface to hydroxymethyl CH_2OH [44]. Further explanations on the change in the surface chemistry can be found for example in [19]. To be able to make an assertion as to the efficiency of these different pre-treatment processes, one can avail oneself of such expedients as the determination of the surface energy of pre-treated surfaces [45] or even the determination of the strength of specifically produced adhesions [46].

The determination of the surface energy on the samples tested were performed 10 minutes after the pre-treatment on an OCA 15 plus contact angle measuring system by the company dataphysics. 10 drops each of various test fluids per sample were measured whereby 5 samples were tested per pre-treatment.

To determine the tensile strength, peeling tests using an adhesive strength testing system by the company Instron GmbH were specified and for each pre-treatment 10 tests were performed.

A comparison between conventional burners (e.g. Webber) and newly developed slot burners was made using polypropylene surfaces. This material was selected because in its untreated form it is practically not glueable on any permanent basis. The tests showed that in the case of an activation with SiO_x deposition, equally good adhesive strengths after stress tests (water immersion for 5 hours at 95 °C) were achieved as with the conventional burners (Figure 3.4). The energy consumption could, however, be reduced with the newly developed burner to approximately a third which is not sufficient for a pure activation without SiO_x deposition.

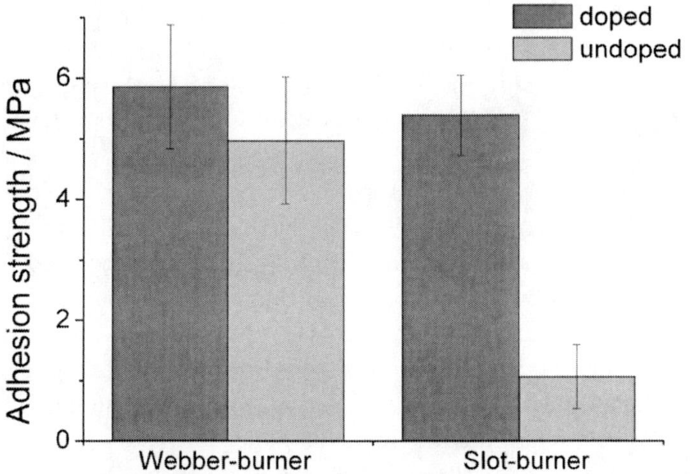

Figure 3.4. Comparison of the adhesive strength of polypropylene CCVD treatments with borehole Webber burner and newly developed slot burner.

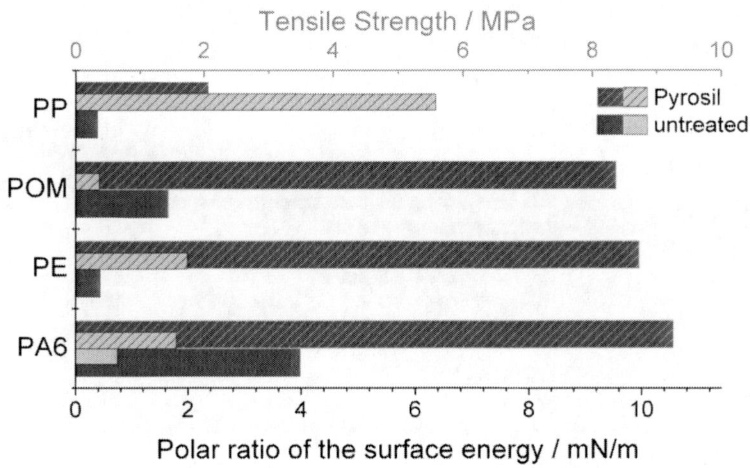

Figure 3.5. Polar ratio of the surface energy and tensile strength of adhesive composites of untreated and PYROSIL® flame treated plastics.

The increase in surface energy in the processes tested is fundamentally based on the rise in the polar component, while the dispersive component remained to the greatest extent unchanged. Figure 3.5 shows the effect of the flaming with PYROSIL® on the polar component of the surface energy for four different plastics tested (polypropylene PP, polyoxymethylene POM, polyethylene PE and polyamide 6 PA6) as well as the measured values of the tensile strength of adhesive bondings with untreated and flamed plastic surfaces.

This demonstrates that polypropylene in particular can be activated well through flaming and a clear improvement in tensile strength is achieved (from 0 to 5.6 MPa). The increase in adhesive strength is accompanied here in the case of polyethylene with the increase of the polar component of the surface energy. That this is not inevitably the case, however, is demonstrated by the two other polymers. Also in the case of POM and PA6 an improvement in the adhesive strength can be observed, even if less so, although the polar component is here reduced through the flaming.

In comparison to other activation processes it appears, based on the tensile strength values (Figure 3.6) which are ultimately relevant for practical application, that the flaming with SiO_x in the case of polypropylene yields a significantly better effect. Only the technologically more complex fluoration achieves a similarly good effect.

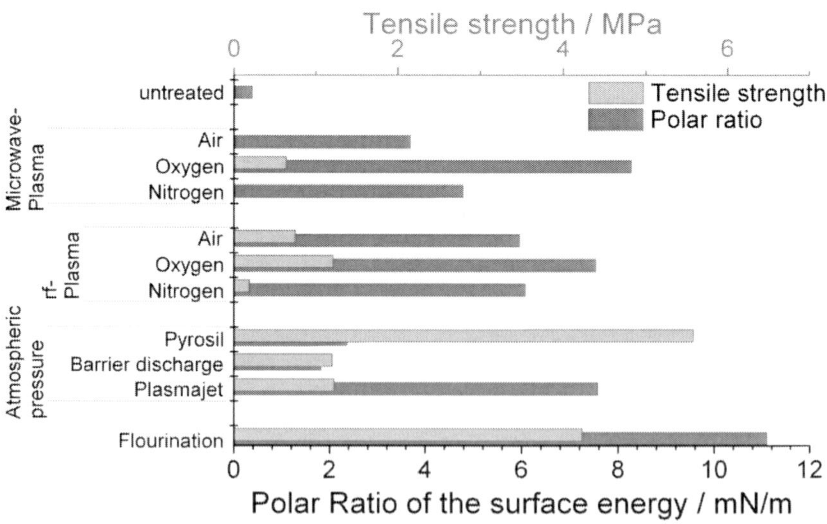

Figure 3.6. Comparison of the polar ratio of the surface energy and tensile strength of adhesive composites for different pre-treatment methods of polypropylene.

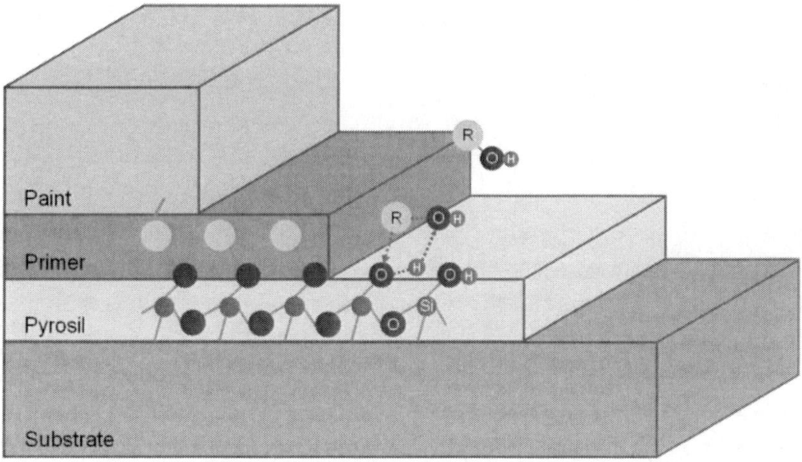

Figure 3.7. Basic molecular structure of the substrate-coating composite using a PYROSIL®/primer adhesion promotion layer.

3.3. Adhesion Promoter

A further application example for the use of flame pyrolytic deposition of SiO_x films (PYROSIL®) is the adhesion improvement of powder paints on glass surfaces.

Until now, ceramic pigments, so-called enamel coatings, have been used for the design of coloured glass. These enamel coatings, after being applied by means of the screen printing method, are burnt in at 600 °C and thus permanently bonded to the glass. The decor applied in this way is indeed very scratch resistant and weatherproof, however the highly energy-intensive process of burning in and the occurring irregularities in the illustration of the image or coloured surface in this coating process has, if anything, a disadvantageous impact.

With the utilisation of the PYROSIL® technology, a process has been developed which allows powder paint coatings to be applied with a permanent bond to the glass surface as an alternative to the enamel coatings. Until now, the use of these powder paint coated glass plates has only been feasible on the interior as the paint did not form a bond with the glass but only lay in a cured form on the glass surface and hence an infiltration of airborne moisture led to rapid detachment of the paint coatings. With the new pre-treatment systems based on the utilisation of PYROSIL® and an adhesion promoter specifically adapted to powder paint, the powder paints can be applied with permanent adhesion to the glass surface and, after curing, form a weathering-stable bond suitable for external use [12].

These cases exploit the high density of active OH groups on the surface of the SiO_x films deposited by flame pyrolysis which adhere very well to previously cleaned glass substrates. Glass surfaces activated in this way are provided with a docking adhesion promoter (primer), targeted at these OH groups, which is sprayed on. This primer is tuned to the binding agent system of the powder paint being used so that, after curing and polymerisation of the paint, an overall permanent and stable bond results (Figure 3.7).

For the production of weatherproof, coloured glass elements, a UV stable, polyester based powder paint is used, as a rule.

To verify the outdoor weather resistance of the coated glass elements, these have been subjected to the applicable quality tests. To evaluate the adhesive strength, significant samples were stressed for 1000 hours in the salt spray mist test in accordance with DIN 50021, 7 days in the water condensate/alternating climate test in accordance with VDA and 500 hours in the artificial weathering test in accordance with DIN EN ISO 11507. Apart from that, the samples were stored in an in-house rapid alkaline test for 72 hours.

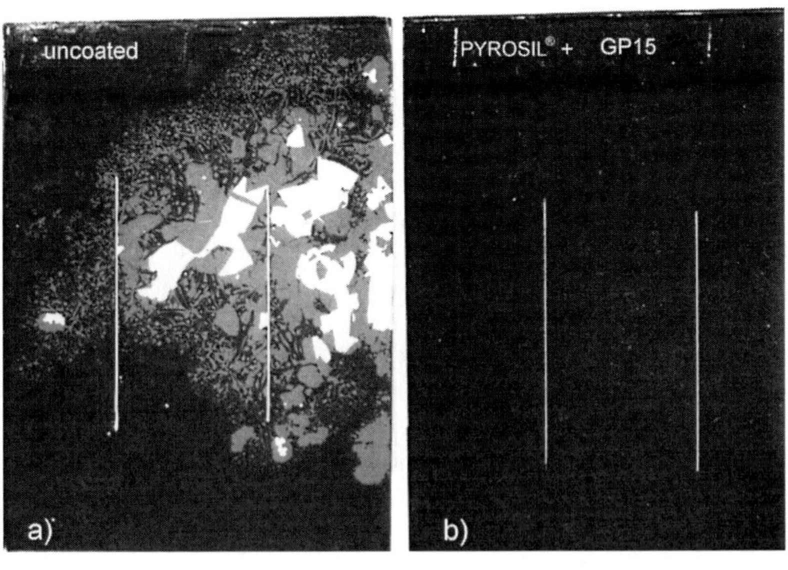

Figure 3.8. Powder coated glass panes on load test without pre-treatment (a) and with PYROSIL®/primer pre-treatment (b).

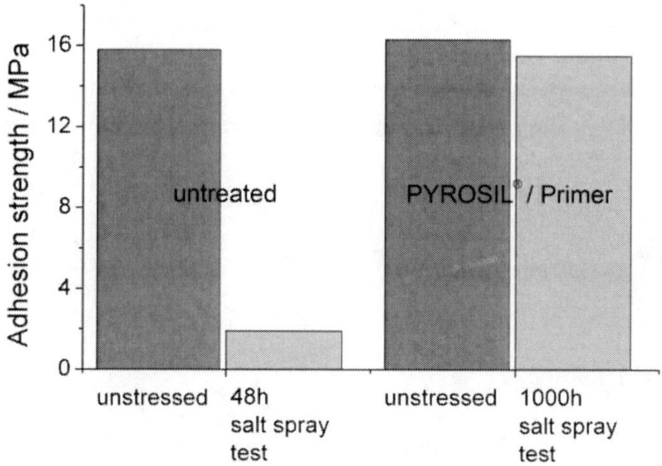

Figure 3.9. Adhesion strength values of a powder coating on untreated and on PYROSIL®/primer pre-treated glass, unloaded and after 1000 h salt spray test.

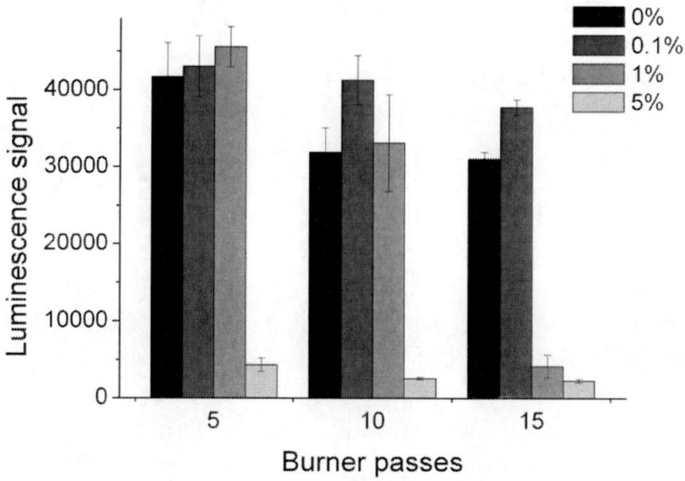

Figure 3.10. Antibacterial activity of SiO_x:Ag layers for different concentrations of $AgNO_3$ solution used for film deposition, in dependence on the burner passes.

The test for adhesive strength was carried out by means of crack detection and cross cut tests as well as quantitatively by means of peeling.

With all stress types, the adhesive strength of the powder paint coat on the glass using surface pre-treatment, and after the repetition of several series of tests, was evaluated as excellent. Infiltration of the powder coating at the scoring was not established even after longer dwell times in the stress test (Figure 3.8).

The peel tests in particular showed that the application of primer without the SiO_x underlayer and also the flaming without subsequent primer application do not deliver the desired adhesive strength values. The best adhesive strength values are only achieved through the combination of PYROSIL® and primer (Figure 3.9).

3.4. Matrix Films

Besides the direct deposition of coating material by means of the chemical conversion of the precursor materials in the flame, is also possible to embed nanoparticles in a matrix film. In so doing, these nanoparticles can either be already available and are, for example, conducted through the flame together with the precursor which forms the matrix film or it is also possible to use a precursor mixture of which one component generates a matrix film and the other the embedded nanoparticle.

To produce antibacterially active films, both the suspensions of silver nanoparticles mixed with HMDSO are sent through the flame as well as a mixture of HMDSO and dissolved silver salts. In both cases it was possible to verify a germ reducing effect of the deposited silver-bearing SiO_x films [47].

An example of the formation of biocide-active silver nanoparticles by means of the addition of dissolved silver nitrate is shown in Figure 3.10. Here the luminescence signal is illustrated, measured in a BacTiterGlo™ test for different silver-bearing SiO_x coatings on glass, whereby the number of burner passes and the concentration of one of the HMDSO precursor admixed $AgNO_3$ solutions in isopropanol is varied. The luminescence signal is hereby a measure for the metabolic activity of the e. coli bacteria used in this test. If the silver concentration is very low, or zero, then no significant effect is recognisable. With increasing silver concentration, the bactericide activity rises initially with higher pass numbers and with even higher concentrations, strong activity is already present for low pass numbers.

3.5. Barrier Films on Float Glass

Float glass corrosion is an on-going and costly problem for all float glass manufacturers. As early as the emergence of the glass from the protective atmosphere shrouded tin bath, the glass surface makes its first contact with the air humidity. During the cooling off process, water accretions occur and the first stages of glass corrosion are initiated [48, 49]. If the glass is yet further transported (different climate zones, sea air) or the storage time exceeded, the goods can become useless for their application or further finishing stages. The formation of deposits and, in the worst case, the degradation of the SiO_2 network [50, 51] can occur.

The idea of the trials was to prevent the initial reaction of the glass surface with the help of the C-CVD process by integrating this technology into a float glass line. In so doing, SiO_x films should be deposited under atmospheric pressure conditions and their influence on the leaching behaviour of the glass was examined. Moreover, film thicknesses and transmission measurements were carried out.

Standard float glass (soda-lime-silicate glass) was coated with SiO_x films by flame pyrolysis and subsequently tested for their corrosion resistance. The area of the glass substrate was (250x150) mm^2. Three different traversing speeds were selected based on normal practice draw rates: 84 mm/s, 147 mm/s, 225 mm/s resulting in three different glass thickness. The number of burner passes (bp = 2; 4; 8; 16) as well as the substrate temperatures were also varied. The latter took place at intervals of 100 °C to 550 °C in 50 K increments. In this way different localities could be simulated in a virtual cooling channel. As constant coating parameters, the volume flow rates of propane (25 l/min) and air (500 l/min), the burner distance from the surface of the glass (30 mm), the precursor concentration (15% HMDSO) and its volume flow rate (3 ml/min) were all determined. The glass sheets were brought to the required temperature on a heating plate with a heat-up rate of 10 K/min. The substrates were covered during the heating. In this way, the destruction of the samples due to stress cracking was prevented. This applies in particular to glass temperatures upwards of 300 °C.

The coating morphology was first documented by SEM and the roughness measured by AFM was recorded for all samples. The transmission of the glasses was then determined in order to check that the coating did not imply a worsening of the glass quality. Lastly, XPS examinations were performed to prove the existence of a SiO_x film on the float glass.

The criterion for the leaching characteristics and destruction of the glass network of the float glass sheets is constituted by the leaching test which makes use of a standardised test [52] in modified form. In so doing, two glass sheets were clamped with the sides to be tested aligned opposite each other and separated by a plastic ring. The volume between the two sheets was then filled with bi-distilled water and tempered for several hours at 95 °C. The liquid was then removed and examined using an ICP/OES measurement for concentrations of the glass constituents Si, Na and Ca. Figure 3.11 shows the principle set up of the leaching test.

Figure 3.12 shows the respective XPS depth profiles of an uncoated (Figure 3.12 a) float glass sheet and a sheet coated at 500 °C, 84 mm/s with 2 bp (Figure 3.12 b). The reference glass sheet showed a constant concentration profile for all constituents. This profile was in accordance with the anticipated composition of float glass. The coated glass sheet exhibited only Si and O as detected elements at the start of the analysis. That is an indication of a deposited SiO_x film. After approximately 80 seconds sputter time, Ca, Mg and Na were detected, whereby the concentration of Si and O decrease slightly, representing the original substrate surface.

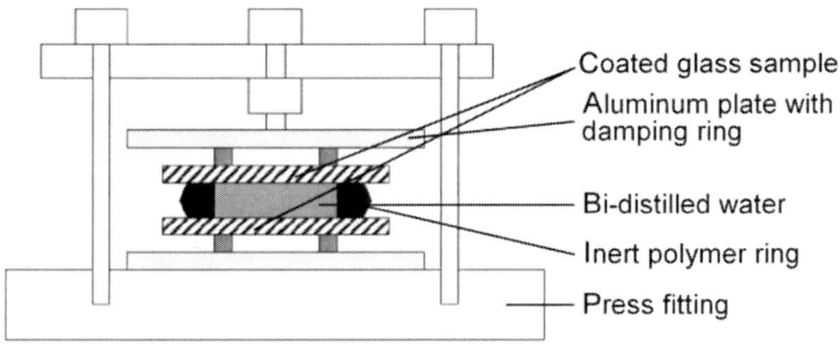

Figure 3.11. Sketch of the leaching test.

The appearance of the Na peak somewhat before the Mg and the Ca peaks points to a slight Na enrichment on the glass surface. That can be explained by the high temperature: the mobile Na diffuses at temperatures close to Tg to the substrate surface. The transmission of float glass represents an outstanding characteristic and should not be decreased by the coating process. The coating experiments were all carried out on float glass with a thickness of 4 mm and a transmission of approximately 90.6 – 90.8 % was measured. The transmission does not decrease within the parameters used (v = 84; 147; 225 mm/s, T = 100 – 550 °C, bp = 2; 4; 8 and 16) or increases slightly.

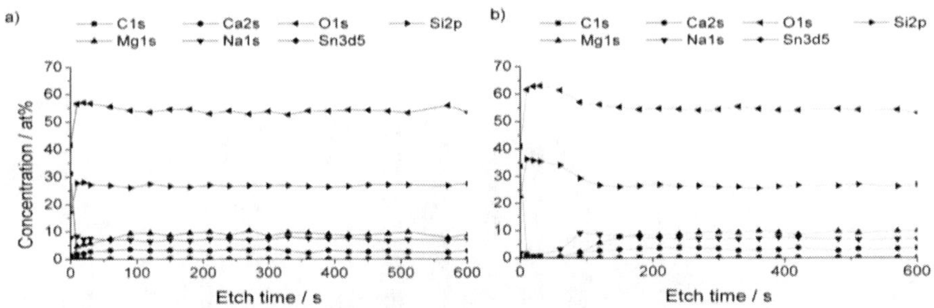

Figure 3.12. XPS depth profile of a float glass as a reference (a) and of a SiO_x barrier coating (500 ° C, 84 mm/s, 2 bp) (b).

Figure 3.13 shows the measured film thicknesses for the parameters v = 84 mm/s, T = 100 – 400 °C with bp = 2; 4; 8 and 16 respectively. It was measured using the profile contact scanning method. It was therefore not possible to ascertain the film thicknesses of the SiO_x coatings deposited over 450 ° C. The value for the column marked white in the diagram could also not be determined clearly. The film thicknesses rise as expected with the number of passes up to the regions of 100 nm at 100 °C and 150 °C. The temperature here, however, appears to have an influence on the film formation as with 16 bp the film thickness at 200 °C to 400 °C hardly exceeds 60 nm. The reason may be an improved sintering between the film forming and agglomerating SiO_x particles and the higher glass surface temperature. The film thicknesses produced at the velocities of 147 mm/s and 225 mm/s are only between 10 nm (for 2 bp) and 45 nm (16 bp).

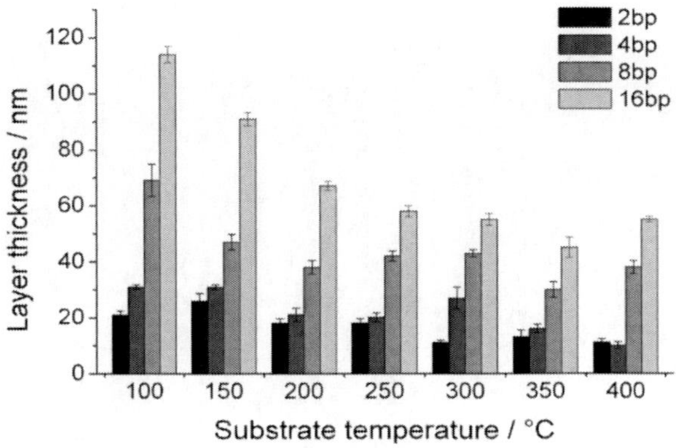

Figure 3.13. Barrier layer thicknesses at v = 84 mm/s as a function of substrate temperature and burner passes.

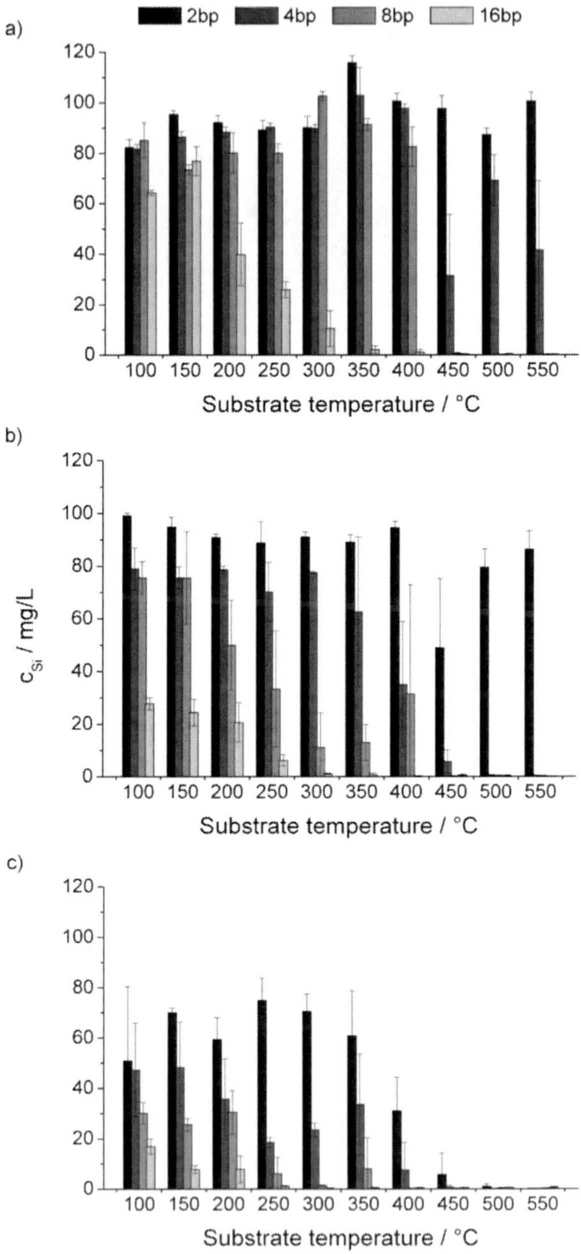

Figure 3.14. Si concentrations as a function of substrate temperature and burner passes for the substrate velocities 225 mm/s (a), 147 mm/s (b) and 84 mm/s (c).

Table 3.1. Measured roughness values Ra for a substrate velocity of 84 mm/s and different substrate temperatures

V = 84 mm/s, bp	T = 100 °C Ra / nm	v = 84 mm/s, bp	T = 550 °C Ra / nm
2	19.9	2	18.8
4	29.1	4	21.2
8	25.4	8	27.2
16	49.7	16	33.8

The Figures 3.14 a – c provide a complete overview of all used coating parameters and their influence on the leaching behaviour of the float glasses. The concentrations of Na and Ca in the aqueous solutions were determined and found to be proportional to the Si concentrations. One can recognise that the variation in the influence magnitudes T, v and bp has a major effect in each case. At v = 84 mm/s, a halving of the Si concentration already results after two passes in comparison to the higher substrate speed. The increase in the number of passes also leads to an improvement in the leaching characteristics of the glasses as expected. At v = 225 mm/s a minimum of 8 bp are necessary to obtain a dense surface even at 450 °C. If one assumes that the films become thinner with a fewer number of passes, it is surprising to ascertain that for instance at v = 147 mm/s and T = 500 °C a completely corrosion resistant and water resistant film is already formed after only 4 bp at a film thickness of probably 20 nm. The same passivating effect occurs at v = 84 mm/s and T = 500 °C after just 2 bp. These results implicate at least 8 burners are necessary for their integration in a float glass line producing a standard float glass at the respective line speed.

Concluding the roughness depending on the successive increase of bp is discussed. The AFM measurements of the samples with the varied parameters v = 84 mm/s, T = 100 °C and 550 °C and also bp = 2; 4; 8; 16 are shown. The agglomeration of the SiO_x particles in the flame and on the glass surface results in an enhanced surface roughness attributed to the particle size depending on the bp (Table 3.1).

3.6. Photocatalytic Films

The deposition of photocatalytic films based on titanium dioxide (TiO_2) is an example of the application of the r-CCVD process. A motivating factor for the development of this specially modified CCVD process is that up to now no TiO_2 films could be produced with this method. The exploitation of hydrolytic reactions of suitable precursors in the direct vicinity of the flame plasma (remote zone) with its reactive species nevertheless opened up the possibility of depositing this type of film using a flame supported process [53, 54]. The precondition for this is a stable precursor showing no pyrolytical degradation in the hot zone of the flame in order not to loose its hydrolysis capability.

The temperature stressing of the substrate, irrespective of the immediate surface, remains really low in the case of the r-CCVD coating, in contrast to films based on Sol-Gel processes or normal pressure CVD processes in which the entire substrate has to be brought to high temperatures.

For the deposition of TiO_2 films, an arrangement in accordance with the principle shown in Figure 2.2 is used applying a hydrolysable organic titanium precursor. It is vaporised and injected into the carrier gas.

The deposited films exhibit a very porous structure (Figure 3.15). The photocatalytic effect was determined based on the decomposition of organic substances (stearic acid). TiO_2 films were deposited on float glass coated with SiO_x. The SiO_x film acts here as a barrier against the diffusion of alkali ions from the glass. The rates of decomposition of the vapour-deposited 100 nm thick stearic acid films were determined based on IR transmission measurements in the range of the CH absorption bands at approximately 2916 cm^{-1} and 2850 cm^{-1} as a function of the duration of the UV-A radiation exposure. The rates achieved thereby were 25% to 70%, compared to the values of commercially available and similarly irradiated photocatalytic reference materials (Pilkington Activ^{TM}). A correlation between the thickness of the TiO_2 film and the decomposition rate is found which increases with an increase in film thickness. This is caused by the larger effective surface of the thicker porous films. The higher decomposition values were reached with films of a thickness of 200 nm.

By means of a subsequent 6 minute tempering at 650 °C, the decomposition rate was increased to values of 100% up to 200% in relation to the reference materials.

The samples also exhibit pronounced photo-induced hydrophilic properties. The contact angles of water drops decrease after irradiation to UV-A light to values of less than 10°. Depending on the surface roughness of the porous films, the values of contact angles are between 30° to over 60° after storage in the dark. The higher angles are observed in the case of the less rough surfaces.

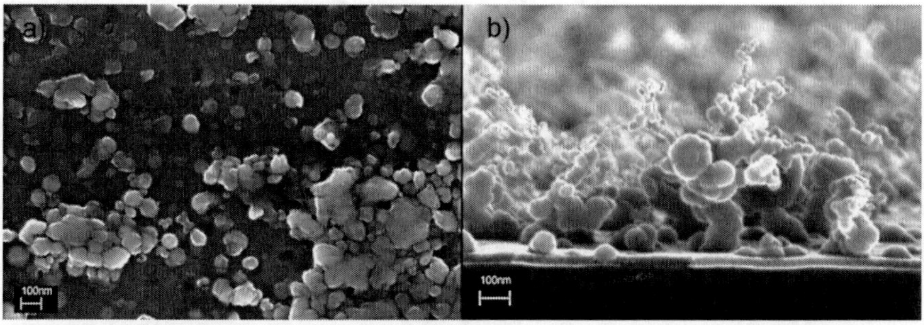

Figure 3.15. SEM images of the top surface (a) and cross sectional view (b) of r-CCVD deposited TiO_2 layer.

The photocatalytic effect and the photo-induced hydrophilic properties of the porous TiO_2 films produced by this process accompany a lower effective refractive index as opposed to compact TiO_2 films. Refractive indices of 1.6 to 2.0 are obtained instead of 2.3 to 2.5, so that significantly reduced reflection is achieved. The consequently higher transmission values make these types of films interesting for applications in the solar energy field for instance.

3.7. Electrically Conductive Films

As an example of the deposition of electrically conductive coatings, film systems will be presented which are based on flame pyrolytically deposited zinc oxide, tungsten oxide and also elementary silver and are the subject of current further development of the CCVD process.

3.7.1. Zinc Oxide

The deposition of zinc oxide (ZnO) as a transparent conductive electrode material (TCO) is very interesting for technical applications as solar cells [55] or other opto-electronic components. Thus doped ZnO is a good candidate for replacing the indium-tin oxide (ITO) normally used for TCOs in solar cells or displays. ZnO is a II-VI n-type semiconductor with a band gap of 3.37 eV [56]. Both doped and undoped ZnO coatings with specific resistances in the range of 10^{-4} to 10^{7} Ωcm are known in literature and have been produced using various processes such as sputtering [57, 58], aerosol assisted CVD (AACVD) [59, 60], atmospheric pressure CVD [61], plasma enhanced CVD [62], pulsed laser deposition [63, 64], spray pyrolysis [65] or Sol-Gel [66, 67].

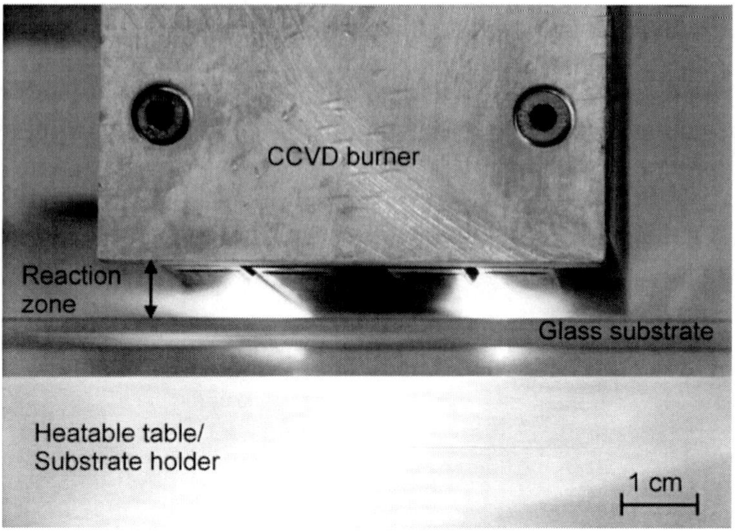

Figure 3.16. Photograph of the CCVD system in operation with the main components labelled. The reaction zone equals the burner-to-substrate distance bsd.

The coatings presented below have been produced by means of the flame pyrolysis mentioned at the outset and offer thereby a cost-effective alternative to energy intensive vacuum processes.

A 10 cm wide slot burner (Figure 3.16) specifically developed for film deposition was used. Silicon wafer pieces and 4 mm float glass (air side) were used as substrates. Zinc nitrate, dissolved in a mixture of water and isopropanol, was used and injected into the gas stream as an aerosol precursor.

Typical film thicknesses are in the range of 20 – 200 nm. The deposition rates of 0.3 – 2 nm/bp can be tightly controlled by means of the substrate temperature and the burner-to-substrate distance. A comprehensive characterisation of these films is given in [14, 68]. The most important results are presented here.

The growth of the ZnO film in the CCVD process is strongly influenced by the substrate temperature. The films deposited at 100 °C exhibited the lowest specific resistance in the range of $2.0 \cdot 10^{-1}\,\Omega$cm. In addition, the specific resistance changed with the film thickness pointing to changes in morphology. The deposition rate is reduced significantly with the increase of substrate temperature. ZnO films produced by a static AACVD system exhibit similar resistances [59]. Films produced in an atmospheric pressure laminar flow reactor [61] deliver values of $4 - 8 \cdot 10^{-4}\,\Omega$cm. Specific resistances of

Sol-Gel films are in the range of $2 \cdot 10^{-2}$ to $2 \cdot 10^{2}$ Ωcm [66] up to values of 10^{7} to 10^{8} Ωcm [60].

Using SEM evidence of columnar phase growth can be found (Figure 3.17 a). The surface is not smooth and exhibits the typical structure of the CCVD process with agglomerated particles (Figure 3.17 b). Even the films deposited at room temperature are very dense and exhibit no porosity or any holes at all.

Figure 3.17. Typical morphology of atmospheric pressure ZnO thin films coated by CCVD as observed by SEM. A top surface view (a) and a cross sectional view (b) is shown.

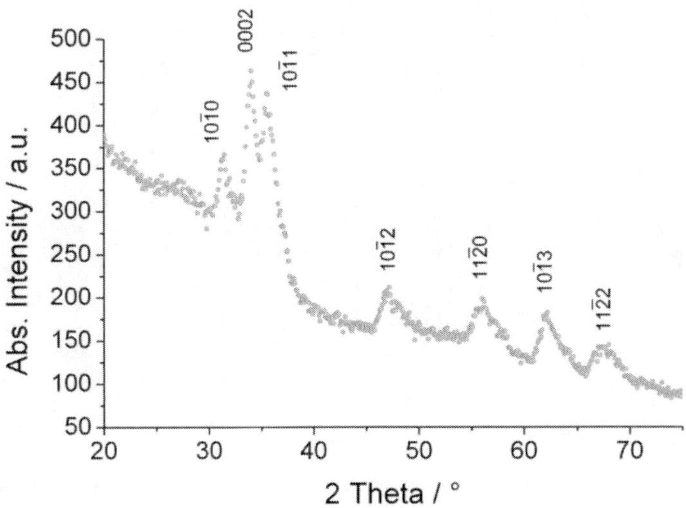

Figure 3.18. XRD pattern of a sample coated with a burner-to-substrate distance of 4 mm at 20 °C.

Examinations of the structure by X-ray diffraction clearly show the presence of crystalline components in the film. In Figure 3.18 the corresponding plains are identified and clearly show the hexagonal structure of the films (space group P63mc). The lattice parameters were determined to be a = (3.254 ± 0.004) Å and c = (5.217 ± 0.008) Å. These values agree well with literature data [69].

The optical properties of the deposited ZnO films are similar to sputtered films [70]. The transmission is temperature-dependant and values of 82% to 84% (at 550 nm) were measured. Thus ZnO is a direct semiconductor, the optical absorption coefficient α can be calculated as

$$\alpha(\mathrm{E}) \sim (E_{Ph} - E_g)^{1/2} \tag{3.1}$$

where $E_{Ph} = h\nu$ can occur is photon energy and E_g is band gap. The band gap energy can be determined by plotting α^2 against E_{Ph} and extending the linear component of the measurement series to the energy axis. The absorption coefficient is thus based on the Beer-Lambert law.

$$\alpha = \ln(100 / T) / d_f \tag{3.2}$$

with T - transmission gained from the transmission spectra.

In Figure 3.19, corresponding transmission spectra and the plot for determining the associated band gap energies are shown.

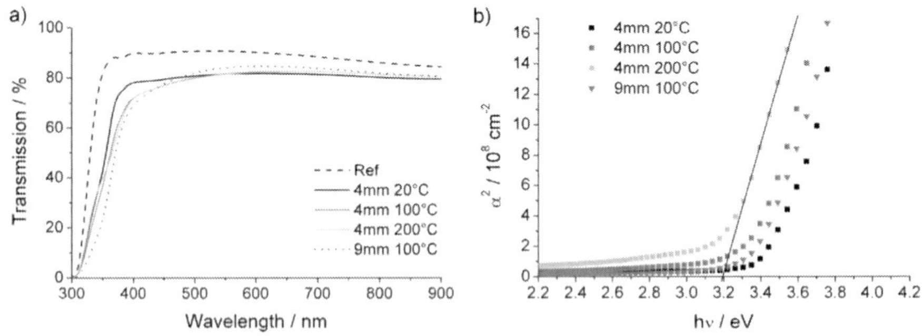

Figure 3.19. Transmission spectra of ZnO samples in the range of 190 to 1100 nm (a). The reference is uncoated float glass. Additionally the calculated data for the band gap determination of the samples is shown (b). The values are obtained from the transmission spectra shown in a).

The highest values of (3.72 ± 0.05) eV are achieved for films which are produced at room temperature and large burner-to-substrate distances. Heating the substrates to 100 °C reduces the energy gaps to values of 3.3 eV. Further heating to 200 °C and 300 °C results in a reduction of the band gap to values of 3.2 eV. Under the same process conditions the band gap increases with increasing film thickness. It is assumed [68] that the crystallite size is an important factor which decisively influences the optical and electrical properties by reducing the grain boundaries and the accompanying scattering of electrons. A shorter burner-to-substrate distance leads to smaller band gaps. With greater burner-to-substrate distances, the particles in the flame have more time to agglomerate and potentially more grain boundaries will be generated.

Infrared transmission spectra of 300 – 700 cm^{-1} also show this trend. An absorption band of 408 – 413 cm^{-1} can clearly be seen in Figure 3.20. For comparison, all spectra have been normalised to the respective film thickness and base line corrected. The profiles were fitted using Voigt profiles (inset Figure 3.20). The derivation leads to the exact peak positions. Both identified modes belong to E_1–TO from ZnO at 410 cm^{-1} [69, 71] and A_1–TO at 380 cm^{-1} [69, 71], respectively.

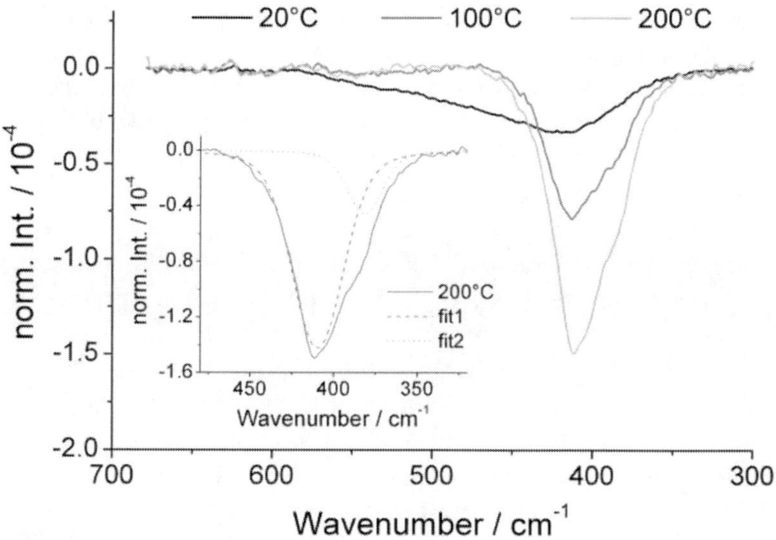

Figure 3.20. IR spectra in the range of 300 – 700 cm^{-1} of ZnO films grown at 20 °C, 100 °C and 200 °C. The inset shows an example fit of two Voigt profile shaped curves for the spectrum of the 200 °C sample.

The increase in the substrate temperature to 100 °C exhibits a significant rise in the intensities and a decrease of the full width at half maximum (FWHM), which indicates the crystallinity of the film. Furthermore, both modes shift with increasing substrate temperature to lower wave numbers.

In addition, precursor residues could be verified in a higher wave number range of 500 – 4000 cm^{-1}. For coatings which were deposited at room temperature, besides OH (stretching vibration, 3000 – 3500 cm^{-1}) [72, 73], nitrate bands $\nu_a NO_3^-$ (1389 cm^{-1} [73]), $\nu_s NO_2$ (1250 – 1300 cm^{-1}) and $\nu_{as} NO_2$ (1480 – 1530 cm¯1 [72]) could also be found. An increase in the substrate temperature with film growth to 100 °C or 200 °C caused these bands to disappear.

3.7.2. Tungsten Oxide

As an example for deposition of electrochromic films, a system is presented which is based on the flame pyrolytic deposited tungsten oxide WO_x, which is the subject of current development of the CCVD process.

The electrochromic effect describes the property of a system reducing light transmission by applying an electric current. Some transition metal oxides exhibit electrochromic properties [74]. The crystal structures of the transition metal oxides consist of more or less distorted oxygen octahedrons with a metal atom in its centre. If these octahedrons are not densely packed, there is free space where the ions can easily move. The incorporation of hydrogen or alkali metal ions is thereby possible. There are two different types of electrochromic transition metal oxides. The cathodic electrochromes are optically transparent and can be coloured by means of incorporating ions. The anodic electrochromes absorb light and become transparent by incorporation. To establish such a system, tungsten oxide was utilised, because of its electrochromic properties [75]. It has a perovskite-similar crystal structure with a missing atom in the centre. It occurs, depending on temperature, as monoclinic, orthorhombic or tetragonal [76]. It is optically transparent and can be coloured dark blue by incorporation of ions. Responsible for this is the reduction of W^{6+} ions from the WO_x film to W^{5+} ions in conjunction with the emplacement of Li^+ ions which leads to blue coloured lithium-tungsten bronze Li_xWO_3 bonds [77]:

$$WO_3 + x\,Li^+ + x\,e^- \rightarrow Li_xWO_3. \tag{3.3}$$

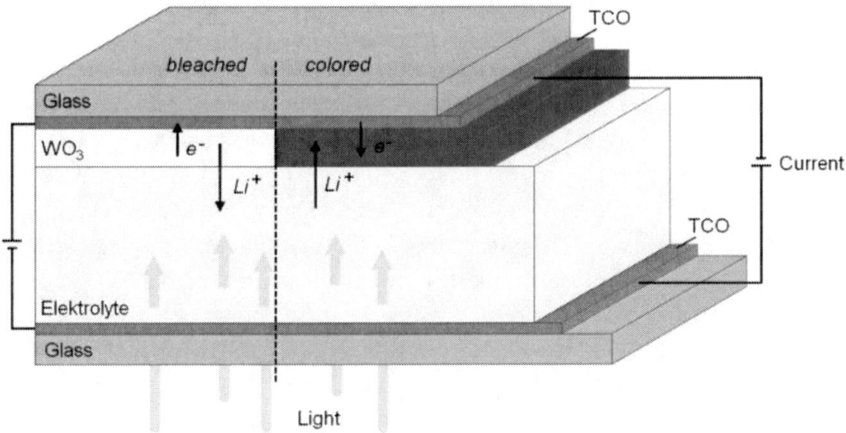

Figure 3.21. Schematic construction and function of a electrochromic glass unit.

This process operates by applying an electrical current and is reversible. By switching this current off the WO_3 film becomes transparent again (maybe the change of current direction is necessary).

An electrolyte which provides the lithium ions, as well as a TCO for contacting is necessary to build an electrochromic cell. Figure 3.21 shows the schematic setup of such a cell. TCOs, in particular ITO, are used for the transparent electrodes. Solid lithium iodide is used for the electrolyte.

Until now tungsten oxide films have been applied to the glass surface by means of PVD processes (sputtering, etc.) which require complex vacuum equipment. Deposition by means of cost-effective atmospheric pressure processes offering interesting perspectives for economical and large-area production of this type of system. In modern architectural glazing they are indeed gaining more importance in terms of the energy balance of buildings but have never gone beyond the stage of pilot production due to high costs and hitherto technical problems [78].

The process currently under development at INNOVENT uses water or alcoholic-soluble tungsten compounds as precursor for the film deposition [79]. As dissolved substances can not be vaporised, the precursor fluid is introduced into the burner stream in the form of aerosol droplets via a peristaltic pump system.

Decisive for the manufacture of a functional electrochromic cell is that the ITO films on the flamed glass substrates are not destroyed by the effect of the flames and the electrical and optical parameters remain fully intact.

Furthermore, the deposited film must contain a high component of WO_x so that a visible effect can occur.

Fully functional cells have been obtained with WO_x films which range from 50 nm to a few 100 nm thickness and deposited on ITO coated glass. Lithium iodide dissolved in gelatine was used for the electrolyte.

A laboratory prototype of these cells is shown in Figure 3.22. The switched and again decoloured states can clearly be seen, whereby the decolouring effect is well visible. Besides the manufacture of electrochromic switching systems, the use of these types of deposited active WO_x films in photoelectrochromic or gaschromic cells is also conceivable.

In contrast to the deposition of SiO_x films, the deposition rates for tungsten oxide are significantly lower. Increasing the deposition rates and increasing the coating width with a simultaneous improvement in the film homogeneity are the subjects of the current research work.

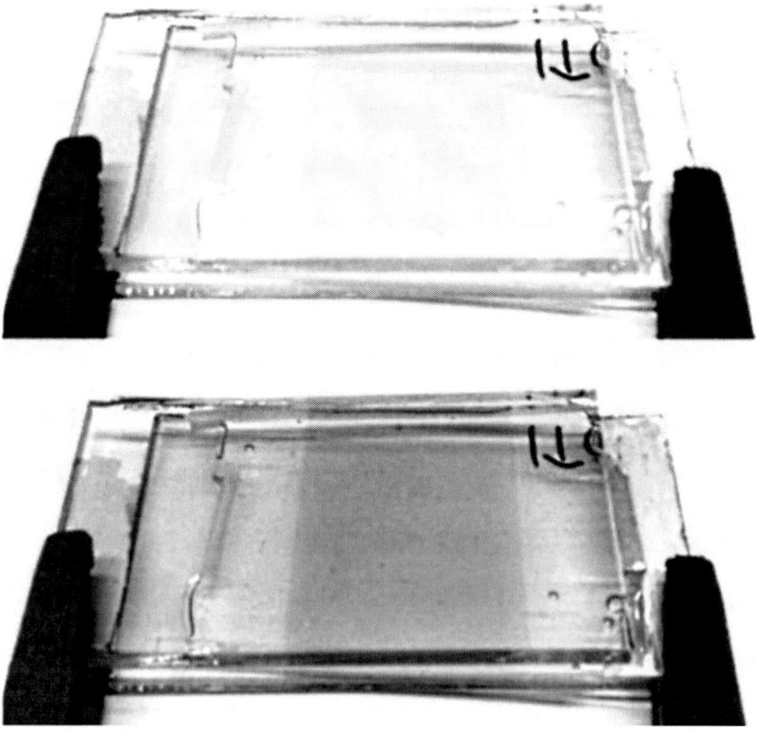

Figure 3.22. Laboratory prototypes of electrochromic units with WO_x layers deposited by CCVD in coloured state (a) and in bleached state (b).

3.7.3. Silver Films

Aside from oxidic films, it is also possible to deposit pure metallic films. This circumstance can be envisaged based on silver films. Besides the deposition of Ag nanoparticles embedded in an SiO_x matrix film already mentioned in section 3.4, compact metallic silver films can also be deposited using the CCVD process.

In contrast to metal oxides, the depositing of thin metal films under the conditions of CCVD is limited to precious metals. Thus it is possible, for example, to deposit elementary silver films if the substrate temperature is maintained above that at which the stable silver oxide occurs [80]. As the precursors for the CVD films, organometallic and metal-organic silver(I) compounds [81, 82] were used. Alongside this, structured silver films can also be printed [83] (direct ink-jet printing) and specific resistances of $1.6 \cdot 10^{-8}$ Ωm close to the solid material are reached. The thin silver films can be used in such varied applications as mirrors [84, 85], high temperature superconductors [86, 87], low-E coatings [88], heat-protective glass [89] and heatable glazing [90].

As a precursor for the deposition by means of CCVD, silver(I)-2-[2-(2-methoxyethoxy)ethoxy]acetat $[AgO_2C(CH_2OCH_2)_3H]$ was used and a solution with ethanol and water in the ratio of 5:1 was used. Apart from that, corresponding trials have been carried out with silver salts. The solution was introduced in the form of aerosol droplets and deposited on float glass. For high quality silver films, a multiple film system (Figure 3.23) consisting of SiO_x for adhesion improvement and tungsten oxide WO_x as a nucleation film for silver [91] is applied. If this is not done and the silver is deposited directly on Si or Si compounds, the formation of silver nanoparticles can be observed, leading to a brownish and electrically non-conducting film.

By means of this setup, an adhesion-fast and highly reflective film is obtained. Deposited film thicknesses vary in the range of 20 nm to 150 nm. A maximum silver content of 95% in the uppermost layer was determined by X-ray photoelectron spectroscopy, confirming the presence of a compact silver film.

To determine the transmission UV-vis spectra were recorded as a comparison to uncoated glass (Figure 3.24 a). A transmission of approximately 10% is observed for the coated glass (70 nm silver) and the roughness is R_a = 3 nm. The reflectivity in the infrared range (Figure 3.24 b) is very constant at ≥ 70% over a broad range of 2500 – 25000 nm.

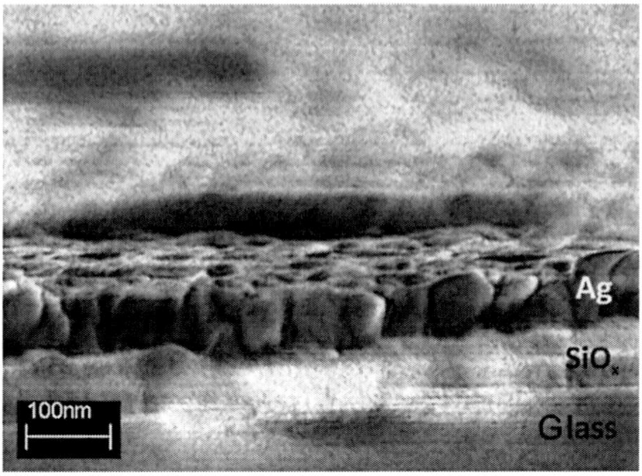

Figure 3.23. Cross section SEM image of a silver-coated float glass (the WO_x layer between SiO_x and silver is not visible in this micrograph due to its low thickness).

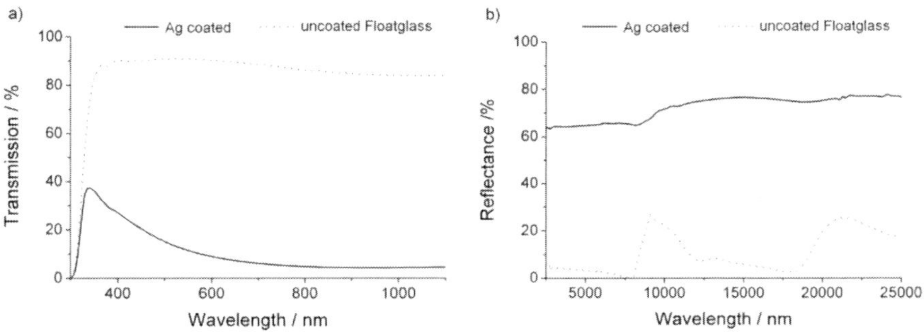

Figure 3.24. UV-vis spectra (a) of silver-coated float glass and uncoated float glass (directed transmission). IR reflection (against gold as standard) spectra (b) of silver-coated and uncoated float glass.

A striking characteristic of these CCVD-produced silver films is the conductivity achieved with a specific resistance of $5 \cdot 10^{-8}$ Ωm, which is very close to the conductivity of solid material.

Conclusion and Outlook

The application examples presented show the versatility as well as the development potential of the CCVD and r-CCVD processes. In addition to the film systems presented, a range of other oxides have been deposited and examined. Thus in addition to WO_x , MoO_x can also be used as an electrochromic film. As protective coatings, ZrO_2 or even ZrO_2 particles embedded in a matrix are also a possibility. Furthermore, for corrosion protection, Al_2O_3 and phosphate films are suitable. The oxides of rare earths can be deposited as superhydrophobic films.

The range of depositable substances can be extended. Suitable precursor compounds are sought and even newly synthesised through further intensive research. The combination of the CCVD and r-CCVD processes in particular with other atmospheric pressure processes promises a broad research field. Thus following and previous Sol-Gel processes are of high interest as the coupling with atmospheric pressure plasma processes.

Potential new fields of applications are seen above all in solar energy generation, both for photovoltaic and solar thermal products. Thus, for example, the combination of film properties which increase transmission and reduce soiling into a single process concept offers an interesting alternative to established and cost-intensive technologies.

One course of development which will be intensively pursued over the near future is the energy reduction of the CCVD and r-CCVD processes. To this end, besides the new gas-saving burner designs, comprehensive adaptation of the process parameters is also inevitable. At the same time, these energy reducing solutions also open new fields of applications with the coating of temperature-sensitive materials and/or the application potentials of new precursor substances.

REFERENCES

[1] Farris, S.; Pozzoli, S.; Biagioni, P.; Duó, L.; Mancinelli, S.; Piergiovanni, L. *Polymer* 2010, 51, pp 3591–3605.

[2] Sutherland, I.; Brewis, D. M.; Health, R. J.; Sheng, E. *Surf. Interface Anal.* 1991, 17, pp 507–510.

[3] Musil, R.; Tiller, H.-J. Der Kunststoff-Metall-Verbund: Silicoater-Verfahren; Hüthig: Heidelberg, Germany, 1989, pp 49–53.

[4] Tiller, H.-J.; Göbel, R.; Magnus, B.; Garschke, A.; Musil, R. *Thin Solid Films* 1989, 169, pp 159–168.

[5] Tiller, H-J.; Richter, T. *Vak. Forsch. Prax.* 2004, 16, pp 85–87.

[6] Hou, X.; Choy, K.-L. *Chem. Vap. Deposition* 2006, 12, pp 583–596.

[7] Struppert, T.; Heft, A.; Grünler, B. *Thin Solid Films* 2012, 520, pp 4106–4109.

[8] Struppert, T.; Jakob, A.; Heft, A.; Grünler, B.; Lang, H. *Thin Solid Films* 2010, 518, pp 5741–5744.

[9] Burkert, A.; Müller, D.; Paa, W. *J. Quant. Spectrosc. Radiat. Transfer* 2013, 114, pp 101–108.

[10] Polley, T. A.; Carter, W. B.; Poker, D. B. *Thin Solid Films* 1999, 357, pp 132–136.

[11] Struppert, T. *Galvanotechnik* 2009, 8, pp 1864–1869.

[12] Heft, A.; Hädrich, T.; Struppert, T.; Pfuch, A.; Homuth, M.; Grünler, B. In Jahrbuch; Ed. Suchentrunk, R.; Oberflächentechnik Band 64; Leuze: Bad Saulgau, Germany, 2008, pp 137–149.

[13] Tiller, H.-J.; Schimanski, A.; Grünler, B.; Glock-Jäger, K.; Weidl, R.; Zobel, B. *Glass Coatings* 2006, 2, pp 37–41.

[14] Zunke, I.; Heft, A.; Schäfer, P.; Haidu, F.; Lehmann, D.; Grünler, B.; Schimanski, A.; Zahn, D. R. T. *Thin Solid Films* 2013, 532, pp 50–55.

[15] Joos, F. Technische Verbrennung: Verbrennungstechnik, Verbrennungsmodellierung, Emissionen; Springer: Berlin Heidelberg New York, Germany, 2006.

[16] Turns, S. An introduction to combustion; McGraw-Hill: New York, NY, 1996.

[17] Dagaut, P.; Boettner, J.; Cathonnet, M. *Combust. Sci. Technol.* 1991, 77, pp 127–148.

[18] Dagaut, P.; Cathonnet, M.; Boettner, J. *Int. J. Chem. Kinet.* 1992, 24, pp 813–837.

[19] Farris, S.; Pozzoli, S.; Biagioni, P.; Duó, L.; Mancinelli, S.; Piergiovanni, L. *Polymer* 2010, 51, pp 3591–3605.

[20] Vandooren, J.; Branch, M. C.; Van Tiggelen, P. *J. Combustion and Flame* 1992, 90, pp 247–58.

[21] Galloway, J. N.; Dentener, F. J.; Capone, D. G.; Boyer, E. W.; Howarth, R. W.; Seitzinger S. P.; Asner, G. P.; Cleveland, C. C.; Green, P. A.; Holland E. A.; Karl D. M.; Michaels, A. F.; Porter, J. H.; Townsend, A. R.; Voosmarty, C. J. *Biogeochemistry* 2004, 70, pp 153–226.

[22] Richter, G. N.; Wiese, H. C.; Sage, B. H. *Combust. Flame* 1962, 6, pp 1–8.

[23] Pijpers, A. P.; Meier, R. J. J. *Electron. Spectrosc. Relat. Phenom.* 2001, 121 pp 299–313.

[24] Glassman, I. Combustion; Academic Press: Orlando, FL, 2nd ed., 1987.

[25] Warnatz, J.; Maas, U.; Dibble, R. Combustion; Springer: Berlin, Germany, 1999.

[26] Struppert, T.; Heft, A. Kunststoffe 2009, 99, pp 14–16.

[27] Volmer, M.; Weber A. *Z. phys. Chemie* 1926, 119, pp 277–301.

[28] Becker, R.; Döring, W. *Ann. Phys.* 1935, 24, pp 719–752.

[29] Frenkel, J. *J. Chem. Phys.* 1939, 7, pp 538–547.

[30] Zeldovich, J. B. *Acta Physicochim.* USSR 1943, 18, pp 1-22.

[31] Park, S. H.; Rogak, S. N. *J. Aerosol. Sci.* 2004, 35, pp 1385–1404.

[32] Pratsinis, S. E.; Vemury, S. *Powder Technol.* 1996, 88, pp 267–273.

[33] Roth, P. Topical Review; Oral presentation, 31st International Symposium on Combustion, University of Heidelberg, Germany, 2006

[34] Ulrich, G. D. *Combust. Sci. Technol.* 1971, 4, pp 47–57.

[35] Wilson, S. J.; Hutley, M. C. *Optica Acta* 1982, 29, pp 993–1009.

[36] Glaubitt, W.; Gombert, A.; Hofmann, T.; Kursawe, M; Patent DE 10146687, 2003.

[37] Giesche, H.; Kinkel, J.; Unger, K.; Patent US 4775520A, 1988.

[38] Walther, M. Proc. Conf. „Die industrielle Anwendung der Sol-Gel-Technologie", Würzburg, 2006.

[39] Chiou, B. S.; Tsai, J. H. *J. Mater. Sci. - Mater. Electron.* 1999, 10, pp 491–495.

[40] laboratory report ift Rosenheim, Nr. 410 32391, 26.10.2006.

[41] laboratory report ift Rosenheim, Nr. 603 32974, 12.04.2007.

[42] laboratory report of the Instituto Tecnologico de Canarias, ITC-EERR-FV- 008.2008, 20.05.2009.

[43] Strobel, M.; Jones, V.; Lyons, C. S. Ulsh, M.; Kushner, M. J. ; Dorai, R.; Branch, N. *Plasmas Polym.* 2003, 8, 61–95.

[44] Garbassi, F.; Occhiello, E.; Polato, F.; Brown, A. *J. Mat. Sci.* 1987, 2, pp 1450-6.

[45] Pfuch, A.; Heft, A.; Ertel, M.; Schiemann, S.; Schimanski, A. *Kunststoffe* 2006, 3, pp 147–150.

[46] Pfuch, A.; Schiemann, S.; Erler, I.; Heft, A.; Schimanski, A. *Kunststoffe* 2007, 3, pp 30–34.

[47] Zimmermann, R.; Pfuch, A.; Horn, K.; Weisser, J.; Heft, A.; Röder, M.; Linke, R.; Schnabelrauch, M.; Schimanski, A. *Plasma Process. Polym.* 2011, 8, pp 295–304.

[48] Marwedel, H. J.; Brückner, R. Glastechnische Fabrikationsfehler; Springer: Berlin Heidelberg New York, Germany, 1980.

[49] Conrad, R. *J. Am. Ceram. Soc.* 2008, 91, pp 728–735.

[50] Clark, D. E.; Zoitos, B. K. Corrosion of Glass, Ceramics and Ceramic Superconductors; Noyes Publications: Park Ridge, NJ, 1992.

[51] Bange, K.; Anderson, O.; Rauch, F.; Lehuédé, P.; Rädlein, E.; Tadokoro, N.; Mazzoldi, P.; Rigato, V.; Matsumoto, K.; Farnworth, M. Part 1: Analysis techniques and corrosion in liquid water Glass Science and Technology – Glastechnische Berichte 74, 2001, 5, pp 127-141. (Report of the International Commission on Glass (ICG), Subcommittee of Technical Committee 19 "Glass Surface Diagnostics").

[52] DIN 52296, method of test and classification, 1989.

[53] Heft, A.; Tiller, H.-J.; Weidl, R.; Patent DE 102004029911, 2006.

[54] Heft, A.; Pfuch, A.; Grünler, B. Conference speech, Congress Industrielle Oberflächentechnik: Braunschweig, 2005.

[55] Aranovich, J.; Ortiz, A.; Bube, R. H. *J. Vac. Sci. Tech.* 1979, 16, pp 994–1003.

[56] Janotti, A.; Van de Walle, C. G. *Rep. Prog. Phys.* 2009, 72, pp pp.126501-1–126501-29.

[57] Shih, W. C.; Wu, M. S. *J. Cryst. Growth* 1994, 137, pp 319–325.

[58] Wang, Q. P.; Zhang, X. J.; Wang, G. Q.; Chen, S. H.; Wu, X. H.; Ma, H. L. *Appl. Surf. Sci.* 2008, 254, pp 5100–5104.

[59] Walters, G.; Parkin, I. P. *Appl. Surf. Sci.* 2009, 255, pp 6555–6560.

[60] O'Brien, S.; Nolan, M. G.; Çopuroglu, M.; Hamilton, J. A.; Povey, I.; Pereira, L.; Martins, R.; Fortunato, E.; Pemble, M. *Thin Solid Films* 2010, 518, pp 4515–4519.

[61] Hu, J.; Gordon, R. G. *J. Appl. Phys.* 1992, 71, pp 880–890.

[62] Fujita, D. C.; Fujita, S.; Park, S. *Phys. Status Solidi* A 1999, 176, pp 579–582.

[63] Jin, B. J.; Woo, H. S.; Im, S.; Bae, S. H.; Lee, S. Y. *Appl. Surf. Sci.* 2001, 169–170, pp 521–524.

[64] Suchea, M.; Christoulakis, S.; Tibeica, C.; Katharakis, M.; Kornilios, N.; Efthimiopoulos, T.; Koudoumas, E. *Appl. Surf. Sci.* 2008, 254, pp 5475–5480.

[65] Rozati, S. M.; Akeste, S. *Cryst. Res. Technol.* 2008, 43, pp 273–275.
[66] Tseng, Y.-K.; Gao, G.-J.; Chien, S.-C. *Thin Solid Films* 2010, 518, pp 6259–6263.
[67] Sagar, P.; Shishodia, P. K.; Mehra, R.M.; Okada,H.; Wakahara, A.; Yoshida, A. *J. Lumin.* 2007, 126, pp 800–806.
[68] Zunke, I.; Wolf, S; Heft, A.; Schimanski, A.; Grünler, B.; Ronning, C.; Seidel, P. *Appl. Surf. Sci*, 2013, script submitted.
[69] Morkoç, H.; Özgür, Ü. Zinc Oxide - Fundamentals,Materials and Device Technology; Wiley: Weinheim, Germany, 2009.
[70] El Amrani, A.; Hijazi, F.; Lucas, B.; Bouclé, J.; Aldissi, M. *Thin Solid Films* 2010, 518, pp 4582–4585.
[71] Ghosh, M.; Dilawar, N.; Bandyopadhyay, A. K.; Raychaudhuri, A. K. *J. Appl. Phys.* 2009, 106, pp 084306.
[72] Yang, L.; Su, Y.; Xu, Y.; Zhang, S.; Wu, J.; Zhao, K. *J. Inorg. Biochem.* 2004, 98, pp 1251–1260.
[73] Srinivasan, S. K.; Ganguly, S. *Catal. Lett.* 1991, 10, pp 279–287.
[74] Granqvist, C. G. *Appl. Phys.* A 1993, 57, pp 3–12.
[75] Granqvist, C. G. *Sol. Energ. Mat. Sol.* C. 1994, 32, pp 369–382.
[76] de Wijs, G. A.; de Boer, P. K.; de Groot, R. A., Kresse, G. *Phys. Rev.* B 1999, 59, pp 2684–2693.
[77] Sivakumar, R.; Moses Ezhil Raj, A.; Subramanian B.; Jayachandran, M.; Trivedi, D. C.; Sanjeeviraja, C. *Mater. Res. Bull.* 2004, 39, pp 1479–1489.
[78] Granqvist, C. G. Handbook of Inorganic Electrochromic Materials; Elsevier: Amsterdam, 1995.
[79] Davis, M. J.; Benito, G.; Sheel, D. W.; Pemble, M. E. *Chem. Vap. Deposition* 2004, 10, pp 29–34.
[80] Hunt, A. T.; Cochran, J. K.; Carter, W. B.; Patent US 5652021, 1997.
[81] Grodzicki, A.; Łakomska, I.; Piszczek, P.; Szymánska, I.; Szłyk, E. *Coord. Chem. Rev.* 2005, 249, pp 2232–2258.
[82] Haase, T.; Kohse-Höinghaus, K.; Bahlawane, N.; Djiele, P.; Jakob, A.; Lang, H. *Chem. Vap. Deposition* 2005, 11, pp 195–205.
[83] Smith, P. J.; Shin, D.-Y.; Stringer, J. E.; Derby, B.; Reis, N. *J. Mater. Sci.* 2006, 41, pp 4153–4158.
[84] Shen, L.; Ji, J.; Shen, J. *Langmuir* 2008, 24, pp 9962–9965.
[85] Saito, Y.; Wang, J. J.; Batchelder, D. N.; Smith, D. A. *Langmuir* 2003, 19, pp 6857–6861.

[86] Aslanoglu, Z.; Akin, Y.; El-Kawni, M. I.; Arda, L.; Sigmund, W.; Hascicek, Y. S. *IEEE Trans. Appl. Supercond.* 2003, 13, pp 2755–2757.

[87] Pan, A. V.; Pysarenko, S. V.; Wexler, D.; Rubanov, S.; Dou, S. X. *IEEE Trans. Appl. Supercond.* 2007, 17, pp 3585–3588.

[88] Arbab, M. *Thin Solid Films* 2001, 381, pp 15–21.

[89] Richardson, C.; Thomsen, C. V. U.S. Pat. Appl. Publ. 2006, US

[90] 2006065350 A1 20060330.

[91] Siddle, J.; Ridealgh, J.; Unquera, J.; Patent WO2007080428, 2007.

[92] Al-Kuhaili, M. F.; Al-Aswad, A. H.; Durrani, S. M. A.; Bakhtiari, I. A. *Sol. Energy* 83 (2009) 1571–1577.

In: Combustion
Editor: Joseph M. Grier

ISBN: 978-1-62948-967-4

Chapter 4

FUNDAMENTALS OF OXY-FUEL CARBON CAPTURE TECHNOLOGY FOR PULVERIZED FUEL BOILERS

Lucía Álvarez, María V. Gil, Juan Riaza, Cova Pevida and Fernando Rubiera*

Instituto Nacional del Carbón, INCAR-CSIC, Oviedo, Spain

ABSTRACT

Coal is expected to continue to contribute to electricity production in the near future. However, it generates more CO_2 per unit of produced energy than other fossil fuels and its use is associated with the formation of contaminants such as NO_x, SO_x, Hg, etc. In recent years, oxy-fuel combustion has attracted great interest as a promising Carbon Capture and Storage (CCS) technology, since it can be easily adapted to both existing and newly built coal-fired power plants. During oxy-fuel combustion coal is burnt in a mixture of oxygen and recycled flue gas (RFG) (mainly CO_2 and water vapor), that is used to lower the flame temperature and yield a rich stream of CO_2. However, the successful implementation of oxy-fuel combustion technology depends on having a full understanding of the difficulties that arise as a result of replacing inert N_2 by reactive CO_2 in the oxidizer stream.

* Corresponding author. Tel.: +34 985118975; Fax: +34 985297662. E-mail address: frubiera@incar.csic.es.

Biomass is a renewable fuel which can be used to reduce CO_2 emissions, as it is considered carbon-neutral. The combination of oxy-coal combustion with biomass co-firing could provide an effective method of CO_2 disposal which has only partially been studied. Furthermore, biomass co-firing is presented as an option for increasing capture efficiency, as the costs and efficiency penalties would be reduced. This chapter presents a technical review of the oxy-fuel combustion process, covering the most recent experimental studies at both laboratory and pilot scale. The impact of replacing N_2 by CO_2 upon heat transfer, devolatilization, ignition, combustion kinetics, pollutant formation (mainly NO_x and SO_x) and fly ash and slag formation is also evaluated. Furthermore, the effect of blending coal and biomass under oxy-fuel conditions on combustion properties is assessed.

1. INTRODUCTION

1.1. Carbon capture Technologies for Coal-fired Power Plants

Coal has been, and will continue to be in the medium term, one of the major sources of energy due to its abundant reserves and competitively low prices. The share of coal in world energy consumption was 29.9% in 2012, compared to 33.1% for oil and 23.9% for natural gas [1]. A diverse power generation portfolio including Carbon Capture and Storage (CCS) and renewable energies is needed to reduce atmospheric CO_2 to below 1990 levels [2]. CCS technologies involve three stages: i) capture of the CO_2 produced in a large industrial plant, ii) transportation of the CO_2 to a suitable storage site and (iii) permanent storage away from the atmosphere. Two types of CO_2 disposal are generally employed: sequestration (permanent disposal) or storage (disposal for a significant period of time). Both terms are often inter-changeable since time periods of more than 10,000 years are considered as permanent. Possible storage sites include depleted oil and gas reservoirs, deep saline aquifers, etc. [3].

The three main technologies for CO_2 capture, which are briefly described below, are post-combustion, pre-combustion and oxy-fuel combustion. The CO_2 concentration in the gas flow, the gas pressure and the fuel type (solid or gas) are important factors when selecting the most appropriate capture system.

- *Post-combustion.* This technology involves the separation of diluted CO_2 (3-15% by volume on a dry basis) from other exhaust gases (typically nitrogen) after the combustion of fossil fuels. The separation is mainly performed via chemical absorption with amines (typically monoethanolamine, MEA). However, the net efficiency of the power plant is reduced by about 10-14% points. Some projects are currently investigating the development of more efficient absorbents [4], and the development of solid sorbent materials for CO_2 capture by adsorption [5].
- *Pre-combustion.* This technology is adequate for application in Integrated Gasification Combined Cycle (IGCC) processes. It involves the primary conversion of the fuel with steam and air/oxygen to produce a mixture of CO and hydrogen (*"synthesis gas"* or *"syngas"*). Extra hydrogen and CO_2 are produced by reacting CO with steam via the water gas shift (WGS) reaction. Although the previous steps for fuel conversion are more elaborated and costly than in post-combustion systems, the high CO_2 concentration in the final gas stream (typically 15-60% by volume on a dry basis), together with the high pressure that often results from the process, are more favorable for CO_2 capture.
- *Oxy-fuel combustion.* This technology involves the combustion of fuel under an oxygen atmosphere instead of air. By avoiding the introduction of nitrogen into the combustion chamber, a rich CO_2 flue gas stream is obtained (higher than 95% by volume on a dry basis). The plant design for the oxy-fuel combustion process comprises the recirculation of flue gases to the burners in order to keep the flame temperature within acceptable limits for the boiler materials. The main disadvantage of this technology is that it needs oxygen of great purity, which may result in an efficiency drop of 7-11% points.

Because of the significant changes required in a power plant by the implementation of the oxy-fuel combustion process, i.e., replacing N_2 (inert) by CO_2 (reactive), an exhaustive research study is needed to assess the impact of the process. The amount of literature on oxy-fuel combustion technology has drastically increased in recent years and significant progress has been made towards the implementation of oxy-fuel combustion in power plants. The objective of the present review is to summarize the current status of oxy-fuel combustion technology.

1.2. Historical Development of Oxy-fuel Combustion

Oxy-fuel combustion technology along with flue gas recirculation were proposed, almost simultaneously, by Abraham et al. [6] and Horn and Steinberg [7] in the 1980s. Whereas Abraham et al. [6] proposed the process as a possible method to produce CO_2 for Enhanced Oil Recovery (EOR), Horn and Steinberg [7] were concerned about the need to reduce the environmental impact caused by the use of fossil fuels in energy generation. Interest in oxy-fuel technology was revived in the mid-90s by the widespread concern about global warming caused by the increasing levels of CO_2 in the atmosphere. Figure 1 shows the historical progression of oxy-fuel combustion projects, including pilot-scale, industrial-scale and demonstration plants.

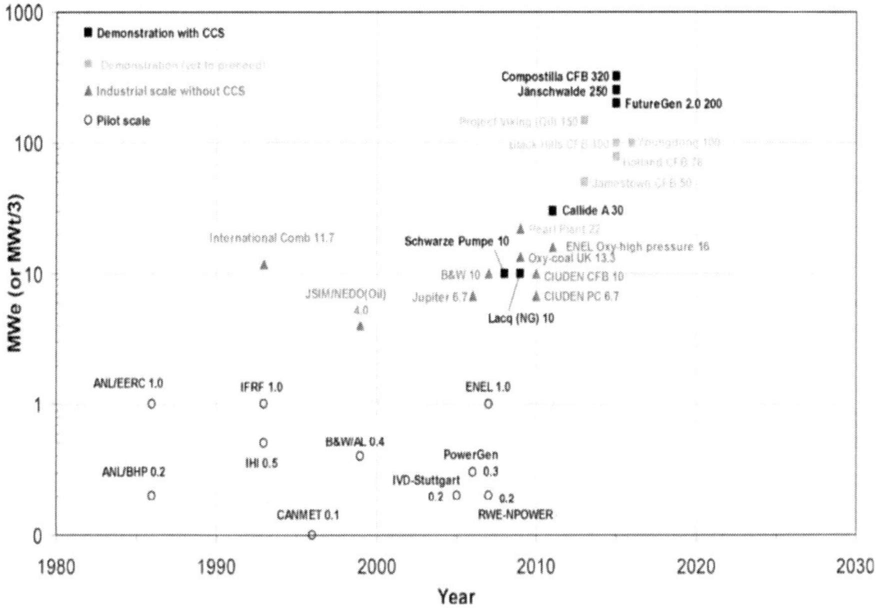

Figure 1. Historical development of oxy-fuel projects (reprinted from [8], with permission from Elsevier).

Oxy-fuel combustion can be applied to any type of fuel for thermal power generation. Until now research has been mainly focused on coal (oxy-coal combustion) and natural gas (oxy-natural gas combustion), although this technology can also be applied to biomass or coal/biomass blends. There are several oxy-fuel pilot plants currently in operation and some industrial-scale power plants are expected to start operating in the near future. Some of these plants are listed below:

- *Vattenfall (2008):* This is the world's first full chain oxy-fuel pilot demonstration plant. It is located in Spremberg (Germany) and it has a 30 MWt pf boiler which is fed with lignite.
- *Total Lacq project (2009):* This is the world's first integrated industrial natural gas-fired oxy-fuel plant and it includes the world's first pipeline for injecting oxy-fuel flue gas. It is located in Lacq (France).
- *CIUDEN (2011):* This is located in Ponferrada (Spain) and it has a 20 MWt pf boiler and a 30 MWt fluidized-bed boiler, which are able to operate in both air and oxy-fuel conditions. Different fuels are employed: anthracitic, bituminous and sub-bituminous coals, coke and biomass.
- *Callide Oxy-fuel Project (2011):* This is the world's first full chain retrofit demonstration plant. It is located in Biloela (Australia) and it is coal-fired to generate electricity.
- *Commercial demonstration (2015-2016):* This is included in the project Future Gen 2.0 (200 MWe) in the United States. This plant is an oil-fired process which will be converted to a coal-fired utility.

These projects and milestones show the trend for the development of oxy-fuel combustion technology in the future. However, the cost of failure of an operating project greatly increases as a function of its technological scale. For this reason, the European Union first made a risk assessment of the CCS technologies in 2008 [9], as part of the European Technology Platform for Zero Emission Fossil Fuel Power Plants (ZEP). In 2010, it outlined the research and development needs of each one of the CCS technologies applied to the supply of electricity, cement and iron [10]. Furthermore, the objective presented by the G8 group to build 20 commercial scale (>1 Mtpa CO_2) CCS plants which will begin operation in 2020 requires that a number of oxy-fuel plants be built, although the exact number is still uncertain [8]. The success of large scale oxy-fuel projects such as Callide and Future Gen 2.0 over the next 5-10 years will therefore be decisive for the development of this technology.

2. Oxy-fuel Combustion Fundamental Research

2.1. Heat Transfer

2.1.1. Radiative Heat Transfer

The major contributor to the transfer of heat from a flame produced by conventional fuels is thermal radiation from water vapor, CO_2, CO, char, particles and soot. In oxy-fuel combustion with RFG, the CO_2 and water vapor concentrations in the combustion atmosphere are significantly increased. Therefore, under oxy-fuel conditions radiative heat transfer in the boiler can be expected to be higher than in conventional air-firing.

Radiative heat transfer in lignite- and propane-fired oxy-fuel flames was studied by Andersson et al. [11-13] in the Chalmers University 100 kW oxy-fuel test facility. In the lignite experiments the flue gas recycle rate was varied in order to ensure the same stoichiometry in all cases, while the oxygen concentration in the oxidizer stream ranged from 25% to 29% (v/v). The temperature and in turn the total radiation intensity of the oxy-fuel flames increased as the flue gas recycle rate decreased, i.e., as the oxygen concentration increased. However, when similar flame temperatures were kept during the air and oxy-fuel combustion with 25% oxygen conditions and dry recycling, the total radiation intensity of both flames remained the same. In fact, the ratio of gas and total radiation intensities was higher under oxy-fuel conditions compared to air-firing, while the particle radiation remained unchanged from air to oxy-fuel combustion. However, when radiation overlap between gas and particles was considered, the ratios of gas/particle-to-total radiation intensities for air-firing and oxy-fuel conditions became similar, since the gas-particle radiation overlap was found to be higher in the CO_2-rich atmosphere. The impact of gas radiation was therefore reduced due to gas-particle radiation overlap. Furthermore, particle radiation, which did not change from air to oxy-fuel conditions, constituted a significant fraction (about 60-70%) of the total radiation emitted by the lignite flames. It can be concluded that no increase in gas radiation due to a higher CO_2 concentration was therefore evident because of the background particle radiation and the fact that the total radiation intensities were similar in air-fuel and oxy-fuel operation under similar temperature distributions. A change in flame temperature caused by changes in the recycling rate, on the other hand, could significantly influence the radiation intensity. When wet recycling was used in oxy-fuel combustion of lignite, the gas radiation became stronger and approached the contribution by the particles. However, the influence of H_2O

on particle radiation characteristics was not studied and requires further experimental studies.

For propane flames, the oxygen concentration in the oxidizer stream ranged from 21% to 45% (v/v). This study revealed that the soot formation in the oxy-fuel flames varied strongly depending on the recycling conditions and this significantly affected the radiation intensity emitted by the flame (Figure 2). It was concluded that the contribution of increased gas radiation intensity by CO_2 was of minor importance. For oxy-fuel conditions with 27% oxygen and dry recycling, the temperature was slightly lower compared to air-fired conditions, but the radiation intensity was significantly higher. It was mainly caused by a significantly increased soot radiation. Images of the flame under oxy-fuel conditions with 45% oxygen clearly indicated a further increase in radiation intensity (Figure 2). Thus, due to its critical role in flame radiation, further research is needed on soot formation in oxy-fuel combustion.

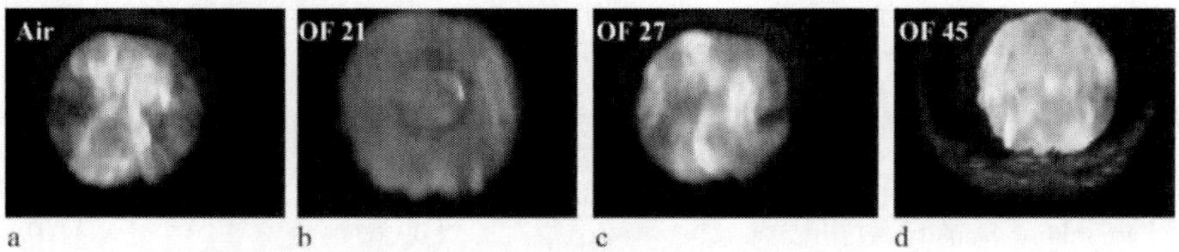

Figure 2. Photographs of the Chalmers flame with propane as a fuel: (a) air flame, (b) oxy-fuel flame with 21% oxygen in the oxidizer, (c) oxy-fuel flame with 27% oxygen in the oxidizer, and (d) oxy-fuel flame with 45% oxygen in the oxidizer; 215 mm from the burner inlet (reprinted from [13], with permission from Elsevier). Images (a)–(c) had previously been published in [12] (reprinted with permission from [12], Copyright © 2008 American Chemical Society).

2.1.2. Convective Heat Transfer

The convective flux of the flue gas, q^{n}_{conv}, can be evaluated as follows:

$$q^{n}_{conv} = h\Delta T \tag{1}$$

where *h* is the convective heat transfer coefficient (which is influenced by the flow velocity and by gas properties such as viscosity, thermal conductivity, heat capacity and density), and ΔT is the temperature difference between the bulk gas and the heated object. The ratio of convective heat transfer coefficient of oxy-fuel combustion (h_{oxy}) to that of air-firing conditions (h_{air}) can be expressed as follows:

$$\frac{h_{oxy}}{h_{air}} = \left(\frac{\mathrm{Re}_{oxy}}{\mathrm{Re}_{air}}\right)^m \left(\frac{\mathrm{Pr}_{oxy}}{\mathrm{Pr}_{air}}\right)^n \left(\frac{k_{oxy}}{k_{air}}\right) \tag{2}$$

where Re and Pr are the Reynolds number and the Prandtl number, respectively; *m* and *n* are empirical factors which vary for different geometries; and *k* is the fluid thermal conductivity. Table 1 summarizes the physical properties of the principal gases used in the air and oxy-fuel conditions. CO_2 has a slightly higher thermal conductivity than N_2, although it does not significantly change the conductive heat transfer. On the other hand, the lower kinematic viscosity of CO_2 may result in a higher Reynolds number, and thus a higher convective heat transfer coefficient.

Table 1. Property table for gases at 1400 K and atmospheric pressure

	$H_2O_{(v)}$	O_2	N_2	CO_2
Density (ρ) (kg/m^3)	0.157	0.278	0.244	0.383
Thermal conductivity (k) (W/mK)	0.136	0.087	0.082	0.097
Specific heat capacity (c_p) (kJ/kmol K)	45.67	36.08	34.18	57.83
Dynamic viscosity (μ) (kg/m s)	5.02 10^{-5}	5.81 10^{-5}	4.88 10^{-5}	5.02 10^{-5}
Kinematic viscosity (ν) (m^2/s)	3.20 10^{-4}	2.09 10^{-4}	2.00 10^{-4}	1.31 10^{-4}

2.1.3. Matching Temperature of Combustion

Furnace heat transfer in an industrial boiler depends on the flame temperature, the properties of the gas and particulates, water cooling, wall temperature and the aerodynamic fluid field. The differences in the radiative and thermo-physical properties of N_2 and CO_2 lead to differences in the radiative and convective heat transfer. When retrofitting existing coal-fired power plants for oxy-fuel combustion, the main task is to maintain the same heat transfer characteristics in the furnace as in the air-fired combustion. Matching the heat transfer values in the furnace can contribute to establishing similar combustion stabilities, carbon burnouts, slagging and fouling tendencies. For new oxy-fuel combustion power plants, it is important to evaluate the effect of operating parameters such as oxidizer composition and oxygen excess levels on flame temperature and heat transfer in the boiler.

Wall et al. [14] calculated the theoretical oxygen mole fraction required at the burner inlet to achieve similar adiabatic flame temperatures (AFT) under air- and oxy-firing conditions. The AFT for air combustion was estimated for 20% excess air and oxygen levels in the flue gas of 3.3% (v/v). The computational results showed that to achieve a similar AFT to that of air combustion, the oxygen fraction at the burner inlet for oxy-fuel combustion using wet and dry flue gas recirculation would have to be 28% and 35%, respectively. The excess oxygen required to maintain oxygen levels in the flue gas similar to those of air combustion was in the range of 3-5% for both oxy-dry and oxy-wet combustion. Hjärtsman et al. [15] obtained temperature profiles during the combustion of lignite in air- and oxy-firing conditions in the Chalmers University 100 kW oxy-fuel test facility. Their measurements showed that under oxy-fuel conditions the oxygen concentration in the oxidizer must be around 25-27% in order to get similar temperatures to those under air.

The effect of oxy-fuel conditions on radiative and convective heat transfer was studied by Smart et al. [16,17], who obtained radiative and convective heat transfer measurements in the RWEn 0.5 MWth combustion test facility located in Didcot UK fired by a semianthracite and two bituminous coals. Recycle ratios varied between 65% and 75% while furnace exit O_2 values were maintained at 3%. Their results showed that peak radiative heat flux values were inversely related to the recycle ratio. Conversely, the convective heat flux values tended to increase with an increasing recycle ratio. Radiative heat flux values in oxy-fuel conditions similar to those in air were obtained for the coals studied in the recycle ratio range of 72-73%. The equivalent air convective heat flux values were achieved at a higher recycle ratio of around 75%.

2.2. Coal Devolatilization and Char Formation

Far from being just another step in the overall reaction, coal particle devolatilization has a strong impact on the overall combustion process since it determines the subsequent steps (e.g., char reactivity and NO formation). Devolatilization is an endothermic process and its kinetics is strongly affected by the operation conditions (i.e., gas temperature and heating rate) and fuel type. Devolatilization kinetics and yield are usually studied under inert gases such as N_2 (typical of air-fuel systems). For this reason, devolatilization characteristics in CO_2 environments (typical of oxy-fuel systems) are of great interest for design purposes.

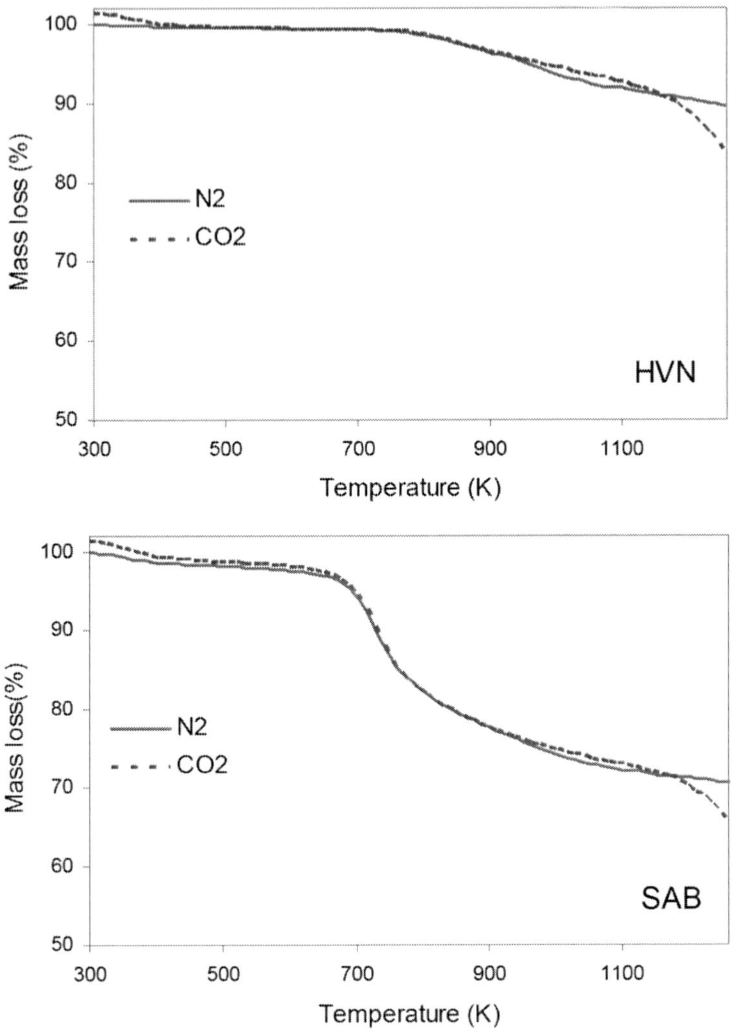

Figure 3. Mass loss rate for two different rank coals (HVN: semianthracite; SAB: high-volatile bituminous coal) during devolatilization tests under N_2 and CO_2 (TG analyzer, heating rate of 15 K/min) (reprinted from [20], with permission from Elsevier).

Rathnam et al. [18] carried out devolatilization experiments on a bituminous coal in N_2 and CO_2 environments in a TGA analyzer. They found that the weight loss histories were similar for both environments below 1050 K. However, in the CO_2 environment the weight loss was higher at temperatures above 1050 K, which they attributed to char-CO_2 gasification.

These authors also carried out devolatilization experiments on four bituminous coals in a drop tube furnace (DTF) in N_2 and CO_2 environments at a nominal wall temperature of 1673 K. SEM images showed partly reacted surfaces in the CO_2 char and significant swelling compared to the N_2 char particles. The volatile yields in the CO_2 environment were ~5-25% higher than those in the N_2 environment. Naredi et al. [19] reached the same conclusion in their pyrolysis experiments on high and low-volatile bituminous coals using a DTF under N_2 and CO_2 environments at temperatures ranging from 1173 K to 1673 K. Likewise, Álvarez et al. [20] obtained similar results when carrying out a comparison of coal devolatilization in N_2 and CO_2 atmospheres conducted in a thermogravimetric apparatus (heating rate of 15 K/min), as illustrated in Figure 3. Before devolatilization was concluded at around 1150 K, the mass loss curves in the N_2 and CO_2 atmospheres were showing a similar trend. The additional mass loss in the CO_2 atmosphere above 1200 K was mainly due to the reaction between char and CO_2.

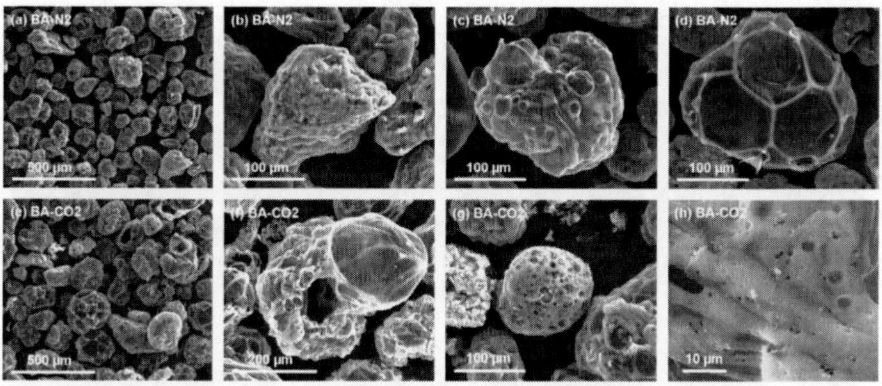

Figure 4. SEM images of the high-volatile bituminous coal (BA) char particles obtained under N_2 (a-d) and CO_2 (e-h) in an entrained flow reactor (EFR) at 1273 K (reprinted from [21], with permission from Elsevier).

Gil et al. [21,22] performed devolatilization experiments on anthracitic and bituminous coals in an entrained flow reactor at 1273 K in N_2 and CO_2 environments. They observed higher volatile yields in the CO_2 environments. An example is presented in Figure 4, where SEM images of the chars reveal a greater degree of swelling in the chars obtained in CO_2. The reactivities of the chars obtained in N_2 and CO_2 in a typical oxy-fuel atmosphere were compared using kinetic parameters but, as can be seen in Table 2, no significant differences were found between both atmospheres. Brix et al. [23] also carried out coal pyrolysis experiments in an entrained flow reactor in N_2 and CO_2 environments, but found no differences in char morphology and volatile yield between both environments. However, these authors have pointed out the need to be careful when attributing an increased volatile yield in CO_2 to the gasification of the fuel because of the short residence times needed for devolatilization at high temperatures. Factors such as mixing cold and hot gas streams, the particle heating rate and total particle residence times need to be included when comparing the data.

Table 2. Kinetic parameters that provide the best fit of the experimental data from oxy-fuel combustion (30%O_2-70%CO_2) to the VM, GM and RPM models (reprinted from [21], with permission from Elsevier)

Char	Volumetric model (VM)		Grain model (GM)		Random pore model (RPM)		
	E (kJ/mol)	k_0 (s^{-1})	*E* (kJ/mol)	k_0 (s^{-1})	*E* (kJ/mol)	k_0 (s^{-1})	ψ
HVN-N2	128	1.64E+05	128	1.21E+05	127	5.09E+04	15.9
HVN-CO2	118	3.55E+04	118	2.51E+04	117	8.10E+03	29.4
UM-N2	125	1.90E+05	123	1.12E+05	123	5.89E+04	6.7
UM-CO2	121	9.10E+04	120	5.76E+04	119	2.14E+04	16.4
SAB-N2	121	1.36E+05	120	8.15E+04	120	9.48E+04	0.7
SAB-CO2	125	2.50E+05	125	1.88E+05	125	2.31E+05	0.2
BA-N2	127	4.09E+05	126	2.66E+05	126	2.82E+05	0.9
BA-CO2	122	1.57E+05	121	1.10E+05	121	1.11E+05	1.0

HVN: semianthracite; UM: medium-volatile bituminous coal; SAB: high-volatile bituminous coal; BA: high-volatile bituminous coal.

2.3. Ignition in Oxy-firing Conditions

The ignition of solid fuel particles is an important preliminary step in the overall combustion process due to its influence on the stability, shape and length of the flame. In practice, the ignition behavior of solid fuels is crucial to identify the optimal location for injecting them into pulverized fuel burners, and flame stability is the main point of issue in furnace operation. In air-fuel combustion, flames are commonly Type-2 swirling flames that involve both internal as well as external recirculation. As observed by Khare et al. [24] for a retrofit of oxy-fuel application, when the burner flows (and hence flow velocities) are reduced, there is the possibility that flames of Type-2 become Type-0 flames (a forward-flowing low swirling flame without internal recirculation). Changes to the burner dimensions and/or swirl levels will be needed to re-establish the operation of the furnace with the Type-2 flame for which many industrial operations are designed.

In the oxy-fuel combustion of pulverized coal, poor ignition quality has been noted during pilot-scale burning trials when operating with substantial flue gas recirculation [25]. After the first operation in oxy-fuel mode, it was clear that adjustments to the swirl and the oxygen concentration at the burner were required to ensure a safe ignition and a stable flame. For this reason efforts have been recently made to comprehend the fundamentals of ignition. Taniguchi et al. [26] developed a model to predict the lean flammability limit and flame propagation velocity for air and oxy-fuel combustion systems. The model could be used to analyze the effect of coal properties, particle diameter, and composition of the surrounding combustion gas. The model was extended by Taniguchi et al. [27] with the development of an engineering design for existing and pilot-scale burner systems using oxy-fuel combustion. A DS®T-burner was developed by Hitachi Power Europe and installed at the Schwarze Pumpe pilot plant. Smart el. [28] evaluated the impact of oxy-fuel combustion on the flame characteristics through digital imaging in a 0.5 MWth coal combustion test facility. They found that the presence of high levels of CO_2 at high recycle ratios resulted in delayed combustion and therefore had a detrimental effect on flame stability, i.e., lower flame oscillation frequencies.

Khatami et al. [29] measured the ignition delay times of two lignites, a sub-bituminous coal and a bituminous coal, in O_2/N_2 and O_2/CO_2 atmospheres with oxygen mole fractions in the range of 20-100%. The experiments were performed in an electrically heated drop tube furnace with a furnace wall temperature of 1400 K. As can be seen in Figure 5, a longer coal particle ignition delay was observed for all the coals in the O_2/CO_2 environment than in O_2/N_2 with an identical oxygen concentration. An increase in the oxygen mole fraction under either O_2/N_2 or O_2/CO_2 reduced the delay in ignition. Similar results were obtained by Zhang et al. [30] when they burned a bituminous coal in air and O_2/CO_2 (21-27% O_2) atmospheres in a drop tube furnace at 1073 K and 1273 K. They observed that ignition was delayed in the presence of CO_2, and that the volatiles preferentially remained unburned in the vicinity of the particle surface and formed a thick protective sheath. Partial oxidation and gasification of the chars to CO was also favored under oxy-fuel conditions.

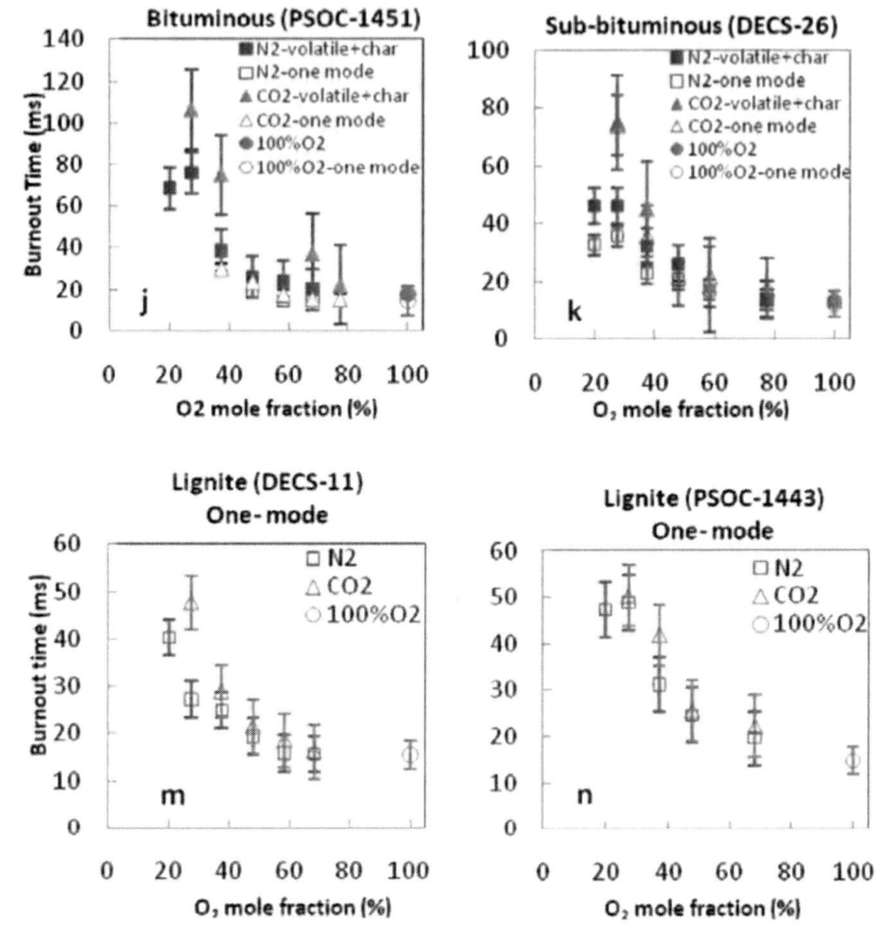

Figure 5. Burnout time for particles (75-90 μm) from Pittsburgh # 8 bituminous coal (PSOC-1451), Wyodak sub-bituminous coal (DECS-26), Beulah lignite coal (DECS-11) and Wilcox lignite coal (PSOC-1443) burning in O_2/N_2 and O_2/CO_2 at 1400 K (reprinted from [29], with permission from Elsevier).

Riaza et al. [31,32] carried out ignition experiments in an entrained flow reactor in air and oxy-fuel conditions for anthracitic and bituminous coals. Their results showed a worsening of the ignition properties when N_2 was replaced by CO_2 with the same oxygen concentration, especially in the case of the anthracitic coals. As can be seen in Figure 6, the anthracite coal ignited in a heterogeneous mode in both air and oxy-firing conditions, the semianthracite coal partially ignited heterogeneously, whereas the bituminous coal ignited in a homogeneous mode. Similar ignition properties to those attained in air were obtained in an oxy-fuel environment when the oxygen concentration was ~30%. In addition, these authors found that the addition of water vapor as a substitute for N_2 or CO_2 caused a worsening of the ignition properties, especially in the atmospheres with a lower oxygen content. Binner at al. [33] also observed that the presence of water vapor in the oxy-fuel atmosphere caused a delay in ignition, an extended combustion time and a lower flame stability.

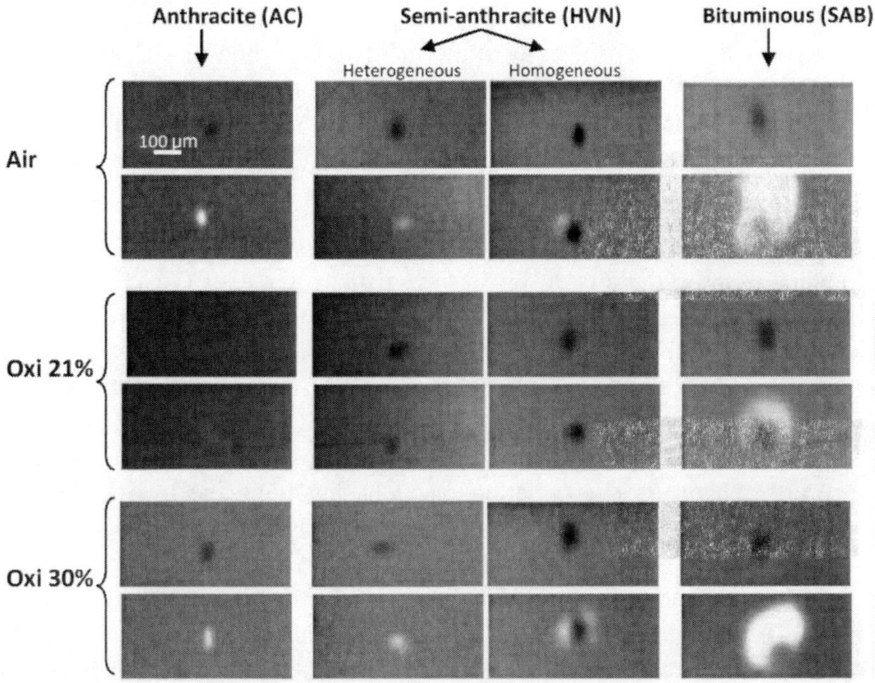

Figure 6. High-speed, high-magnification cinematographic images of single particles (75-150 µm) of AC, HVN and SAB coals in air and in two oxy-fuel conditions ($21\%O_2$-$79\%CO_2$ and $30\%O_2$-$70\%CO_2$). In each case, a particle is shown prior to and after ignition (reprinted from [32], with permission from John Wiley and Sons).

2.4. Combustion in Oxy-firing Conditions

In pf coal boilers, combustion usually takes place at high temperatures in the diffusion control regime (Zone III). In the case of oxy-fuel combustion in which temperatures are lower, combustion occurs in the diffusion/chemical control regime (Zone II). The lower diffusivity of oxygen in CO_2 than in N_2 affects the flux of oxygen towards the coal particle, reducing the burning and heat generation rates of the coal particles [34].

2.4.1. Combustion of Volatiles

Although volatile combustion kinetics is relatively quick and is not a limiting step in the overall coal combustion process, some researchers have observed differences in oxy-fuel conditions and an air-fuel environment [30,35]. For example, they have observed that the consumption kinetics are reduced by the lower diffusivity of the small hydrocarbons in CO_2 than in N_2.

Glarborg and Bentzen [36] in their study of the chemical effects of high CO_2 concentrations on the oxidation of methane found that high CO_2 concentrations competed with oxygen for atomic hydrogen and led to the formation of CO through the reaction $CO_2 + H \leftrightarrow CO + OH$. In addition, the reactions of CO_2 with hydrocarbons could also contribute to CO formation. Mendiara and Glarborg [37] observed that higher CO_2 concentrations increased the OH/H ratio, lowered the O/H radical pool and increased the amount of CO. In the light of these findings, Andersen et al. [38] tested and refined two global multistep mechanisms for oxy-fuel conditions, the two-step mechanism by Westbrook and Dryer and the four-step mechanism by Jones and Lindstedt, on the basis of model predictions using a detailed chemical kinetic mechanism. The initiation reactions involving the hydrocarbon fuel remained unaltered in the modified model and changes were made to the H_2-CO-CO_2 reactions to obtain an improved fit for CO levels and steady state emissions predicted by the detailed chemical kinetic mechanism.

Bejarano and Levendis [39] performed combustion experiments on lignite and bituminous coal particles in a 4.2 kW electrically heated laminar-flow drop-tube furnace in air and oxy-fuel conditions with high temperatures (1400-1600 K) and different oxygen contents (20-100%). The combustion temperatures of the coal volatiles were measured and found to be higher in the O_2/N_2 environment than in O_2/CO_2 conditions, whereas their corresponding combustion times were shorter. Khatami et al. [29] performed similar particle combustion experiments on coals of different rank and they observed that on replacing N_2 by CO_2 but maintaining the same oxygen mole fraction, particle combustion appeared dim and blurry (indicative of slow oxidation), as can be seen in Figure 7. The combustion of volatiles in envelope flames was suppressed in the presence of CO_2. However, increasing the oxygen mole fraction in CO_2 atmospheres led to an increase in the intensity of combustion and temperature.

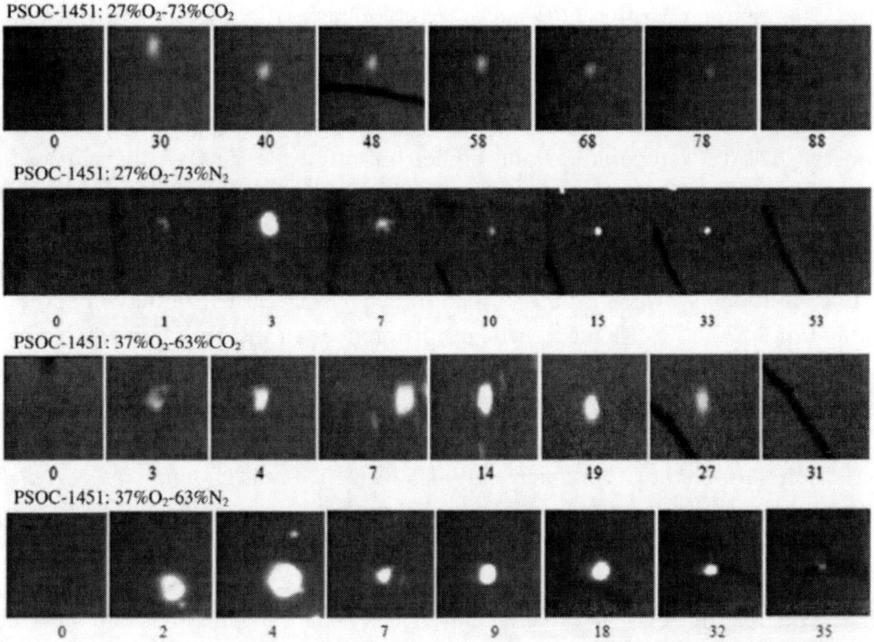

Figure 7. High-speed, high-magnification cinematography of typical combustion events of single particles of Pittsburgh # 8 bituminous coal (PSOC-1451) in different O_2/N_2 or O_2/CO_2 atmospheres. The numbers displayed under each frame denote milliseconds, zero representing the beginning of each depicted sequence. Nominal particle sizes were in the range of 75-90 μm, and the furnace wall temperature was set to 1400 K (reprinted from [29], with permission from Elsevier).

2.4.2. Char Oxidation

An increase in the oxygen concentration in the oxidant stream under oxy-fuel combustion can compensate for the larger specific heat capacity of CO_2 compared to N_2 and for the lower diffusivity of oxygen and hydrocarbons in CO_2 than in N_2. Bejarano and Levendis [39] found that the combustion of bituminous char particles was hotter and faster in N_2 than in CO_2 background gases. In the case of the bituminous coals, similar char burning times and temperatures to those attained in air were obtained in the O_2/CO_2 atmospheres with ~30-35% O_2. However, in the case of the lignite coal particles the corresponding oxygen required in the O_2/CO_2 atmosphere was around 25%. Similar conclusions were drawn by Kathami et al. [29], who found that the burning time differences under N_2 and CO_2 were considerable for low oxygen contents (~21%), whereas they were not significant for high oxygen contents (higher than 35%).

Shaddix and Molina [40] measured the char particle temperature reached during the combustion (gas temperature of 1700 K) of a sub-bituminous coal and a high-volatile bituminous coal in an entrained flow reactor. The oxygen concentrations ranged between 12-36% in N_2 and CO_2 atmospheres. They observed lower temperatures and higher burning times in CO_2 than in N_2 for the same oxygen content. Brix et al. [23] carried out char combustion experiments in an entrained flow reactor at temperatures ranging between 1173 K and 1673 K, using N_2 and CO_2 as carrier gases with inlet oxygen concentrations between 5-28%. When the char was burned in the temperature range of 1173-1573 K, no significant differences in the char combustion rate between the two environments were observed. However, when char was burned at 1573 K and 1673 K, a higher combustion rate was observed in the N_2 environment. This was attributed to the lower molecular diffusion coefficient of O_2 in CO_2, which only has influence at high temperatures when the external mass transfer may affect the combustion process.

2.4.3. Effect of Gasification Reactions

Considering that the CO_2 and H_2O(v) partial pressures in oxy-fuel combustion atmospheres are much higher than in the case of air-fuel, gasification should become more prominent in the later stages of the oxy-fuel combustion process. As long as the oxygen concentration is higher, the contribution of the gasification reactions is smaller than that of the oxidation reactions. Gasification reactions are highly endothermic, and the external energy required for such reactions is usually provided by the heat released from the oxidation reactions. In any case due to their endothermicity, the temperature of the particles and their reaction rate would be reduced [41]. Várhegyi et al. [42] measured the char consumption rate in Ar/O_2 and CO_2/O_2 mixtures and they found that the reaction rate was proportional to the oxygen concentration, but it was not influenced by the high concentration of CO_2 at 1223 K.

2.5. Pollutant Formation

Coal combustion results in the formation of gaseous pollutants such as NO_x, SO_x, mercury, and particulate matter including fly ash and soot. Recent studies on the impact of replacing N_2 by CO_2 upon pollutant formation have been carried out, the main findings of which are as follows.

2.5.1. Nitrogen Oxides

An extensive review on NO_x emissions and their control in oxy-fuel combustion has been compiled by Normann et al. [43]. In comparison with conventional combustion, lower NO_x emission intensities were observed in oxy-fuel conditions since the lower N_2 concentration inhibits the formation of thermal and prompt NO. Furthermore the NO contained in the recycled flue gas is reburned in the combustion zone to yield N_2. Also, due to the higher CO_2 concentrations in the oxy-fuel atmosphere, the radical pool may be altered, thereby influencing NO formation.

A higher CO content can be expected under oxy-fuel conditions due to the dissociation of CO_2 and char gasification reactions, which could contribute to the reduction in the amount of NO formed. Shaddix and Molina [44] measured NO concentrations during the combustion of pulverized coals and coal chars at 1323 K in O_2/N_2 and O_2/CO_2 environments with different oxygen concentrations (12-36% O_2). They found that N-char had slightly lower conversion values in O_2/CO_2 environments, although the fuel nitrogen conversion increased with the oxygen concentration (Figure 8). These authors also performed experiments with high background NO_x concentrations to investigate the importance of NO reburning. The reburn reactions were found to be of great importance (between 20-40%) in high background NO_x environments, net NO_x production being observed only in the environment with 36% oxygen. Similar results were obtained by Ndibe et al. [45] when they burned a lignite and a high-volatile bituminous coal in an entrained flow reactor at a wall temperature of 1373 K in air and O_2/CO_2 conditions (21-40%O_2). They found that NO emissions were significantly lower in the oxy-fuel experiments than in air and that they increased with the increase in inlet oxygen concentrations under oxy-fuel conditions.

Álvarez et al. [46] reached similar conclusions in their oxy-fuel combustion experiments with a semianthracite and a high-volatile bituminous coal in an entrained flow reactor at a wall temperature of 1273 K (Figure 9). These authors also observed that the substitution of steam for N_2 or CO_2 decreased the formation of NO. However, the efficiency of steam in the reduction of NO decreased with increasing inlet oxygen concentrations.

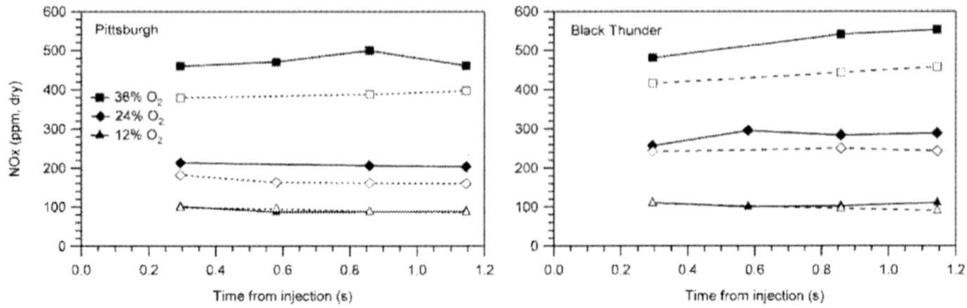

Figure 8. NO_x concentration measured during coal combustion at different oxygen levels. The solid lines and symbols indicate the N_2 diluent, whereas the dashed lines and open symbols indicate the CO_2 diluent (reprinted from [44], with permission from Elsevier).

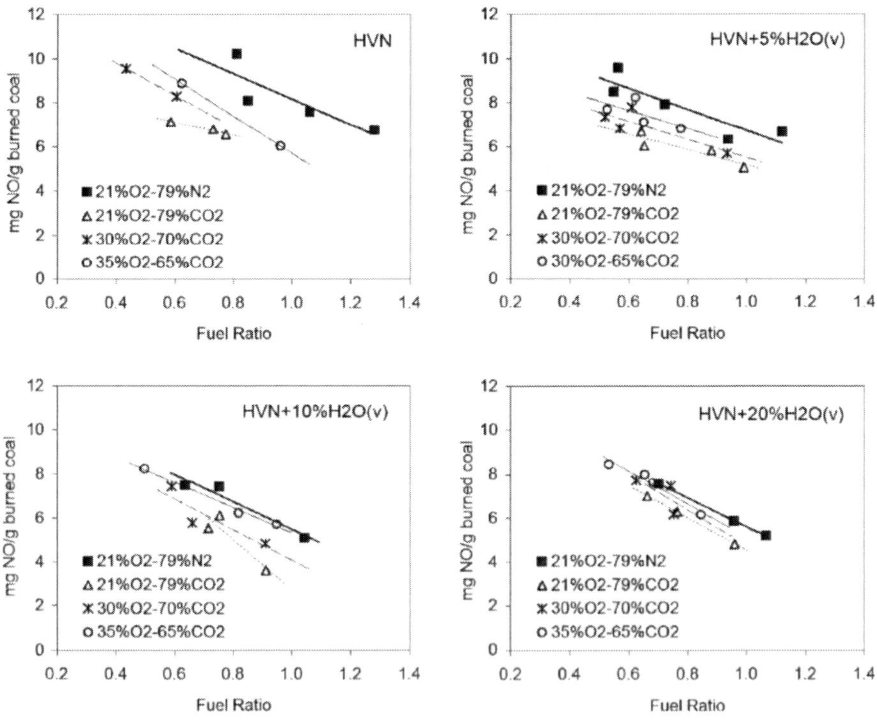

Figure 9. NO concentration from the combustion of a semianthracite (HVN) under air and oxy-fuel conditions with the addition of steam (reprinted from [46], with permission from John Wiley and Sons).

The recirculation of nitric oxides along with the flue gas through the burner will lead to a decrease in NO emissions. The reduction of NO occurs via one of three different mechanisms. Okazaki and Ando [47] evaluated the relative importance of each of these mechanisms in oxy-fuel conditions in a small-scale laboratory reactor, where they found that: (i) the reduction of NO to cyanide and amine intermediates by hydrocarbons accounts for 50-80% of the decrease in NO, (ii) the reduction of NO to N_2 through reactions with other nitrogen species (mainly NH_3 and HCN according to De Soete's scheme [48]) accounts for 10-50% of the total NO reduction, and (iii) the reduction of NO to N_2 by heterogeneous reactions on the char was of only minor importance in oxy-fuel conditions.

2.5.2. Sulfur Oxides

Stanger et al. [49] reported the fate of sulfur under oxy-coal combustion and found that the fuel sulfur conversion ratio did not significantly change under air-fuel and oxy-fuel combustion conditions. However, due to the lower volume of flue gases and the recirculation of some of the gases (and therefore of NO_x and SO_2), the NO_x and SO_2 concentrations measured in the furnace may be significantly higher under oxy-fuel conditions than under air-firing.

Due to the higher SO_x and $H_2O_{(v)}$ concentrations, the dew point is bound to increase [50] and increases of about 20-40 K from ~403 to 433 K have been reported [51]. This may cause corrosion problems in the boiler and at the end of the flue gas duct if the temperature decreases below the gas dew point [52]. The effect of SO_2 on fuel oxidation behavior has been studied by Giménez-López et al. [53], who showed that SO_2 can prevent CO oxidation, although this effect is more pronounced under air-firing conditions than under oxy-fuel conditions.

2.5.3. Fly Ash and Slag

Stam et al. [54] investigated the thermodynamic equilibrium composition of fly ash and slag in oxy-fuel conditions. These authors found similar fly ash compositions in both an air and oxy-fuel environment. However, due to the lower particle temperature in oxy-fuel conditions (and hence a lower coal burning rate), the vaporization of both volatile metals and metal sub-oxides is significantly smaller, and therefore the formation of submicrometer ash particles is prevented [55]. Suriyawong et al. [56] compared two combustion cases, air and $20\%O_2/80\%CO_2$, and they found that the ratio between fine and coarse particles shifted significantly towards a smaller number of sub-micrometer particles in the oxy-fuel case.

According to the research carried out on slag formation, changes in slag deposition in an oxy-fuel plant compared to an air-fired one will not greatly affect the operation of the plant. Studies on ash deposition, slagging and fouling at the ANL 3 MWth pilot scale furnace plant showed no significant differences between air and oxy-firing conditions [57]. Neither did the ash samples from the deposits in the radiative and convective section of the IHI 1.2 MW test facility show any notable differences between air and oxy-fuel conditions [14]. However, the furnace deposits under oxy-coal combustion contained more sulfur than those formed under air-firing.

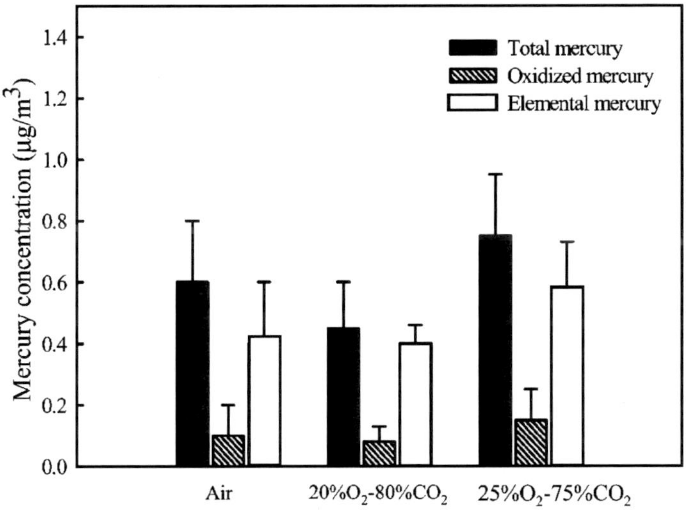

Figure 10. Mercury concentration measured at the furnace exit. Experiments were performed with a sub-bituminous coal in air and two O_2/CO_2 atmospheres at a reactor temperature of 1473 K (reprinted with permission from Suriyawong et al. [56], Copyright © 2006 American Chemical Society).

2.5.4. Trace Elements: Mercury

Although oxy-coal combustion has been extensively studied at a laboratory and pilot scale, very few data have been reported on the oxidation of mercury or its retention in fly ashes. In a series of tests in the 30 MW pilot scale facility of Babcock and Wilcox, mercury concentration in the flue gas and its oxidation were found to be higher under oxy-coal combustion than under air. Ongoing research on mercury behavior in oxy-fuel systems is being performed by the US Department of Energy/National Energy Technology Laboratories (DOE/NETL) [58] and CANMET [59]. Suriyawoong et al. [56]

measured mercury emissions from the air- and oxy-fuel combustion of a sub-bituminous coal in a laminar reactor and determined the ratio between the elemental and oxidized forms (Figure 10). Some attempts at modeling have also been made and Gharebaghi et al. [60] developed a model for predicting the mercury speciation of flue gases in air and oxy-coal combustion.

2.6. Biomass Co-firing

Biomass is a renewable fuel with a limited greenhouse effect, since the carbon dioxide released during its combustion is recycled as an integral part of the carbon cycle. Biomass absorbs during its growth the carbon dioxide which is emitted during its combustion. It is therefore considered as carbon-neutral, although this is not strictly true, since fossil fuels are typically consumed in the cultivation, irrigation and transportation of biomass. Biomass can be classified into the following different types: residues (e.g., sewage sludge, agricultural and algae-based residues), herbaceous (e.g., grasses and straw), aquatic (e.g., kelp), woody (e.g., sugarcane and sawdust), and derivatives (e.g., paper) [61,62]. Biomass is being increasingly implemented as an energy source for residential and industrial use, for energy production in modern biomass power plants and for co-firing with other fuels such as coal. It also has an important advantage as a fuel due to its high reactivity and high volatiles content. However, in comparison with solid fossil fuels, biomass contains much less carbon and so it has a lower heating value.

Direct air-combustion is the traditional way of using biomass and it is responsible for over 97% of the world's bio-energy production in energy units [61]. The same challenges to the use of biomass for reducing CO_2 emissions in coal-fired power plants also apply to CCS power plants. The abbreviation BECS is employed to denote the concept of Biomass Energy for Carbon Capture and Sequestration [63]. In principle, all the CCS technologies proposed for fossil-fired power systems can be applied to systems using biomass fuels. However, the same problems that affect biomass and coal co-firing, such as biomass handling and management and problems associated with the lower ash melting point and the Cl and alkali contents can also be expected to appear in oxy-fuel power plants.

Biomass produces high volatile matter, besides a high concentration of CO in the fuel-rich zone of the flames, which is even higher in O_2/CO_2 environments. Riaza et al. [64] burned two coals, a semianthracite and a high-volatile bituminous coal, and their blend with up to 20% weight of biomass (olive residue) in air and three O_2/CO_2 atmospheres (21-35% O_2). In the case of the individual coals, a worsening in coal burnout was observed when N_2 was replaced by CO_2 at the same oxygen concentration (21%). In order to achieve similar burnout values to those in air, the oxygen concentration under oxy-fuel conditions must be of the order of 30%. Biomass shows a higher reactivity than coal and improves combustion behavior when it is blended with coal under both air and oxy-fuel firing conditions (Figure 11). However, the effect of blending coal and biomass is more marked with high-rank coal (semianthracite) than with low-rank coal (hvb coal).

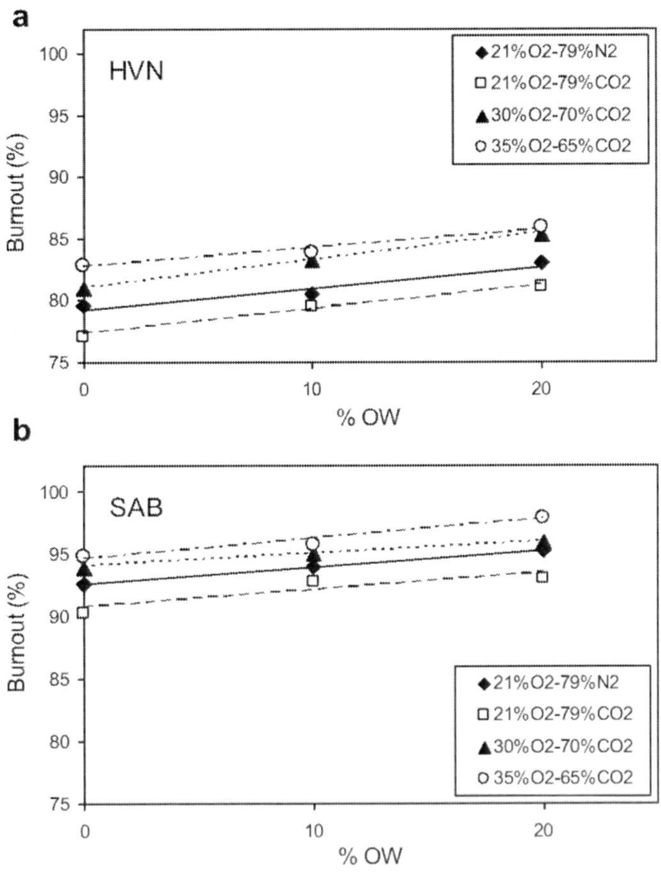

Figure 11. Burnout values of: (a) a semianthracite, HVN and (b) a high-volatile bituminous coal, SAB, as well as their blends with olive waste (OW) under different atmospheres at a fuel equivalence ratio of 0.8 (25% excess air) (reprinted from [64], with permission from Elsevier).

Borrego et al. [65] studied the characteristics of wood chips, rice husk and forest residue chars obtained by pyrolysis in air and oxy-fuel atmospheres. They observed that the characteristics of the char obtained under air- and oxy-fuel combustion conditions, such as its morphology, pore volume, optical texture, specific surface area and reactivity, showed no significant differences. Farrow et al. [66] obtained chars from biomass (sawdust and pinewood) and coal (South African bituminous) in a drop tube furnace in pure N_2 and CO_2 at temperatures ranging from 1173 to 1573 K. The biomass/coal char blends burned off significantly faster than the unblended coal chars in both air and CO_2, highlighting the catalytic effect of biomass char blending on coal combustion reactivity. They also found that the catalytic effect of alkali and alkaline metals present in the biomass fuels increased with the proportion of biomass char included in the blends.

Arias et al. [67] investigated the combustion behavior of blends of a bituminous coal with biomass (olive residue) in an electrically heated entrained-flow reactor (EFR) in air and three oxy-fuel atmospheres (21%O_2/79%N_2, 30%O_2/70%O_2 and 35%O_2/65%O_2). Riaza et al. [64] performed similar tests in the same EFR but with blends of a semianthracite with biomass (olive residue). In the case of the coal/biomass blends, the same ignition behavior under oxy-fuel conditions was observed as in the case of the individual coals. The ignition behavior of coals improved as the addition of biomass increased in both air and oxy-fuel conditions. The effect of adding biomass was especially noticeable in the ignition of high rank coals, such as the semianthracite.

Smart et al. [17] burned two bituminous coals (Russian and South African) with 20% biomass (Shea Meal) in a 0.5 MWt Combustion Test Facility. When the coal was cofired with biomass, the peak radiative flux values were inversely related to the recycle ratio, although these radiative fluxes were lower than in the case of the individual coals. Black et al. [68] numerically investigated the effect of replacing coal by biomass in both air and oxy-fuel conditions. In the case of oxy-coal firing, they pointed out that the optimum oxygen concentration for attaining a similar heat transfer to that of air-coal combustion was between 25% and 30%, whereas for oxy-biomass firing an oxygen concentration of more than 30% would be required to achieve similar heat transfer characteristics to those obtained with air-coal combustion.

Some works on oxy-fuel combustion of coal/biomass blends focus on the ash depositions, which could represent the most serious problem associated with the addition of biomass. Depending on type of biomass used, some problems can be attributed to the composition of the biomass ash. Fryda et al.

[69] compared the deposition and fouling behavior of coal/biomass blends in oxy-firing and air-firing conditions. The fuels studied were two bituminous coals and a biomass (Shea meal). A horizontal deposition probe, fitted with thermocouples and heat transfer sensors, was placed at a fixed distance from the burner to simulate ash deposition on heat transfer surfaces. The ash analysis did not reveal differences in the ash chemistry between the oxy- and air-fuel cases. The results indicated that temperature and the composition of the fuels/blends were the major factors affecting gaseous compounds and ash composition rather than the combustion environment, which seems to affect the ash deposition mechanisms. However, this needs to be confirmed by data obtained at a larger scale. Finally, it remains to mention an indirect way of using biomass in oxy-fuel power plants by means of the biomass gasification. The indirect co-firing of biomass in oxy-fuel combustion has been planned in Ciuden's project [70].

Acknowledgments

J. Riaza acknowledges funding from the Government of the Principado de Asturias (Severo Ochoa program). Financial support from the CSIC (Project PIE 201080E09) is gratefully acknowledged.

References

[1] BP (2013). *Statistical Review of World Energy 2013.* http://www.bp.com/en/global/corporate/about-bp/statistical-review-of-world-energy-2013.html.

[2] IEA (2011). *World Energy Outlook; IEA: Paris, FR, 2011.* http://www.worldenergyoutlook.org.

[3] Bachu, S. *Prog Energy Combust Sci.* 2008, 34, 254-273.

[4] Oexmann, J., Kather, A., Linnenberg, S., Liebenthal, U. *Greenhouse Gas Sci Tecnol.* 2012, 2, 80-98.

[5] Olivares-Marín, M., Maroto-Valer, M. *Greenhouse Gas Sci Technol.* 2012, 2, 20-35.

[6] Abraham, B. M., Asbury, J. G., Lynch, E. P., Teotia, A. P. S. *Oil Gas J.* 1982, 80, 68-70.

[7] Horn, F. L., Steinberg, M. *Fuel.* 1982, 61, 415-422.

[8] Wall, T., Stanger, R., Santos, S. *Int J Greenhouse Gas Control.* 2011, 5S, S5-S15.

[9] ZEP, EU (2008). EU Zero Emissions Platform – Matrix of Technologies. http://www.zeroemissionsplatform.eu.

[10] ZEP, EU (2010). EU Zero Emissions Platform – Recommendations for Research to Support the Deployment of CCS in Europe Beyond 2020. http://www.zeroemissionsplatform.eu.

[11] Andersson, K., Johansson, R., Hjärtstam, S., Johnsson F, Leckner, B. *Exp Therm Fluid Sci.* 2008, 33, 67-76.

[12] Andersson, K., Johansson, R., Johnsson, F., Leckner, B. *Energy Fuels.* 2008, 22, 1535-1541.

[13] Andersson, K., Johansson, R. Johnsson, F. *Int J Greenhouse Gas Control.* 2011, 5S, S58-65.

[14] Wall, T. F., Liu, Y., Spero, C., Elliott, L., Khare, S., Rathman, R., Zeenathal, F., Moghtaderi, B., Buhre, B., Sheng, C., Gupta, R., Yamada, T., Makino, K., Yu, J. *Chem Eng Res Des.* 2009, 87, 1003-1016.

[15] Hjärtstam, S., Johnsson, F., Andersson, K., Leckner, B. *Fuel.* 2009, 88, 2216-2224.

[16] Smart, J. P., Patel, R., Riley, G. S. *Combust Flame.* 2010, 157, 2230-2240.

[17] Smart, J. P., Riley, G. S. *Fuel.* 2011, 90, 2812-2816.

[18] Rathnam, R. K, Elliot, L. K., Wall, T. F., Liu, Y., Moghtaderi, B. *Fuel Proc Technol.* 2009, 90, 797-802.

[19] Naredi, P., Pisupati, S. *Energy Fuels.* 2011, 25, 2452-2459.

[20] Álvarez, L., Yin, C., Riaza, J., Pevida, C., Pis, J. J., Rubiera, F. *Energy.* 2013, 62, 255-268.

[21] Gil, M. V., Riaza, J., Álvarez, L., Pevida, C., Pis, J. J., Rubiera, F. *Appl Energy.* 2012, 91, 67-74.

[22] Gil, M. V., Riaza, J., Álvarez, L., Pevida, C., Pis, J. J., Rubiera, F. *Energy.* 2012, 48, 510-518.

[23] Brix, J., Jensen, P. A., Jensen, A. D. *Fuel.* 2010, 89, 3373-3380.

[24] Khare, S. P., Wall, T. F., Farida, A. Z., Liu, Y., Moghtaderi, B., Gupta, R. P. *Fuel.* 2008, 87, 1042-1049.

[25] Strömberg, L., Lindgren, G., Jacoby, J., Giering, R., Anheden, M., Burchhardt, U., Altmann, H., Kluger, F., Stamatelopoulos, G.-N. *Energy Procedia.* 2009, 1, 581-589.

[26] Taniguchi, M., Shibata, T., Kobayashi, H. *Proc Combust Inst.* 2011, 33, 3391-3398.

[27] Taniguchi, M., Yamamoto, K., Okazabi, T., Rehfeldt, S., Kuhr, C. *Int J Greenhouse Gas Control.* 2011, 5S, 111-119.
[28] Smart, J., Lu, G., Yan, Y., Riley, G. *Combust Flame.* 2010, 157, 1132-1139.
[29] Khatami, R., Stivers, C., Joshi, K., Levendis, Y. A., Sarofim, A. F. *Combust Flame.* 2012, 159, 1253-1271.
[30] Zhang, L., Binner, E., Li, C. H. *Fuel.* 2010, 89, 2703-2712.
[31] Riaza, J., Álvarez, L., Gil, M. V., Pevida, C., Rubiera, F., Pis, J. J. *Energy.* 2011, 36, 5314-5319.
[32] Riaza, J., Álvarez, L., Gil, M. V., Khatami, R., Levendis, Y., Pis, J. J., Pevida, C., Rubiera, F. *Greenhouse Gas Sci Technol.* 2013, 3, 397-414.
[33] Binner, E., Zhang, L., Bhattacharya, S. *International Pittsburgh Coal Conference,* Pittsburgh, EEUU. 2009.
[34] Li, X., Rathman, R. K., Yu, J., Wang, Q., Wall, T., Lin, B. *Energy Fuels.* 2010, 24, 160-164.
[35] Molina, A., Shaddix, C. R. Proc Combust Inst. 2007, 31, 1905-1912.
[36] Glarborg, P., Bentzen, L. B. *Energy Fuels.* 2008, 22, 291-298.
[37] Mendiara, T., Glarborg, P. *Energy Fuels.* 2009, 23, 3565-3572.
[38] Andersen, J., Rasmussen, C. L., Giselsson, T., Glarborg, P. *Energy Fuels.* 2009, 23, 1379-1389.
[39] Bejarano, P. A, Levendis, Y. A. *Combust Flame.* 2008, 153, 270-287.
[40] Shaddix, C. R., Molina, A. *Proc Combust Inst.* 2009, 32, 2091-2098.
[41] Hecht, E. S., Shaddix, C. R., Molina, A., Haynes, B. S. *Proc Combust Inst.* 2011, 33, 1699-1706.
[42] Várhegyi, G., Szabó, P., Jakab, E., Till, F. *Energy Fuels.* 1996, 10, 1208-1214.
[43] Normann, F., Andersson, K., Leckner, B., Johnsson, F. *Prog Energy Combust Sci.* 2009, 35, 385-397.
[44] Shaddix, C. R., Molina, A. *Proc Combust Inst.* 2011, 32, 2091-2098.
[45] Ndibe, C., Spörl, R., Maier, J., Scheffknecht, G. *Fuel.* 2013, 107, 749-756.
[46] Álvarez, L., Riaza, J., Gil, M. V., Pevida, C., Pis, J. J., Rubiera, F. *Greenhouse Gases: Sci Technol.* 2011, 1, 180-190.
[47] Okazaki, K., Ando, T. *Energy.* 1997, 22, 207-215.
[48] De Soete, G. G. *Symp (Int) Combust.* 1975, 15, 1093-1102.
[49] Stanger, R., Wall, T. *Prog Energy Combust Sci.* 2011, 37, 69-88.
[50] Fleig, D., Normann, F., Andersson, K., Johnsson, F., Leckner, B. *Energy Procedia.* 2009, 1, 383-390.

[51] Toftegaard, M. B., Brix, J., Jensen, P. A., Glarborg, P., Jensen, A. D. *Prog Energy Combust Sci.* 2010, 36, 581-625.
[52] Syed, A. U., Simms, N. J., Oakey, J. E. *Fuel.* 2012, 101, 62-73.
[53] Giménez-López, J., Martínez, M., Millera, A., Bilbao, R., Alzueta, M. U. *Combust Flame.* 2011, 158, 48-56.
[54] Stam, A., Ploumen, P., Brem, G. *The 34th international technical conference on coal utilization and fuel systems,* Clearwater, Florida, USA. 2009.
[55] Sheng, C., Lu, Y., Gao, X., Yao, H. *Energy Fuels.* 2007, 21, 435-440.
[56] Suriyawong, A., Gamble, M., Lee, M. -H., Axelbaum, R., Biswas, P. *Energy Fuels.* 2006, 20, 2357-2353.
[57] Payne, R., Chen, S. L., Wolsky, A. M., Richter, W. F. *Combust Sci Technol.* 1989, 67, 1-16.
[58] Fout, T., Li, Z. S., Ciferno, J., Brickett, L., Matuszewski, M. IEA GHG *1st Oxyfuel Combustion Conference,* Cottbus, Germany. 2009.
[59] Zheng, L., Zhuang, Q., Tan, Y., Clements, B., Pomalis, R. *IEA GHG 1st Oxyfuel Combustion Conference,* Cottbus, Germany. 2009.
[60] Gharebaghi, M., Hughes, K. J., Porter, R. T. J., Pourkashanian, M., Williams, A. *Proc Combust Inst.* 2011, 33, 1779-1786.
[61] Demirbas, A. *Prog Energy Combust Sci.* 2004, 30, 219-230.
[62] Williams, A., Pourkashanian, M., Jones, J. M. *Prog Energy Combust Sci.* 2001, 27, 587-610.
[63] Kraxner, F., Nilsson, S., Obersteiner, M. *Biomass Bioenergy.* 2003, 24, 285-296.
[64] Riaza, J., Gil, M. V., Álvarez, L., Pevida, C., Pis, J. J., Rubiera, F. *Energy.* 2012, 41, 429-435.
[65] Borrego, A. G., Garavaglia, L., Kalkreuth, W. D. *Int J Coal Geol.* 2009, 77, 409-415.
[66] Farrow, T. S., Sun, C., Snape, C. E. *Fuel.* 2013, 114, 128-134.
[67] Arias, B., Pevida, C., Rubiera, F., Pis, J. J. *Fuel.* 2008, 87, 2753-2759.
[68] Black, S., Szuhánszki, J., Pranzitelli, A., Ma, L., Stanger, P. J., Ingham, M., Pourkashanian, M. *Fuel.* 2013, 113, 780-786.
[69] Fryda, L., Sobrino. C., Cieplik, M., van de Kamp, W. L. *Fuel.* 2010, 89, 1889-1902.
[70] Lupion, M., Diego, R., Loubeau, L., Navarrete, B. *Energy Procedia.* 2011, 4, 5639-5646.

In: Combustion
Editor: Joseph M. Grier

ISBN: 978-1-62948-967-4

Chapter 5

Combustion of Lignocellulosic Biomass and Marine Biomass by Means of Thermal Analyses: Kinetic and Gas Evolution Analysis

Diego López González, José Luis Valverde and María Luz Sánchez Silva*

Departamento de Ingeniería Química, Universidad de Castilla-La Mancha (UCLM), Ciudad Real, Spain

Abstract

Biomass fuels are gaining particular attention as a potential alternative to increase energy independence of fossil fuels and reduce environmental pollution. The main ones refer to lignocellulosic biomass and marine biomass (especially microalgae). There is a growing interest in the cultivation of lignocellulosic biomass for energy production as they typically have less capital-intensive conversion technologies, attractive opportunity for local and regional self-sufficiency, reduction in greenhouse gas emissions and viable alternative to fossil fuel use. On the other hand, microalgae have received increasing attention due to the fact that they can be cultured in ponds or photobioreactors with supply of nutrients or waste water. Moreover, the production of microalgae does

* Tel.: +34-926295300; Fax: +34-926295256; *E-mail address: marialuz.sanchez@uclm.es.

not require of high quality arable land and therefore it does not compete with food crops.

Direct combustion is the oldest and still the most commonly used route for converting biomass to heat, power and combined heat and power. Combustion can be defined as the conversion of biomass fuels to several forms of useful energy in the presence of air or oxygen. Despite its inherent potential as a biofuel resource, the commercial viability of biomass combustion technology has not been achieved yet. In this regard, thermogravimetric analysis is one of the most common techniques used to rapidly investigate and compare thermal events and kinetics during combustion of biomass. TGA provides useful information concerning the temperatures at which combustion or decomposition reactions in the sample start. Biomass characteristics and kinetics of biomass combustion are essential for modeling the combustion in industrial processes. Therefore, from TGA data suitable mathematical models can be derived for a better comprehension of the oxidation behavior of these complex feedstock that allow to perform economic analysis and develop technology for a more efficient biomass conversion.

Additionally, during the combustion of biomass, the composition of the gas emissions should be determined before industrial application. In spite of the environmental advantages, some aspects concerning the release of contaminants during biomass combustion must be taken into account. In this regard, NO_x and SO_x emissions depend on raw biomass composition, which usually is variable. Furthermore, the chloride amount in biomass might turn into operational problems such as corrosion. Other organic compounds such as benzene and toluene are considered to be part of the most dangerous emissions from biomass combustion causing diseases as lung infection and leukemia. Therefore, the knowledge of pollutant release during biomass conversion is truly important in order to reassure the use of biomass from the environmental point of view. In this regard, the coupling of thermogravimetric analysis with mass spectrometry (TGA-MS) stands out as the only experimental technique able to afford real-time and sensitive detection of evolved gases from the thermochemical conversion of a very small sample.

In this chapter, the combustion behavior and the gas evolution analysis of different biomass feedstock (lignocellulosic and microalgae) by means of TGA-MS will be addressed. Finally, the kinetic analysis of the combustion process will be presented.

1. INTRODUCTION

The global energy demand has been increased exponentially in the last two decades being mainly due to the increase in world population. This fact, together with the depletion of fossil fuel resources and the recent global awareness of environment pollution are the main reasons that have been proposed to make a change towards a global sustainability. In this scenario, the potential of biomass as an energy source has been widely recognized. It is estimated that biomass provides approximately the 14 % of the total world-wide energy needs [1].

Biomass is a term for all organic material that stems from plants including algae, trees and crops that are susceptible to be converted into energy [2]. One of the most controversial points in the use of biomass as an energy source is its possible competition with human food supply. Nevertheless, there are different types of biomass that fit into this definition without compromising the world food supply. The main ones refer to lignocellulosic biomass and marine biomass (especially algae) [3].

A major part of biomass is ligno-cellulose [3]. Lignocellulosic material is the non-starch, fibrous part of plant materials. It is comprised primarily of cellulose, hemicellulose and lignin with lesser amounts of extractives [4]. Unlike carbohydrate or starch, lignocellulose is not easily digestible by humans, thus it does not threaten the world´s food supply. There is a growing interest in the cultivation of lignocellulosic biomass for energy production as they typically have short growing period and high yields, and require little or no fertilizer, so they provide quick return on investment [3].

On the other hand, algae can be divided according to their size as: microalgae and macroalgae. Microalgae have received more attention than macroalgae for biofuels production, due to the fact that they can be cultured in ponds or photobioreactors with supply of nutrients or waste water. Moreover, the production of microalgae does not require of high quality arable land and therefore it does not compete with food crops [5]. The generic term microalgae refers to a large group of very diverse photosynthetic micro-organisms of microscopic dimensions. They are a very promising biomass for the following reasons: a high growth rate, high yield per area and high efficiency in CO_2 capture and solar energy conversion. Generally, microalgae varied in their proportions of protein (6-52 wt.%), carbohydrate (5-23 wt.%) and lipid (7-23 wt.%). According to Ross et al. (2009) [6], microalgae with high lipid content could be a future source of third generation biofuels and chemicals. The oil content itself can be estimated to be 64.4 % of the total lipid component.

Thermochemical conversion of biomass is the most promising route for biomass utilization [7]. These processes mainly include the direct combustion to generate heat and electricity, pyrolysis and gasification, to produce liquid and gaseous fuels that are suitable for feeding efficient gas engines and gas turbines [8]. Pyrolysis plays an important role in these processes, being the first chemical step in both gasification and combustion processes. Generally, pyrolysis can be considered as a two-stage process involving the devolatilization of biomass and the slow heterogeneous conversion of char [8]. The char generated during pyrolysis is a high energy-density solid fuel suitable for combustion and gasification processes. Among the different thermo-chemical conversion technologies, the direct combustion is the oldest and still the most commonly used route for converting biomass to heat, power and combined heat and power [9]. Combustion can be defined as the conversion of biomass organic matter to several forms of useful energy in the presence of air or oxygen [10]. Thermogravimetric analysis is one of the most common techniques used to rapidly investigate and compare thermal processes and kinetics during combustion of biomass [11]. Furthermore, complementary techniques such as differential scanning calorimetry (DSC) can be used to evaluate the amount of heat released during the combustion of biomass. The combustion of biomass using thermogravimetric analysis has not been studied intensively yet [12]. Recently, the combustion behavior of different types of wood has been performed [12-16]. Furthermore, Kay et al. (2011) [17] studied the effect of biomass components on the co-combustion of biomass and coal.

TGA provides useful information concerning the temperatures at which combustion or decomposition reactions occur. Moreover, quantitative methods can be applied to TGA curves in order to obtain kinetic parameters [18, 19]. As reported by Naik et al. (2010), suitable mathematical models can be derived for a better comprehension of the thermal behavior of these complex feedstock that allow to perform economic analysis and develop technology for a more efficient biomass conversion [20].

2. Combustion of Biomass

The combustion of biomass is used to convert the chemical energy stored in biomass into heat, mechanical power or electricity [2]. Eq. 1 shows a basic scheme of the biomass combustion process:

$$\overbrace{C_xH_yO_z+N+S+\text{Inorganic constituents (Ca, K, Mg, \ldots)}}^{\textbf{Biomass Fuel}} + \text{Air} \rightarrow CO_2 + H_2O + C_xH_y + NO_x + SO_x + \text{Ash} + \text{Heat} \quad (1)$$

The first term of the equation represents the biomass fuel. It can be seen that it is a highly heterogeneous material which is mainly composed of C, H and O and in a lesser extent of N and S. The amount of these elements in the biomass fuel varies from different species of biomass. Furthermore, it also contains a small but important percentage of inorganic constituents which are the main components of the biomass ash. The main products obtained in this process are CO_2 and H_2O, together with a wide range of sub-products such as hydrocarbons, nitrogen oxides (NO_x) or sulfur oxides (SO_x). There is also a solid residue (Ash), which might influence the process in many ways, contributing to the occurence of operational problems such as fouling, slagging or corrosion. Furthermore, they can also affect the development of the process.

Despite its apparent simplicity, biomass combustion is a complex process that involves simultaneous decomposition reactions, coupled heat and mass transfer effects with chemical reaction and fluid flow [21]. Generally, the combustion process is generally described by three stages:

- Drying. In this stage, the moisture of the biomass fuel is removed.
- Pyrolysis. The second stage of conversion involves the devolatilization of the biomass fuel. Also, most of volatiles are released and a solid fuel (char) is formed. This stage usually takes place between 400 and 600 ºC.
- Combustion of the char. The last stage represents the oxidation of the formed char. This stage is the more exothermic one. This stage takes place at temperatures above 600 ºC.

The composition of biomass plays an important role on the conversion of biomass into energy, determining both the potential of the feedstock for energy processing and the operational problems that may arise [2]. Thus, a comprehensive evaluation of biomass composition must be performed before their use in energy conversion processes. The main analyses of interest are: elemental analysis, proximate analysis and ash composition.

Ultimate Analysis

The ultimate analysis determines the amount of C, H, N, S and O in the biomass fuel. This analyses provide an insight of the energy value of a fuel by measuring the significance of O:C and H:C ratios. High contents of O and H implies a reduction of the energy value of the fuel due to the lower energy contained in carbon-oxygen and carbon-hydrogen bonds, than that contained in carbon-carbon bonds [2].

Table 1 shows the ultimate analysis for different types of biomass. Carbon and oxygen are the main elements that constitute the biomass together with hydrogen. Nitrogen content was much higher for microalgae than that for ligenocellulosic biomass. This is mainly due to the high protein content in microalgae. Finally, the sulfur content was lower than that of nitrogen being slightly higher for microalgae than for lignocellulosic biomass. The presence of N and S implies the emission of NO_x and SO_x during the oxidation process.

Table 1. Ultimate analysis of different types of lignocellulosic and marine biomass

	Ultimate analysis (%wt.)					
Biomass sample	C	H	O	N	S	Reference
Eucalyptus wood	41.6	4.9	53.1	0.38	0.03	[22]
Fir wood	50.1	6.1	43.4	0.44	0.01	
Pine bark	52.7	5.5	41.7	0.01	0.08	
Chlorella vulgaris	42.8	6.5	43.1	6.7	1.1	[23]
Nannochloropsis gaditana	47.7	7.4	37.0	6.8	1.1	
Scenedesmus almeriensis	40.7	6.5	46.3	5.7	0.8	

Proximate Analysis

The proximate analysis determines the way in which the chemical energy of biomass is stored [2]. The proximate analysis provides the amount of moisture, ash, volatile matter and fixed carbon in the biomass fuel. The two first properties have adverse effects on the quality of the fuel. High values of moisture decreases the calorific value of the fuel to an uneven overall energy balance whereas high values of ashes lead to an increase of operational costs. On the other hand, proportion of fixed carbon and volatile matter determine the available energy that biomass is able to yield.

Table 2 shows the proximate analysis of different types of lignocellulosic and marine biomass obtained from literature. Volatile matter content varied among the different species of biomass. Lignocellulosic biomass contains the higher amount of fixed carbon. Finally, it is remarkable the high ash content of marine biomass compared to that of lignocellulosic biomass. The ash content of biomass affects both the handling and processing costs of the overall biomass conversion process cost [2].

Table 2. Proximate analysis of different types of lignocellulosic and marine biomass. The proximate analyses were carried out according to the technical specifications UNE-EN, UNE-EN 14775:2010, UNE-EN 15148:2010 and UNE-EN 1474-2 for ash, volatile matter and moisture determination, respectively

	Proximate analysis (%wt.)				
Biomass sample	Moisture	Ash	Volatile matter	Fixed carbon	Reference
Eucalyptus wood	2.6	6.8	73.8	16.8	
Fir wood	2.6	3.4	74.4	19.5	
Pine bark	4.4	2.7	61.6	31.3	[22]
Chlorella vulgaris	4.4	15.9	67.2	12.4	
Nannochloropsis gaditana	3.5	6.4	79.8	10.2	
Scenedesmus almeriensis	2.9	19.4	67.9	9.7	[23]

Ash Composition

Finally, the ash composition is important to control or mitigate the possible fouling or slagging phenomena that might take place when processing the biomass fuel. Table 3 summarizes the mineral content of different types of lignocellulosic and marine biomass. It can be observed that all samples contained a high concentration of alkali and alkali-earth metals. Furthermore, other elements such as Si or Al were found in high proportions.

The reaction among these elements produces a sticky, mobile liquid phase, which can lead to blockages of the furnaces and boilers [2].

The detailed chemistry of the process is still far from being understood and the study of the process is usually described on the basis of generalizations and simplifications. Thus, further research must be done in order to obtain a deeper insight into the biomass combustion process. In this regard, thermal analysis techniques provide an excellent tool for studying biomass conversion processes at laboratory scale.

Table 3. Mineral content of different species of lignocellulosic and marine biomass determined by inductively coupled plasma (ICP)

	Mineral content (ppm)							
Biomass sample	K	Ca	Mg	Al	Na	Si	Fe	References
Eucalyptus wood	5078	4116	1062	43	1431	247228	33	[22]
Fir wood	1880	10921	1774	557	1807	353166	717	
Pine bark	1254	2726	776	946	2764	474344	385	
Chlorella vulgaris	3116	10397	3577	107	5277	14817	712	[23]
Nannochloropsis gaditana	2689	7493	3123	433	41620	38394	2415	
Scenedesmus almeriensis	21038	7297	22455	106	4402	11239	2769	

Thermal analysis has been formally defined by the International Confederation for Thermal Analysis and Calorimetry (ICTAC) as a group of techniques in which a property of the sample is monitored against time or temperature while the temperature of the sample, in a specified atmosphere, is programmed [4]. Complementary techniques can be used to analyze the gaseous products evolved during thermal analysis measurements. This group of techniques involves the mass spectrometry (MS), Fourier transform spectrometry (FTIR) and gas chromatography (GC).

This work is aimed to the application of the most common thermal analysis techniques (TGA and DSC) to the study of the combustion process of biomass feedstock. Furthermore, the coupling of mass spectrometry for the study of the gaseous products obtained during the combustion of biomass is also reviewed.

3. Study of Combustion by Thermal Analysis

3.1. Thermogravimetric Analysis (TGA)

Lignocellulosic biomass is mainly composed of three different types of components: cellulose, hemicellulose and lignin. In this regard, the thermal behavior of biomass can be regarded as the addition of the primary decomposition of biomass main components. This study reports the thermal behavior of the hemicellulose component of biomass as the xylan one, which has been widely used as representative of the hemicellulose component in thermochemical conversion processes [12, 24]. This is mainly due to the fact that the hemicellulose component is not easily separated from biomass and thus, it can hardly be purchased commercially. Figure 3.1 shows the thermogravimetric (TGA) and derivative thermogravimetric (DTG) analysis for the combustion process of pine bark and the biomass main components (cellulose, hemicellulose and lignin). It can be observed that the combustion process can be sub-divided into the two abovementioned main stages (the drying stage is not reported here): the devolatilzation stage (pyrolysis) and the char oxidation stage.

There are substantial differences in the thermal behavior of the main components of lignocellulosic biomass, which are mainly attributed to their different chemical structures [17]. Lignin was the first component to decompose (146 ºC) whereas xylan and cellulose started decomposing at 187 and 266 ºC, respectively. This behavior is attributed to the fact that lignin and xylan have methoxy functional groups, which tend to break easily [17].

Lignin was the most thermal stable component decomposing in two steps, starting at 146 and 550 ºC, respectively. DTG curve for lignin oxidation presented the smallest and widest peaks. This fact is due to lignin is polymeric in nature with a three-dimensional structure consisting of phenylpropane coupled with C-C or C-O-C bonds whose activity covers a wide range of temperatures [17]. The DTG curve for lignin combustion shows two main peaks at 397 and 518 ºC (decomposition rate of 0.295 and 0.326 %/ºC, respectively). The burnout temperature was established at 757 ºC.

Xylan is the less thermally stable component of biomass. Two strong decomposition peaks overlapped in the temperature range of 147-371 ºC, having the maximum DTG peak at 264 ºC (0.841 %/ºC). This degradation step is mainly attributed to C-O-C and some pyranose C-C bonds breakdown [12]. A second step was observed between 429 and 615 ºC.

The primary decomposition of cellulose, in which 87 % by weight was lost, took place between 266 and 423 ºC. The DTG peak for cellulose oxidation was found at 354 ºC and presented the highest weight loss rate of all samples (2.74 %/ºC). During this stage, a complex set of reactions as denitration and deacetylation, scission of O-N, C-O, CC and C-H bonds might take place [25]. A smaller peak was found between 441 ºC and 637 ºC (0.10 %/ºC) that can be attributed to char oxidation.

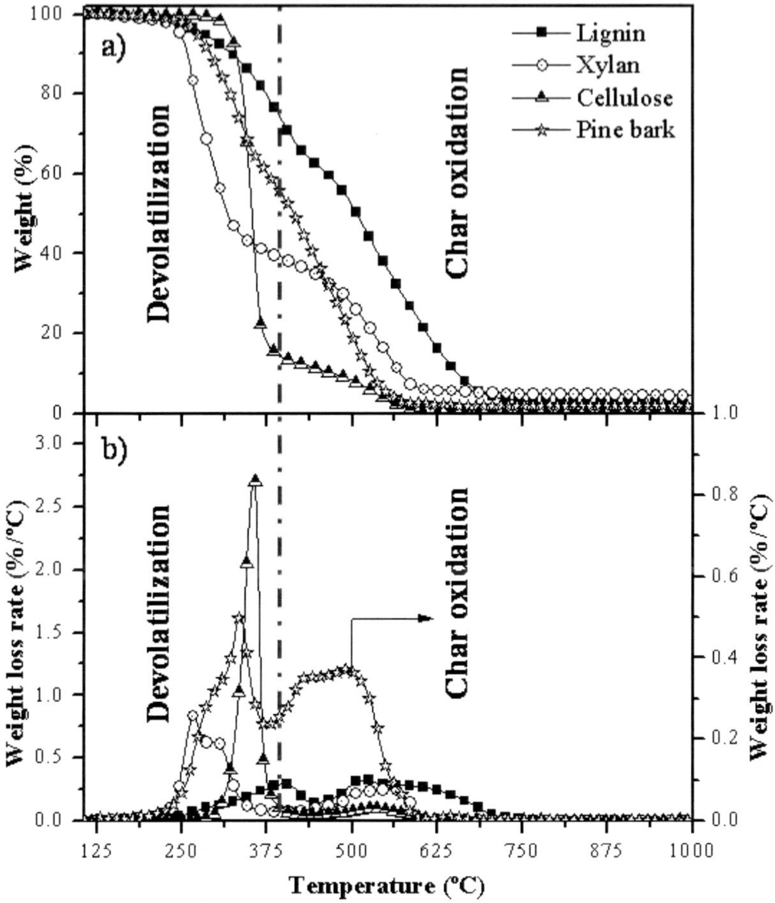

Figure 1. TGA-DTG plots for the combustion process of Pine bark and lignocellulosic biomass main components (cellulose, hemicellulose and lignin) (Temperature range: 125-1000 °C; Heating rate: 40°C/min; Flow rate: 100 Nml/min; Reactive gas: 79 % vol. Ar + 21 % vol. O_2).

Concerning the thermal behavior of the lignocellulosic biomass sample (Pine bark), three decomposition peaks can be clearly observed. These peaks matched the main decomposition steps of biomass main components. In the devolatilization stage, two peaks could be observed. The first one, detected between 250 and 310°C, represented as a shoulder in the DTG plot corresponded to the thermal degradation of the hemicellulose component. The second one, where the maximum weight loss rate took place, was detected at 330 °C which was pretty close to that obtained for the cellulose component

(350 ºC). The oxidation stage (440-610 ºC) was coincidental with the oxidation of the lignin component. Thus, a clear representation of the thermal behavior of lignocellulosic biomass can be done by considering the lignocellulosic biomass main components decomposition.

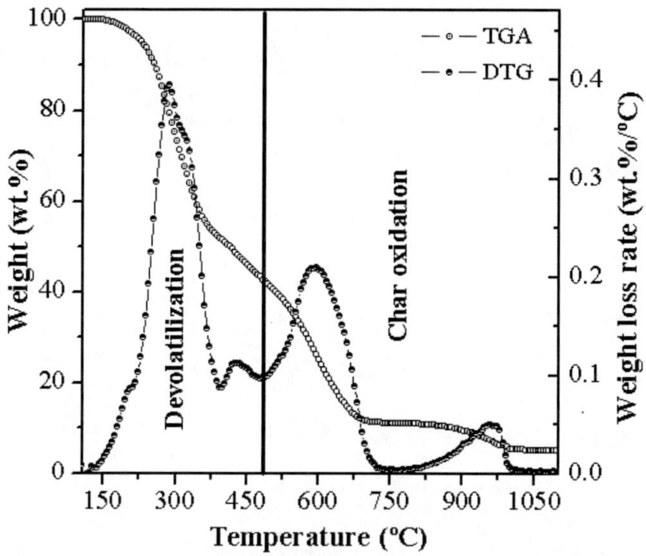

Figure 2. TGA/DTG plots for the combustion process of microalgae *Nannochloropsis Gaditana* (Temperature range: 125-1000 ºC; Heating rate: 40ºC/min; Flow rate: 100 Nml/min; Reactive gas: 79 % vol. Ar + 21 % vol. O_2).

Microalgae composition is different than lignocellulosic biomass. Microalge are mainly composed of lipids, carbohydrates and proteins. In a similar way than for lignocellulosic biomass, the thermal behavior of microalgae can be regarded as a step-wise process where carbohydrates, proteins and lipids decompose. Figure 2 shows the TGA/DTG plot for the combustion process of the microalgae *Nannochloropsis gaditana*. In this case the devolatilization stage could be described by three decomposition peaks. It can be observed that the decomposition of microalgae started at around 170 ºC being ascribed to carbohydrates decomposition. The second sub-step was detected by a peak at 284 ºC and was associated to the degradation of carbohydrates and proteins, obtaining the highest weight loss rate (0.44 wt%/ºC). Finally, a third peak was observed close to the char oxidation stage. This peak appeared at 430 ºC. In this stage, the final decomposition of lipids took place and was mainly associated to the break-down of hydrocarbon chains of fatty acids [26-28]. The char oxidation stage took place between 478 and 725 ºC. The final amount of ashes remaining after the combustion process of the microalgae was around 6.0 wt.%. Additionally, a last decomposition step was observed between 826 and 998 ºC. According to Wang et al. [29], this step is mainly related to volatile metal loss and carbonate decomposition.

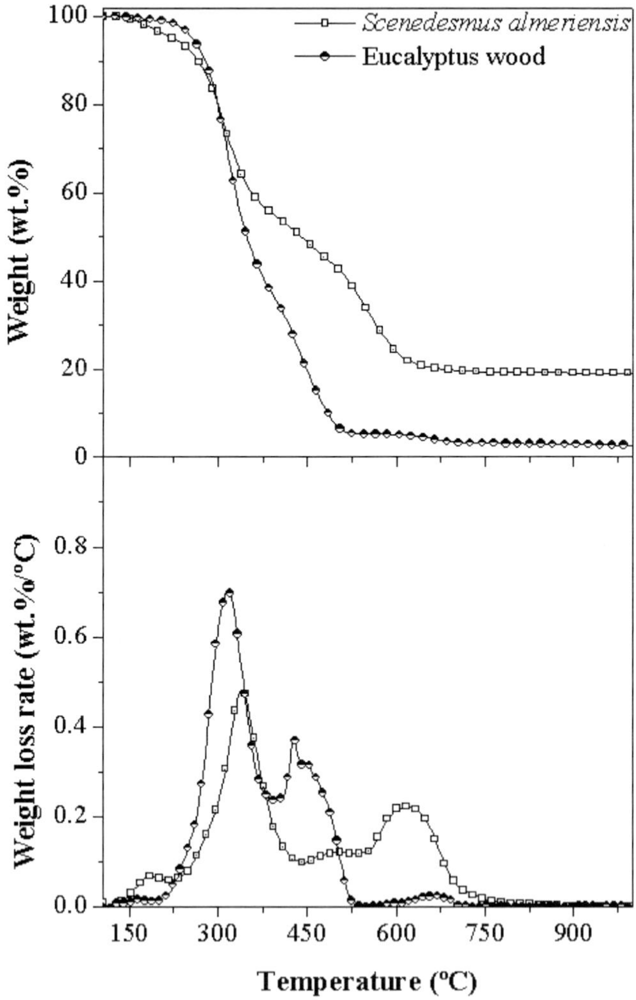

Figure 3. TGA-DTG plots for the combustion process of marine microalgae (*Scenedesmus almeriensis*) vs lignocellulosic biomass (Eucalyptus wood sample) (Temperature range: 125-1000 ºC; Heating rate: 40ºC/min; Flow rate: 100 Nml/min; Reactive gas: 79 % vol. Ar + 21 % vol. O_2).

From the view of point of the pyrolysis of lignocellulosic biomass and marine biomass, Figure 3 shows the TGA-DTG plots for the combustion process of Eucalyptus wood and the specie of microalgae *Scenedesmus almeriensis*. It can be observed that microalgae sample started decomposing at lower temperatures than terrestrial sample although they decomposed in a broader range of temperatures. Furthermore, microalgae samples showed a last decomposition step at temperatures around 1000 ºC. These differences are attributed to the different compositions of these materials. As abovementioned, microalgae are mainly composed carbohydrates, proteins and lipids, which are less thermal resistant than lignocellulosic biomass main components (cellulose, hemicellulose and lignin) [30]. Additionally, the last decomposition step was attributed to the decomposition of inorganic material being in good agreement with the high ash content in microalgae. This fact cannot be overlooked either, as the catalytic effect of ashes can also contribute to the progress of the pyrolysis process [6].

3.2. Differential Scanning Calorimetric Analysis (DSC)

Figure 4 shows the differential scanning calorimetry (DSC) profile for the combustion process of Fir wood and the specie of microalgae *Nannochloropsis gaditana*. The DSC plots showed the existence of two exothermic regions. The first region is associated to the combustion of light volatile matters, which provides reactivity of biomass fuels. This peak is short and lower, so less heat was released. The second one represents the combustion of fixed carbon [31]. Similar conclusions can be derived from this study. It can be observed that the DSC plot for the combustion of the microalgae sample started at lower temperatures than those for the lignocellulosic one. However, the DCS profile of the microalgae covered a wider temperature range. This way, the combustion of fixed carbon peak appeared at higher temperatures (620 ºC) than in the lignocellulosic biomass sample (480 ºC).

The DSC curve allows to compute the combustion heat (H_{comb}) by integrating the area of the positive or negative peaks for exothermic and endothermic processes, respectively. Table 4 shows the calculated heat of combustion obtained for different types of lignocellulosic and marine biomass at a heating rate of 40 ºC/min. It can be observed that the value of this magnitude obtained for the combustion process of lignocellulosic biomass samples was higher than that obtained for the microalgae ones. This fact is due to the higher amount of fixed carbon of lignocellulosic biomass. Furthermore, the high amount of ash in marine biomass can also influence the process as it reduces the energetic density by mass of the sample.

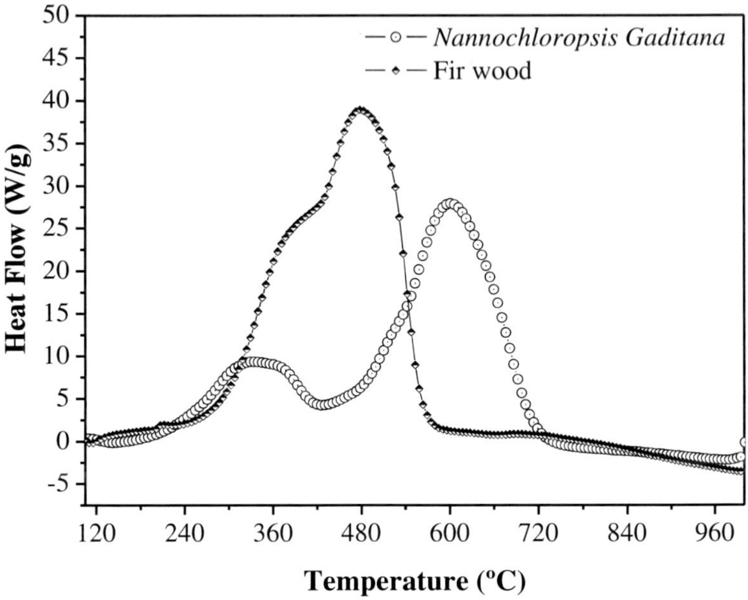

Figure 4. DSC plots for the combustion process of marine microalgae (*Nannochloropsis Gaditana*) and lignocellulosic biomass (Fir wood). (Temperature range: 125-1000 °C; Heating rate: 40°C/min; Flow rate: 100 Nml/min; Reactive gas: 79 % vol. Ar + 21 % vol. O_2).

Table 4. Heat released during the combustion process of lignocellulosic biomass and marine biomass

Biomass feedstock		H_{comb} (kJ/g)	Reference
Lignocellulosic biomass	Fir wood	9.4	
	Eucalyptus wood	8.6	[22]
	Pine bark	10.1	
Microalgae	*Nannochloropsis gaditana*	7.9	
	Scenedesmus almeriensis	7.8	[23]
	Chlorella vulgaris	8.8	

3.3. Mass Spectrometric Analysis

Figure 5 shows the MS spectra of the main products (H_2O, CO and CO_2) obtained in the combustion of lignocellulosic biomass (Eucalyptus wood) and marine biomass (*Chlorella vulgaris*). MS profile of biomass could be divided into different stages related to their degradations steps studied in the TGA/DTG curves and described in previous sections. The gas distribution observed for the combustion process of lignocellulosic biomass and marine biomass was different.

Concerning lignocellulosic biomass, CO and CO_2 evolved over the whole temperature range with a higher proportion of CO_2. The release of these products took place by two emission peaks at temperatures around 320 and 400 ºC. CO is assumed to be produced by the formation of molecular oxide complexes that further rearrange turning into the evolution of CO [32]. However, CO_2 presented a more complex formation process [32], as its production can be catalyzed or inhibited by the formation of carbon intermediates [33]. The maximum emission peak was obtained for CO_2 at 420 ºC corresponding to the oxidation of the char. A peak for CO_2 was detected at 680 ºC which was related to the decomposition of inorganic materials. H_2O was released in two steps. The higher peak for water was formed in the main devolatilization stage, being associated to the evolved aliphatic OH groups [33, 34]. Finally, a shoulder appeared in the MS spectra, which was related to the water formed by the oxidation of H_2 and calcium carbonate decomposition [33].

The mass spectra obtained for the combustion process of marine biomass (*Chlorella vulgaris*) was different from that of the lignocellulosic biomass. The main difference was the marked signal obtained for H_2O during the devolatilization of the sample. CO and CO_2 main peaks took place at temperatures between 555 and 610 ºC being coincident with their DTG peaks. The CO and CO_2 emissions detected in this temperature range were due to the fixed carbon oxidation [29]. Emission peaks at lower temperatures were associated to the decomposition of carboxyl groups in protein and saccharides [35]. Furthermore, an additional peak was found at temperatures above 800 ºC, which was attributed to the decomposition of mineral matter, as carbonates, in the ash [29, 36]. This peak was more prominent than that obtained for the combustion process of eucalyptus wood. This fact agrees well with their ash composition analysis (Table 2). On the other hand, H_2O emissions reached their maximum evolution rate at 298 ºC.

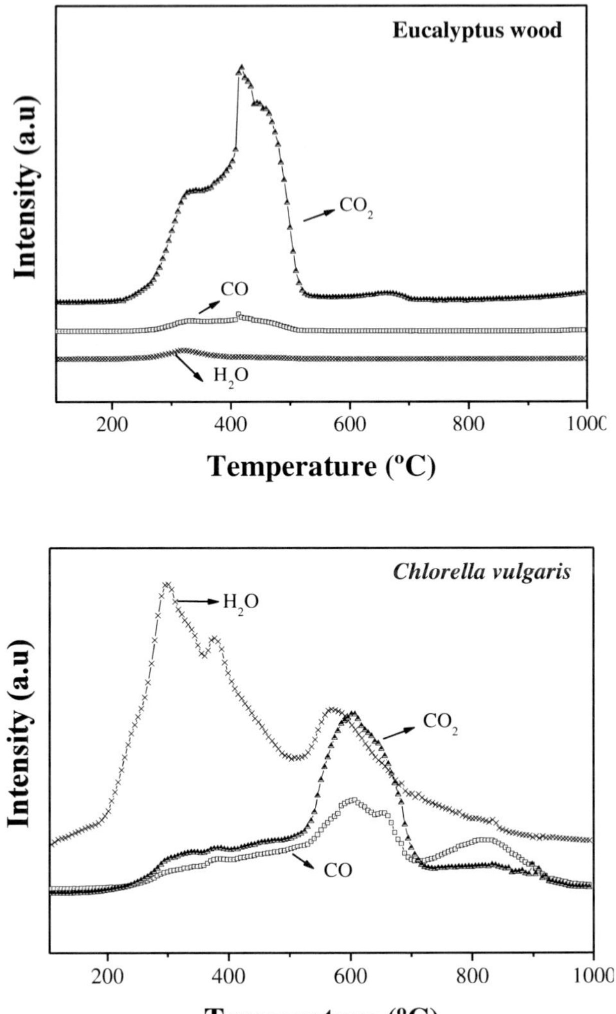

Figure 5. Mass spectra profile of CO, CO_2 and H_2O for the combustion process of lignocellulosic biomass (Eucalyptus wood) and microalgae (*Chlorella vulgaris*) (Temperature range: 125-1000 ºC; Heating rate: 40ºC/min; Flow rate: 100 Nml/min; Reactive gas: 79 % vol. Ar + 21 % vol. O_2).

The H_2O produced at this step was mainly associated to the oxidation of oxygen containing functional groups (especially hydroxyl groups). H_2O peaks at lower temperatures were attributed to the loss of cellular water and external water bounded by surface tension [37].

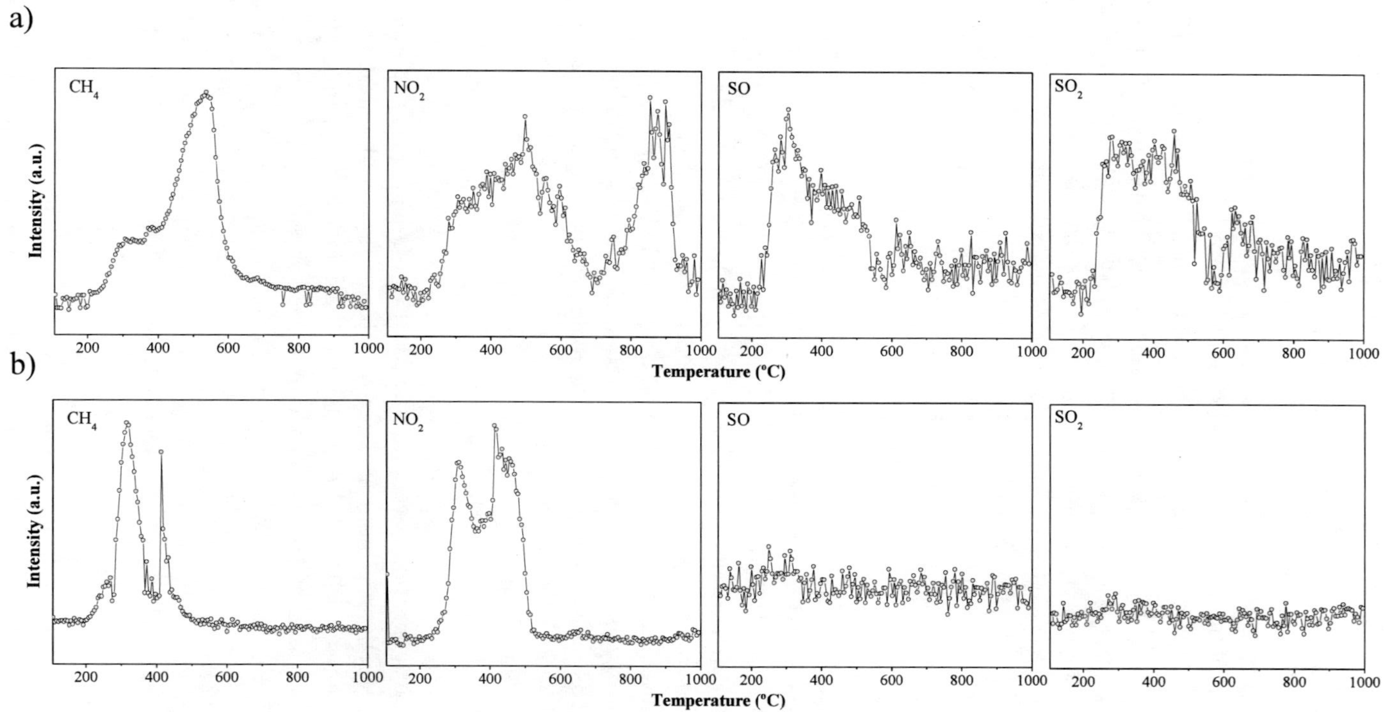

Figure 6. Gas evolution profile of CH_4, NO_2, SO and SO_2 for the combustion process of a) microalgae *Chlorella vulgaris* and b) lignocellulosic biomass eucalyptus wood. (Temperature range: 125-1000 °C; Heating rate: 40°C/min; Flow rate: 100 Nml/min; Reactive gas: 79 % vol. Ar + 21 % vol. O_2).

Finally, the water produced at higher temperatures was associated to the evolution of H_2. H_2O peaks were found at lower temperatures than CO_2 ones, indicating that microalgae samples decomposition started via dehydration of their components followed by their combustion [5].

Figure 6 shows the mass spectra for other compounds which were released in lower proportions. Light hydrocarbons (HC), especially CH_4, were the main secondary products detected. CH_4 had two main origins related to devolatilization and charring processes [14]. It can be observed that the maximum emission peak for the microalgae sample was obtained at higher temperatures than for the Eucalyptus wood one. The origin of HC in microalgae is attributed to the decomposition of carbohydrates and lipids. In this sense, Marcilla et al. [38] reported that the main source of methyl groups was the decomposition of lipids, which agrees well with the maximum CH_4 production observed at 529 ºC.

The evolution of nitrogen compounds took place through different emission peaks. Nitrogen compounds mainly evolved NO_2 although other components such as CH_4N, NO or HCN were also expected to be produced [22]. The first NO_2 emission peak for Eucalyptus wood was detected between 200 and 400 ºC and was mainly associated to the decomposition of proteins. The last peak at temperatures above 400 ºC corresponded to the oxidation of the remaining nitrogen in the char. The maximum yield of NO_2 was reached in this stage. The NO_2 emission peak for the microalgae sample was broader than that for the lignocellulosic one, encompassing the devolatilization and oxidation stages. Furthermore, a peak at temperatures higher than 700 ºC was found and it could be related to the catalytic effect of some compounds in the ash [23].

Finally, sulfur compounds were also detected as SO and SO_2. The signal obtained for lignocellulosic biomass was lower being close to the noise level of the MS equipment. The release of these compounds in the combustion process of microalgae is mainly associated to the decomposition and oxidation of sulphated polysaccharides [35]. SO and SO_2 maximum peaks were found at 300 ºC. Furthermore, a second emission peak was found in the lipid decomposition temperature range (350-480 ºC) for both compounds due to the degradation of organic sulfides in organic residues [35]. These peaks were characterized by the apparition of SO_2 at lower temperatures than SO. The emission of SO and SO_2 in the combustion of lignocellulosic biomass took place in the devolatilization stage at temperatures around 330 ºC. The origin of these compounds in the combustion was associated to the decomposition of sulphoxide and sulphone groups of the lignin component, which facilitates its further depolymerization [39].

3.4. Kinetic Analysis

3.4.1. Theory

Kinetic data from solid-state combustion can be obtained by using thermogravimetric analysis. The thermal devolatilization curve has been usually obtained as a sum of the contributions of the corresponding biomass individual components [11, 40]. However, solid-state reactions consist of more complex processes, involving a superposition of several elementary processes such as nucleation, adsorption, desorption interfacial reaction and surface/bulk diffusion [4].

There are numerous models that have been used for interpreting the kinetics of a process when using thermal analysis, such as single-step global reaction models, multiple-step models, and semi-global models. The model reported in this work is similar to that reported by Gil et al. (2010) [11] for the determination of the combustion kinetic parameters of coal/biomass blends. A several-stage reaction kinetic scheme consisting of several independent reactions was proposed for the thermal decomposition of biomass under an oxidative atmosphere [11, 13], represented by the DTG curves. The kinetic scheme includes the following reactions:

$$\begin{gathered}
A(\text{Biomass}) \rightarrow V_1(\text{Volatiles}_1) + C_1\ (\text{Biomass}_1) \\
C_1(\text{Biomass}) \rightarrow V_2(\text{Volatiles}_2) + C_2\ (\text{Biomass}_2) \\
\cdots\cdots\cdots \\
C_{x-1}(\text{Char}) \rightarrow C_x\ (\text{Volatiles}_x) + D\ (\text{Ash})
\end{gathered} \tag{2}$$

The sub-index (1,2,…,x-1 and x) refers to each of the decomposition stages, determined by the number of DTG peaks observed, of the biomass combustion process. The approach used in the present work for the determination of the combustion kinetic rates of a material was based on the Arrhenius equation, which has been used by other researchers in order to obtain kinetic parameters of thermal events under combustion conditions [11, 14].

$$d\alpha/_{dt} = k \cdot f(\alpha) \tag{3}$$

$$k = K_0 \cdot e^{-E_a/R \cdot T} \tag{4}$$

where f(α) represents the hypothetical model of the reaction mechanism; k is the reaction rate; K_0 represents the pre-exponential factor (min^{-1}); E_a is the activation energy ($kJ\ mol^{-1}$);T is the absolute temperature (K); t represents the time (min), and α is the degree of conversion defined by:

$$\alpha = {(m_o - m_t)}/{(m_o - m_f)} \tag{5}$$

where m_o and m_t represent the mass at t=0 and t=t, respectively, and m_f is the final mass of the sample.

For a constant heating rate β ($K\ min^{-1}$), β= dT/dt, Eq. (3) can be transformed into:

$$dx/_{f(\alpha)} = K/_{\beta}\, dT \tag{6}$$

By integrating Eq. (6) gives:

$$g(\alpha) = \int_0^\alpha dx/_{f(\alpha)} = K_o/_{\beta} \int_{T_o}^{T} e^{-E/_{RT}}\, dT \tag{7}$$

where g(α) is the integral function of conversion.

Eq. (7) is integrated by using the Coats-Redfern method [41]:

$$\ln[g(\alpha)/_{T^2}] = \ln[K_o \cdot R/_{\beta \cdot E \cdot (1 - 2RT/_{E})}] - E/_{R \cdot T} \tag{8}$$

Generally, the term $2RT/E_a$ can be neglected as it is much less than unity [42]. It has been demonstrated that for both the temperatures of combustion range and most values of E_a, the expression $\ln[K_0R/\beta E]$ in Eq. (8) is essentially constant [18]. Thus, if the correct expression of g(α) is used, the plot of $\ln[g(\alpha)/T^2]$ against 1/T should give a straight line with a high correlation coefficient of linear regression analysis, from which the values of E_a and K_0 can be respectively calculated from the slope of the line; and by the intercept term in Eq. (8). The functions f(α) and g(α) referred to the different reaction models and are presented in Table 5 [4].

Table 5. Expressions for the most common reaction mechanisms in solid-state reactions [4]

Reaction model	f(α)	g(α)
Reaction order		
O_0	$(1-\alpha)^n$	α
O_1		$-\ln(1-\alpha)$
O_2		$-(1-\alpha)^{-1}$
O_3		$1/2\ (1-\alpha)^{-2}$
Phase boundary controlled reaction		
R_2	$(1-\alpha)^{(1-1/n)}$	$1-(1-\alpha)^{(1/2)}$
R_3		$1-(1-\alpha)^{(1/3)}$
Power Law		
P_1	$n(\alpha)^{(1-1/n)}$	$\alpha^{1/4}$
P_2		$\alpha^{1/3}$
P_3		$\alpha^{1/2}$
P_4		$\alpha^{3/2}$
Nucleation and growth (Avrami-Erofeev equation)		
N_1	$n(1-\alpha)[-\ln(1-\alpha)]^{(1-1/n)}$	$[-\ln(1-\alpha)]^{(1/1.5)}$
N_2		$[-\ln(1-\alpha)]^{(1/2)}$
N_3		$[-\ln(1-\alpha)]^{(1/3)}$
N_4		$[-\ln(1-\alpha)]^{(1/4)}$
Diffusion		
D_1	$1/2\alpha$	α^2
D_2	$[-\ln(1-\alpha)]^{-1}$	$(1-\alpha)\ln(1-\alpha)+\alpha$
D_3	$3/2(1-\alpha)^{2/3}[1-(1-\alpha)^{1/3}]^{-1}$	$[1-(1-\alpha)^{1/3}]^2$
D_4	$3/2[(1-\alpha)^{1/3}-1]^{-1}$	$1-2/3\alpha-(1-\alpha)^{2/3}$

3.4.2. Kinetic Results

The kinetic model used in this work was derived from the PMSM (*Pseudo Multi-component Separate-stage Models)* approach. In this type of models, the biomass sample is composed of multiple pseudo components [42]. In this regard, the kinetic parameters can be determined assuming single separate reactions for the different stages of thermal conversion. As aforementioned, the DTG plots represented different decomposition peaks dividing the combustion process of biomass samples in different stages. Each stage represents a separate reaction. Biomass combustion was clearly defined by two main stages: devolatilization stage (Dev. stage) and char oxidation stage

(Oxid. stage). However, some samples underwent additional decompositions that should be considered as supplementary separate reactions before using this type of models. This way, Eq. (8) can be separately used in each of the stages commented above.

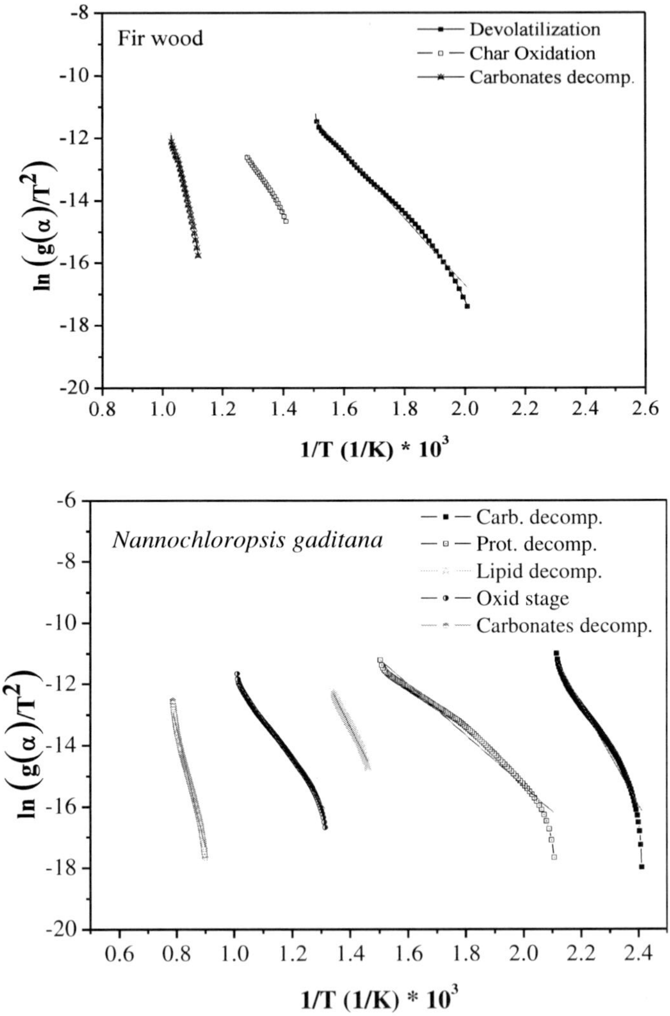

Figure 7. $\ln(g(\alpha))/T^2$ vs 1/T that showed the best linearity for Fir wood lignocellulosic biomass sample and the microalgae *Nannochloropsis gaditana*. (Temperature range: 125-1000 ºC; Heating rate: 40ºC/min; Flow rate: 100 Nml/min; Reactive gas: 79 % vol. Ar + 21 % vol. O_2).

The model as a function of *g(α)* (Table 5), which delivered the highest correlation coefficient, was considered to be the function representing the mass loss kinetics for the samples under study. The function *g(α)* depends on the mechanism controlling the reaction and the size and shape of the reacting particles [43]. Figure 7 shows the typical lineal plots of $ln[g(\alpha)/T^2]$ versus $1/T$ that are obtained with this type of models for the combustion process of the microalgae *Nannochloropsis gaditana* and the lignocellulosic biomass sample Fir wood at 40 °C/min. Table 6 summarizes the main kinetic parameters of different biomass samples gathered from literature. Generally, the models based on reaction order (O_i), nucleation (N_i) and diffusion (D_i) are the most representative ones when evaluating the combustion behavior of biomass samples [11, 42, 44]. The general disadvantage of using dynamic thermal analysis is that in many cases more than one function *g(α)* fits the experimental results. Consequently, selection of the responsible mechanism and estimation of the real kinetic parameters can be difficult as previously reported [44]. Furthermore, separate stage models fail to predict the transition region between two mass loss processes and do not consider chemical processes.

Table 6. Comparison for the kinetic parameters obtained for different types of biomass feedstocks

Biomass		E_a (kJ/mol)		References
	Heating rate (°C/min)	Dev. stage	Oxid. stage	
NG microalgae	40	63	113	[23]
SC microalgae	40	81	125	
CV microalgae	40	71	126	
Pine bark	40	97	188	[22]
Fir wood	40	89	167	
Eucalyptus wood	40	90	135	
Maple wood	Modulated	164	150	[12]
Pine sawdust	15	102	219-236	[11]
Oak wood	30	118	130	[13]
Aspens wood	30	120	95	
Pine wood	30	125	120	
Birch wood	30	115	140	

Table 6 shows the calculated activation energies for the combustion process of different types of biomass feedstock obtained by different authors. It can be observed that marine biomass samples presented lower activation energies than lignocellulosic biomass ones. This fact can be attributed to both the catalytic effect of ashes in the marine biomass and the low thermal resistance of their main constituents.

For validating this type of analysis the reconstruction of the weight loss curves can be performed. Considering *n* separate reactions, the kinetic rates of thermal decomposition of a material can be easily derived from Eq. (6) as follows:

$$\mathbf{d\alpha_i}/\mathbf{dT} = \mathbf{K_o}/\boldsymbol{\beta}\, \mathbf{e}^{-\mathbf{E_{ai}}/\mathbf{RT}} \mathbf{f_i(\alpha_i)} \quad (9)$$

where α_i, K_o, E_{ai} and $f_i(\alpha_i)$ are the degree of conversion, the pre-exponential factor, activation energy and model functions obtained for each stage of the combustion process, respectively.

Figure 8 shows the experimental data (solid line) compared to the predicted one (dotted line) for the combustion process of the Fir wood lignocellulosic biomass sample and the microalgae *Scenedesmus almeriensis*. It can be observed that the used model adequately reproduces the experimental values, obtaining a low error.

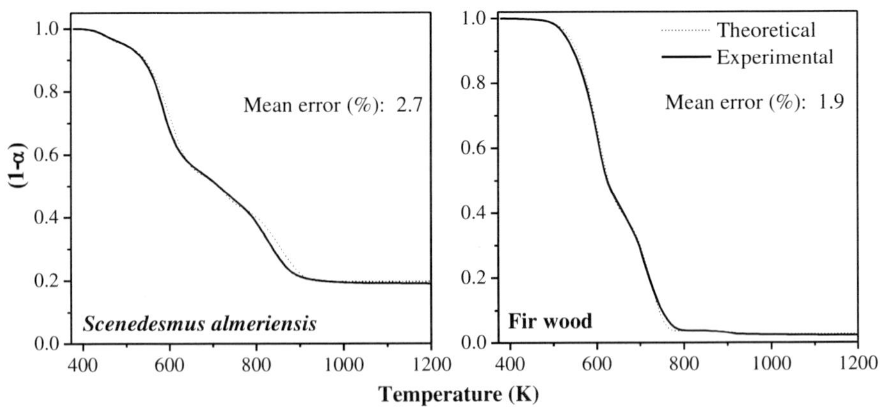

Figure 8. Comparison between experimental and theoretical results for the combustion process of the microalgae *Scenedesmus almeriensis* and the lignocellulosic biomass Fir wood. (Temperature range: 125-1000 °C; Heating rate: 40°C/min; Flow rate: 100 Nml/min; Reactive gas: 79 % vol. Ar + 21 % vol. O_2).

CONCLUSION

Thermal analysis coupled to mass spectrometry is an excellent tool for the characterization of the combustion process of different biomass feedstock. The TGA-DTG figures showed that the oxidation of biomass took place into two main stages. Firstly, the devolatilization of the sample took place leading to char formation (Devolatilization stage). The second stage involved the oxidation of the formed char (Oxidation stage). DSC and MS plots corroborated the existence of two stages. DSC analyses showed that most of the heat was liberated during the char oxidation stage being lignocellulosic biomass samples the ones which released a higher amount of H_{comb}. The mass spectra of the combustion process showed two main emission peaks. The main products obtained were CO, CO_2 and H_2O together with other compounds such as CH_4, NO_x and SO_x. The presence of NO_x during the oxidation stage implies that part of the initial nitrogen in the biomass remained in the char after the oxidation process. According to their ultimate analysis, NO_x and SO_x were expected to be released in higher proportions for microalgae than for lignocellulosic biomass. Other pollutants such as SO_x were detected in lower proportions according to biomass ultimate analysis. Lignocellulosic biomass had a higher thermal resistance than marine biomass samples, being mainly due to the higher resistance of their main constituents. The combustion process can be successfully reproduced by using Pseudo-Multi-component-separate stage models. Models of reaction order and diffusion ones were the ones that led to the best fits. It could be observed that the activation energy of combustion process was lower in marine biomass than in lignocellulosic biomass ones.

REFERENCES

[1] Koçar G, Civaş N. An overview of biofuels from energy crops: Current status and future prospects. *Renewable and Sustainable Energy Reviews*. 2013;28:900-16.

[2] McKendry P. Energy production from biomass (part 1): Overview of biomass. *Bioresource Technology*. 2002;83:37-46.

[3] Basu P. Biomass Gasification and Pyrolysis: *Practical Design and Theory*. 2010.

[4] White JE, Catallo WJ, Legendre BL. Biomass pyrolysis kinetics: A comparative critical review with relevant agricultural residue case studies. *Journal of Analytical and Applied Pyrolysis*. 2011;91:1-33.

[5] Babich IV, van der Hulst M, Lefferts L, Moulijn JA, O'Connor P, Seshan K. Catalytic pyrolysis of microalgae to high-quality liquid bio-fuels. *Biomass and Bioenergy*. 2011;35:3199-207.

[6] Ross AB, Anastasakis K, Kubacki M, Jones JM. Investigation of the pyrolysis behaviour of brown algae before and after pre-treatment using PY-GC/MS and TGA. *Journal of Analytical and Applied Pyrolysis*. 2009;85:3-10.

[7] Shen DK, Gu S, Bridgwater AV. The thermal performance of the polysaccharides extracted from hardwood: Cellulose and hemicellulose. *Carbohydrate Polymers*. 2010;82:39-45.

[8] Di Blasi C. Combustion and gasification rates of lignocellulosic chars. *Progress in Energy and Combustion Science*. 2009;35:121-40.

[9] Hood EE, Nelson P, Powell R. Plant Biomass Conversion: Wiley; 2011.

[10] McKendry P. Energy production from biomass (part 2): Conversion technologies. *Bioresource Technology*. 2002;83:47-54.

[11] Gil MV, Casal D, Pevida C, Pis JJ, Rubiera F. Thermal behaviour and kinetics of coal/biomass blends during co-combustion. *Bioresource Technology*. 2010;101:5601-8.

[12] Cheng K, Winter WT, Stipanovic AJ. A modulated-TGA approach to the kinetics of lignocellulosic biomass pyrolysis/combustion. *Polymer Degradation and Stability*. 2012;97:1606-15.

[13] Shen DK, Gu S, Luo KH, Bridgwater AV, Fang MX. Kinetic study on thermal decomposition of woods in oxidative environment. *Fuel*. 2009;88:1024-30.

[14] Barneto AG, Carmona JA, Alfonso JEM, Blanco JD. Kinetic models based in biomass components for the combustion and pyrolysis of sewage sludge and its compost. *Journal of Analytical and Applied Pyrolysis*. 2009;86:108-14.

[15] Zheng G, Koziński JA. Thermal events occurring during the combustion of biomass residue. *Fuel*. 2000;79:181-92.

[16] Meng X, de Jong W, Badri FS, Benito P, Basile F, Verkooijen AHM. Combustion study of partially gasified willow and DDGS chars using TG analysis and COMSOL modeling. *Biomass and Bioenergy*. 2012;39:356-69.

[17] Kai X, Yang T, Huang Y, Sun Y, He Y, Li R. The effect of biomass components on the co-combustion characteristics of biomass with coal. *Zhangjiajie, Hunan* 2011. p. 1274-8.

[18] Zhou L, Wang Y, Huang Q, Cai J. Thermogravimetric characteristics and kinetic of plastic and biomass blends co-pyrolysis. *Fuel Processing Technology*. 2006;87:963-9.

[19] Wang C, Wang F, Yang Q, Liang R. Thermogravimetric studies of the behavior of wheat straw with added coal during combustion. *Biomass and Bioenergy*. 2009;33:50-6.

[20] Naik S, Goud VV, Rout PK, Jacobson K, Dalai AK. Characterization of Canadian biomass for alternative renewable biofuel. *Renewable Energy*. 2010;35:1624-31.

[21] Jenkins BM, Baxter LL, Miles Jr TR, Miles TR. Combustion properties of biomass. *Fuel Processing Technology*. 1998;54:17-46.

[22] López-González D, Fernandez-Lopez M, Valverde JL, Sanchez-Silva L. Thermogravimetric-mass spectrometric analysis on combustion of lignocellulosic biomass. *Bioresource Technology*. 2013;143:562-74.

[23] López-González D, Fernandez-Lopez M, Valverde JL, Sanchez-Silva L. Kinetic analysis and thermal characterization of the microalgae combustion process by thermal analysis coupled to mass spectrometry. *Applied Energy*. 2014;114:227-37.

[24] Wang G, Li W, Li B, Chen H. TG study on pyrolysis of biomass and its three components under syngas. *Fuel*. 2008;87:552-8.

[25] Li XG. High-Resolution Thermogravimetry of Cellulose Esters. *Journal of Applied Polymer Science*. 1999;71:573-8.

[26] Kebelmann K, Hornung A, Karsten U, Griffiths G. Intermediate pyrolysis and product identification by TGA and Py-GC/MS of green microalgae and their extracted protein and lipid components. *Biomass and Bioenergy*. 2013;49:38-48.

[27] Tang Y, Ma X, Lai Z. Thermogravimetric analysis of the combustion of microalgae and microalgae blended with waste in N2/O2 and CO2/O2 atmospheres. *Bioresource Technology*. 2011;102:1879-85.

[28] Ross AB, Jones JM, Kubacki ML, Bridgeman T. Classification of macroalgae as fuel and its thermochemical behaviour. *Bioresource Technology*. 2008;99:6494-504.

[29] Wang S, Jiang XM, Han XX, Liu JG. Combustion characteristics of seaweed biomass. 1. Combustion characteristics of enteromorpha clathrata and sargassum natans. *Energy and Fuels*. 2009;23:5173-8.

[30] Shuping Z, Yulong W, Mingde Y, Chun L, Junmao T. Pyrolysis characteristics and kinetics of the marine microalgae Dunaliella tertiolecta using thermogravimetric analyzer. *Bioresource Technology*. 2010;101:359-65.

[31] Kok MV, Özgür E. Thermal analysis and kinetics of biomass samples. *Fuel Processing Technology*. 2013;106:739-43.

[32] Li C, Brown TC. Carbon oxidation kinetics from evolved carbon oxide analysis during temperature-programmed oxidation. *Carbon*. 2001;39:725-32.

[33] Shen D, Hu J, Xiao R, Zhang H, Li S, Gu S. Online evolved gas analysis by Thermogravimetric-Mass Spectroscopy for thermal decomposition of biomass and its components under different atmospheres: *Part I. Lignin. Bioresource Technology*. 2013;130:449-56.

[34] Jakab E, Till F, Várhegyi G. Thermogravimetric-mass spectrometric study on the low temperature oxidation of coals. *Fuel Processing Technology*. 1991;28:221-38.

[35] Wang S, Jiang XM, Wang N, Yu LJ, Li Z, He PM. Research on pyrolysis characteristics of seaweed. *Energy and Fuels*. 2007;21:3723-9.

[36] Marcilla A, Gómez-Siurana A, Gomis C, Chápuli E, Catalá MC, Valdés FJ. Characterization of microalgal species through TGA/FTIR analysis: Application to nannochloropsis sp. *Thermochimica Acta*. 2009;484:41-7.

[37] Li D, Chen L, Chen S, Zhang X, Chen F, Ye N. Comparative evaluation of the pyrolytic and kinetic characteristics of a macroalga (Sargassum thunbergii) and a freshwater plant (Potamogeton crispus). *Fuel*. 2012;96:185-91.

[38] Marcilla A, Gómez-Siurana A, Gomis C, Chápuli E, Catalá MC, Valdés FJ. Characterization of microalgal species through TGA/FTIR analysis: Application to nannochloropsis sp. *Thermochimica Acta*. 2009;484:41-7.

[39] Amen-Chen C, Pakdel H, Roy C. Production of monomeric phenols by thermochemical conversion of biomass: a review. *Bioresource Technology*. 2001;79:277-99.

[40] Heikkinen JM, Hordijk JC, De Jong W, Spliethoff H. Thermogravimetry as a tool to classify waste components to be used for energy generation. *Journal of Analytical and Applied Pyrolysis*. 2004;71:883-900.

[41] Coats AW, Redfern JP. Kinetic parameters from thermogravimetric data *Nature*. 1964;201:68-9.

[42] Liu NA, Fan W, Dobashi R, Huang L. Kinetic modeling of thermal decomposition of natural cellulosic materials in air atmosphere. *Journal of Analytical and Applied Pyrolysis*. 2002;63:303-25.

[43] Gadalla AMM. Kinetics of the decomposition of hydrated oxalates of calcium and magnesium in air. *Thermochimica Acta*. 1984;74:255-72.

[44] Yorulmaz SY, Atimtay AT. Investigation of combustion kinetics of treated and untreated waste wood samples with thermogravimetric analysis. *Fuel Processing Technology*. 2009;90:939-46.

In: Combustion
Editor: Joseph M. Grier
ISBN: 978-1-62948-967-4

Chapter 6

METHANE COMBUSTION BY ELECTROCHEMICAL PROMOTION OF CATALYSIS PHENOMENON

Carmen Jiménez Borja*, Nuria Gutiérrez, Fernando Dorado and Jose Luis Valverde
Departamento de Ingeniería Química,
Universidad de Castilla-La Mancha (UCLM), Ciudad Real, Spain

ABSTRACT

Methane is the main component of natural gas and is probably the most abundant organic compound on Earth. In view of the fact that the world petroleum reserves are on the decline, a great deal of emphasis has been placed on developing alternatives for energy production. Significant efforts have been focused on finding effective processes and technology for the optimum utilization of the abundant natural gas inenergy production. During the last decades, the catalytic combustion of methane has been extensively studied as an alternative to conventional thermal combustion. This method allows converting methane at relatively low temperatures and with very high combustion efficiency, thus avoidingemissions. Hence, it is considered an effective route to produce

* Tel.: +34-926295300; Fax: +34-926295256; *Email address: Carmen.JBorja@uclm.es / Carmenjborja@hotmail.com.

power with low environmental impact and then suitable for gas turbine applications.

On the other hand, one of the lastest advances in catalysis is the phenomenon of Electrochemical Promotion Of Catalysis (EPOC), or Non-faradaic Electrochemical Modification of the Catalytic Activity (NEMCA), that has had a strong impact in the fields of electrochemistry, heterogeneous catalysis and surface science. This phenomenon is an important tool to improve the catalytic performance of a metal catalyst interfaced with a solid electrolyte by the application of low current or potentials over the catalytic film. During the last years, it has been demostrated that itis based on the change of the work catalytic function due to the elctrochemical pumping of ions from the solid electrolyte to the catalyst.

In this chapter, the catalytic combustion of methane by this new phenomenon will be addressed. Finally, the application of this tecnhology to the methane combustion over palladium based catalyst-electrodes will be presented.

1. Introduction

Methane, which is referred to as CH_4, is an organic compound made up of one carbon atomand four hydrogen atoms. The hydrogen atoms are spaced equally around the carbon atom, placed at the four verticesof a tetrahedron. The chemical structure of methane molecule is shown in Figure 1.It has a density of 0.717 kgm^{-3}, so methane gas is lighter than air, and a molecular weight of 16.043 g mol^{-1}.

Current estimations by the EPA (Environmental Protection Agency) place global emissions of methane at 3 trillion cubic feet (85 km^3) per year [1]. Methane is the second most prevalent greenhouse gas and constitutes 14% of the global greenhouse gas emissions (Figure 2) [1].

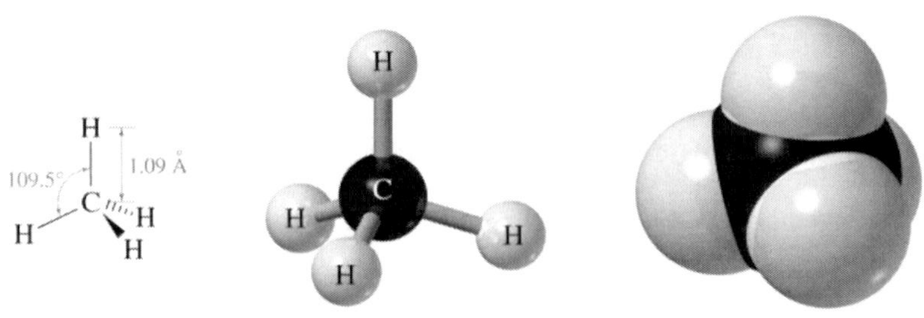

Figure 1. Chemical structure of methane molecule.

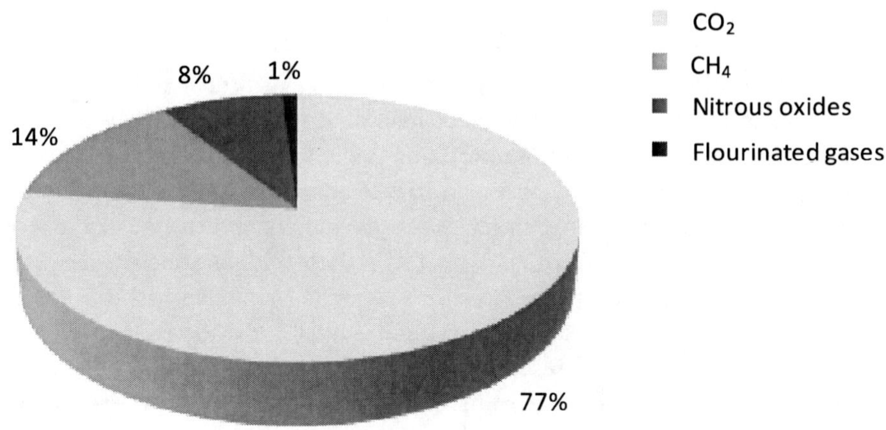

Figure 2. Global greenhouse gas emissions (in terms of CO_2 equivalent) [1-2].

Table 1. Typical composition of natural gas [2]

Component	Chemical formula	Composition
Methane	CH_4	70–90%
Ethane	C_2H_6	
Propane	C_3H_8	0–20%
Butane	C_4H_{10}	
Carbon dioxide	CO_2	0–8%
Oxygen	O_2	0–0.2%
Nitrogen	N_2	0–5%
Hydrogen sulfide	H_2S	0–5%
Rare gases	A, He, Xe, Ne	traces

Furthermore, although the changing rate is less than that of carbon dioxide, methane affects climate change at least as much as carbon dioxide due to its Global Warming Potential (GWP) of21. This means that every kilogram of methane emitted into the atmosphere has an equivalent forcing effect on the Earth's climate that is 21 times greater than that of carbon dioxide over a 100 year period. Taking this into account, methane is a good candidate to mitigate global warming, but also represents a challenge for research into its reuse and maximum exploitation, for example, as an energy source.

Methane is the main component of natural gas and probably the most abundant organic compound on Earth. Although natural gas is formed primarily from methane, it can also include ethane, propane and butane at levels up to 20%. The composition of natural gas can vary widely. Atypical composition of natural gas prior to refining is shown in Table 1.

In view of the fact that the world petroleum reserves are on decline, a great deal of emphasis has been placed on developing alternatives for energy production. Significant effort has been focused on finding effective processes and technology for the optimum utilization of the abundant natural gas for energy production as the supply of natural gas and coal-derived methane is expected to last for the foreseeable future. Furthermore, unlike other fossil fuels, natural gas is clean burning and emits lower levels of potentially harmful byproducts into the air [3].

Methane has the highest hydrogen to carbon ratio of all hydrocarbons (it contains four C–H bonds and only one carbon atom) and, as a consequence, combustion of methane yields the lowest amount of CO_2 per unit of energy produced. For instance, while the combustion of methane generates 890 kJ mol^{-1} of CO_2 produced, the corresponding values for n-decane and coal are 680 and 390 kJ mol^{-1}, respectively [4]. Moreover, the ratio of the heat of combustion (890 kJ mol^{-1}) to the molecular mass (16.0 g mol^{-1}, of which 12.0 g mol^{-1} is carbon) shows that methane, being the simplest hydrocarbon, produces more heat per mass unit (55.7 kJ g^{-1}) than other complex hydrocarbons [5]. Stricter regulations on CO_2 emissions to the atmosphere make natural gas even more attractive as a fuel for the purposes of energy production.

2. Catalytic Combustion of Methane

The overall reaction for methane combustion is given by the following equation:

$$CH_4 + 2O_2 \rightarrow CO_2 + 2H_2O \ \Delta H_{298} = 890 \text{ kJ mol}^{-1}$$

The strength of the C–H bond of the methane molecule is 435 kJ mol^{-1}. This dictates a very high overall activation energy for the homogeneous combustion of methane. On the other hand, the apparent activation energies for the dissociative adsorption of methane on certain crystal faces of transition metals have been reported to be as low as 29–41 kJ mol^{-1} [6, 7]. This large difference in the activation energies for homogeneous and heterogeneous dissociation of the C–H bond in the methane molecule suggests that catalysts can allow the combustion process to have a fast rate. Thus, the main advantage of the catalytic process in comparison to traditional homogenous combustion is the possibility, in principle, of obtaining complete oxidation of the fuel at low temperatures. The importance of this process has resulted in significant effort being applied to investigate a large range of catalytic systems for the combustion of methane. In the subsequent sections the main catalysts used in this reaction and their main applications are described in detail.

i) Catalysts for Methane Oxidation

The complete oxidation of methane can be performed over either noble metals or transition metal oxides. These two families of catalysts have been extensively studied over the last few decades with the aim of developing catalytic combustion applications [8-10].

Metal oxides, especially those belonging togroups 3–12 of the Periodic Table, have been studied as catalysts for methane oxidation. These oxides are characterized by high electron mobility and positive oxidation states. Among the various metal oxides, the perovskiteshavebeen studied extensively [9]. However, despite the fact that metal oxides are relatively inexpensive, they do suffer from disadvantages such us lower catalytic activity and higher light-off temperatures.

The majority of investigations directed towards methane combustion have involved noble metals. The main advantage of noble metal catalysts over metal oxides is their superior specific activity, which makes them the best candidates for the low-temperature combustion of hydrocarbons. This is particularly important in the case of methane, which is the most difficult hydrocarbon to oxidize. Among the noble metals, platinum and palladium have been widely considered due to their high volatility in comparison to the other noble metals. However, Pd has been widely reported as the catalyst of choice for methane oxidation applications [7, 10, 11].

The rate of methane oxidation on Pd catalysts strongly depends on its chemical state. Because oxygen adsorbs much more effectively than methane, the metal surface will be oxidized eitherunder an excess of oxygen in the gas mixture or at low conversions with fuel-rich mixtures [11]. It is commonly accepted that at low temperature the active phase is crystalline PdO, which may exist in more than one form depending on the size of the oxidized particles and on the nature of the support, but at high temperature metallic Pd can also be active for methane oxidation [7].

Another key issue is whether methane oxidation on Pd is structure-sensitive or structure-insensitive. Despite the fact that most research groups involved in this area have reported the methane oxidation reaction to be structure-sensitive [12], there are still some discrepancies [13]. The structure sensitivity of the methane oxidation reaction has been attributed to different reactivities of the adsorbed oxygen on the Pd surface. Oxidation of small crystallites of palladium produces PdO dispersed on the support, while the oxidation of large crystallites yieldsPdO dispersed on small crystallites of palladium. Palladium oxides on small crystallites of palladium are more active than Pd oxides on the support. Therefore, the large crystallites of palladium are more active than the small crystallites [10]. In contrast, Baldwin and Burch [14] reported that the particle sizes of the supported palladium catalysts did not have a clear effect on the activity inmethane oxidation once the rates were measured in terms of specific rate constants.

The phase transformation between the metallic and the oxidized states of the palladium catalysts is accompanied by changes in the particle size and morphology. However, not only phase transformation, but also the method of catalyst preparation, type of support, the precursor and the reaction conditions could affect the performance of the palladium catalyst in the methane oxidation reaction. In addition, the oxygen mobility of the support can have a marked effect on the oxidation/reduction behavior of palladium catalysts [7].

The kinetics and mechanism of the catalytic methane oxidation over Pd-based catalysts have been widely studied. There is general agreement that the reaction rate is close to zero order with respect to oxygen and has a first order dependence with respect to methane except when there is insufficient oxygen for the reaction [7]. However, several studies have been reportedin the literature on CH_4 combustion kinetics, but there is no consensus about the mechanism of this reaction. Thus, Eley–Rideal (ER), Langmuir–Hinshelwood (LH) and Mars van Krevelen (MvK) mechanisms have been proposed [15].

ii) Applications

The catalytic combustion of hydrocarbons is generally considered to be an effective method to generate power and reduce emissions. In particular, interest in the catalytic combustion of methane has increased considerably in recent years and the following main applications can be highlighted:

(1) As an alternative to the conventional thermal process in gas turbine combustors used for power generation.
(2) In compressed natural-gas vehicles (NGVs).

Regarding the first application, the use of a catalyst would minimize the emissions. Catalytic combustion has the potential to reduce the emissions of NO_x to almost zero and achieve ultra-low emissions of both CO and unburned hydrocarbons [3]. The presence of the catalyst means that the combustor can operate at higher air/fuel ratios than those with a flammable mixture. The maximum temperature attainable in the combustor can be controlled by varying the air/fuel ratio. This is a unique feature of the catalytic combustor, since without a catalyst a flame can only be sustained within a narrow air/fuel ratio range [5]. In conventional flame combustion systems, air enters the combustor at the compressor discharge temperature and, upon blending with fuel, the mixture is ignited and burned at temperatures in the range from 1700 to 2000 °C. At these temperatures, nitrogen and oxygen can react to form NO_x. Procedures to reduce NO_x formation, such as reducing the residence time, lead to an increase in both operating costs and emissions of CO and unburned hydrocarbons. However, it is well known that the emissions of nitrogen oxides can be minimized by reducing the average temperature at which the combustion takes place. Catalysis is therefore an alternative approach to avoid the aforementioned problems [16]. In a catalytic flameless combustor, fuel and air are thoroughly mixed before entering the catalyst. The fuel-air mixture is then ignited over the catalyst outside the flammability region at lower temperatures, i.e., below 1100 °C. At temperatures above 1500 °C, the thermal NO_x production is doubled for each 40 °C increase in temperature. Whereas a standard diffusion flame combustion turbine produces exhausts with NO_x concentrations higher than 150 ppm, a flameless catalytic combustor achieves concentrations inthe order of 3 ppm [17]. Furthermore, catalytic combustion reduces the risk of blow-out or instability and dynamics during combustion. The catalytic combustor also does not require the use of expensive clean-up systems and involves negligible efficiency loss compared to gas turbines with conventional flame combustion systems. Thus, catalytic combustion has been shown to be effective in producing energy in gas turbine combustors while reducing emissions.

The second main application of the catalytic total oxidation of methane concerns its use in natural gas combustion devices such as, for example, lean-burn natural gas vehicles (NGVs). Interest is growing in the expanded use of natural gas as an alternative to conventional liquid fuels in automotive internal combustion engines. In addition to the economical and/or political reasons, the use of NGVs offers significant environmental advantages over gasoline or diesel. Natural gas engines can operate under lean conditions, thus increasing their fuel efficiency and minimizing the typical products that arise from incomplete combustion, such as soot, CO and volatile organic compounds. Emissions from properly functioning natural gas vehicles are generally considered to be lower than those produced by vehicles powered by gasoline or diesel [18, 19].

3. Methane Combustion by the Electrochemical Promotion of Catalysis Phenomenon

Since the discovery of the catalytic phenomenon by Berzelius in 1835, many discoveries and advances have been made in the field of heterogeneous catalysis. One of the latest is the phenomenon of Electrochemical Promotion of Catalysis (EPOC), also known as the NEMCA effect (Non-Faradaic Electrochemical Modification of Catalytic Activity), which was discovered and first reported by Stoukides and Vayenas in 1981 [20].

The EPOC phenomenon is an innovative concept that is a tool to improve the behavior of a heterogeneous catalyst by electrochemistry. This effect can be used to modify both the activity and selectivity of catalyst electrode films supported on a solid electrolyte in a very pronounced and controllable manner. The process is based on the control, by applied current or potential, of the work function due to the electrochemical pumping of ions between the solid electrolyte and the catalytic surface deposited on it [21]. The rate and the selectivity can be reversibly and predictably modified by externally applying voltages (± 2.5 V) or currents ($\pm\ 1–10^4\ \mu A$) between a catalyst and a counter electrode [21]. Since its discovery, several research groups throughout the world have studied this effect and it has been widely established that EPOC is a general phenomenon that combines different scientific disciplines such as catalysis, electrochemistry and surface science.

The origin and mechanism have been found to arise from the electrochemically controlled transport of ionic species, which act as promoters, from the solid electrolyte to the metal-gas interface. This electrochemically assisted migration of promoter species together with their image charge in the catalyst create an overall neutral double layer and affect the chemisorptive bond strength of reactants and intermediates, thus influencing the catalytic activity and selectivity in a dramatic manner. Hence, it can be defined as a special type of heterogeneous catalysis in which catalyst properties are deliberately modified by applying an electrical field.

The effect of one type of polarization or another will have a positive or negative effect on the overall kinetics of the reaction process depending on whether the adsorbates involved in the reaction are electronegative or electropositive or the promoter ions are either cations (electropositive promoters) or anions (electronegative promoters) [21].

The magnitude of the Electrochemical Promotion phenomenon for different catalytic systems is commonly described by the rate enhancement ratio, ρ, defined by [21]:

$$\rho = r/r^{o}$$

and the Faradaic efficiency, Λ, defined by:

$$\Lambda = \Delta r/(I/2F)$$

where r and r^0 are the electrochemically promoted and the open-circuit (unpromoted) catalytic rate values, Δr (= r –r^o) is the electrochemically induced change in catalytic rate, I is the applied current and F is Faraday's constant.

Since its discovery, the electrochemical promotion of catalysis phenomenon has been employed to improve the catalytic performance of several processes of industrial and environmental interest. One of the processes that hasbeen investigated is the catalytic combustion of light hydrocarbons, including methane. The reported studies on the electrochemical promotion of methane combustion are summarized in Table 2. It can be seen that only noble metals have been used as catalysts. On the other side, palladium has been selected in most cases due to its high catalytic activity in this reaction. The first two studies were carried out in double chamber reactors and at temperatures above 600 °C. In these studies, Pt films prepared by paste deposition were used as the catalyst-working electrode. Tsiakaras and Vayenas investigated the electrochemical promotion of methane oxidation on Pt deposited on YSZ [22]. The catalytic rate was enhanced by both supplying and removing O^{2-} to or from the Pt catalyst, i.e. the system exhibited volcano behavior [21]. Due to the high operating temperatures, the observed rate enhancements were restricted and the measured Λ values were quite low (up to 5) whereas ρ values were found to be up to 70.

Table 2. NEMCA studies for the methane combustion reaction

Catalyst	Solid Electrolyte	Reactor configuration	Temp (°C)	Ref.
Pt	YSZ	Double chamber	600–750	[22]
Pt	$SrCe_{0.92}Dy_{0.08}O_{3-\alpha}$	Double chamber	600–645	[23]
Pd	YSZ	Single chamber	400–420	[24]
Pd	YSZ	Single chamber	400	[25]
Pd	YSZ	Single chamber	500–600	[26, 27]
Rh	YSZ	Single chamber	350–550	[28]
Pd	YSZ	Single chamber	300–600	[29-36]

Moreover, Belyaev et al. [23] reported studies on the complete oxidation of methane by electrochemical promotion but in this case over a Pt catalyst deposited on a solid proton-conducting electrolyte ($SrCe_{0.92}Dy_{0.08}O_{3-\alpha}$). The induced change in the reaction rate at anodic polarizations was over two orders of magnitude higher than the rate of hydrogen pumping from the solid electrolyte to the catalyst film. However, cathodic polarizations did not affect the catalytic reaction rate, meaning that the system exhibited electrophobic behavior.

Pd catalysts prepared by paste deposition on YSZ were later investigated by Giannikos et al. [37] and Frantzis et al. [25]. They used, for the first time, single chamber reactors and operated at lower temperatures (400–420 °C). These studies were performed only under near stoichiometric or slightly reducing conditions. Electrophobic behavior was observed and Faradaic efficiency values of around 100–150 and catalytic rate increases up to 70–90 times under anodic polarization were reported. Frantzis et al. [25] also observed slightly electrophilic behavior but, in this case, the catalytic rate was enhanced by up to a factor of 10 upon cathodic polarizations.

Roche et al. [26] studied the electrochemical promotion of methane oxidation on sputtered Pd thin films deposited on YSZ. This study was carried out exclusively under fuel-lean conditions. Only at relatively high temperatures of 500 to 600 °C, they found a catalytic rate increase up to 3 and Faradaic efficiency values around 260 upon positive overpotentials, i.e. pure electrophobic behavior was observed. These authors also used YSZ honeycomb monoliths as supports [27]. Only inhibiting non-Faradaic effects were observed under anodic and cathodic polarizations for methane oxidation, with a maximum Faradaic efficiency of 47 (at –2 V). The magnitude of these effects was weak and this resulted in a slight conversion decrease.

Recently, the electrochemical promotion of methane oxidation has been investigated over Rh films prepared by paste deposition and interfaced with YSZ at temperatures from 350 to 550 °C [28]. The NEMCA effect on the catalytic rate was found to decrease on increasing the partial pressure of oxygen and temperature. The maximum effect was observed under reducing or stoichiometric conditions. Under reducing conditions at 430 °C, positive current application caused a three-fold increase in the catalytic rate, while the apparent Faradaic efficiency was 170 and, therefore, electrophobic behavior was exhibited.

In recent years, our group has carried out an in-depth investigation on the EPOC phenomenon in the methane combustion reaction at temperatures between 250 and 600 °C using Pd-based catalyst-electrodes deposited on YSZ discs [29-36]. Firstly, some preliminary studies were carried out in which Pd films were prepared by either organometallic paste deposition or impregnation and these were deposited on either YSZ or CeO_2/YSZ. The aim was to assess the feasibility of the electrochemical promotion of methane oxidation over these palladium films. In all cases, a thermal pre-treatment was necessary in order to activate the catalysts. During this treatment a 'negative activation' was observed and the activity strongly increased upon the formation of PdO. However, all catalysts were rapidly deactivated. Thus, it was only possible to electropromote the catalysts at high temperatures and with the presence of a ceria interlayer. A galvanostatic NEMCA transient is shown in Figure 3. It depicts the response of the reaction rate of methane consumption and the catalyst potential on applying a constant positive current (I = +5 mA) between the Pd film and the counter electrode at 470 °C.

For the first hour, the electrochemical cell was under open circuit conditions (occ) in order to stabilize the catalyst. The corresponding steady-state catalytic rate, r^0, was found to be 1.45×10^{-8}mol O/s. The catalytic reaction rate under anodic polarization, where oxygen ions were transferred to the catalyst at a rate of $I/2F = 2.6 \times 10^{-8}$mol O/s, was found to be more than two times higher than that registered under open circuit potential ($\rho = 2.1$).

One of the main goals of these systems was to achieve electrochemically promoted methane oxidation at low temperatures. As a result, further studies on the pre-treatment of these catalysts were carried out in order to activate and stabilize them at temperatures below 400 °C. It was revealed that a pre-treatment under oxidizing ethylene/methane gas mixtures led to the enhancement and stabilization of the methane consumption rate.

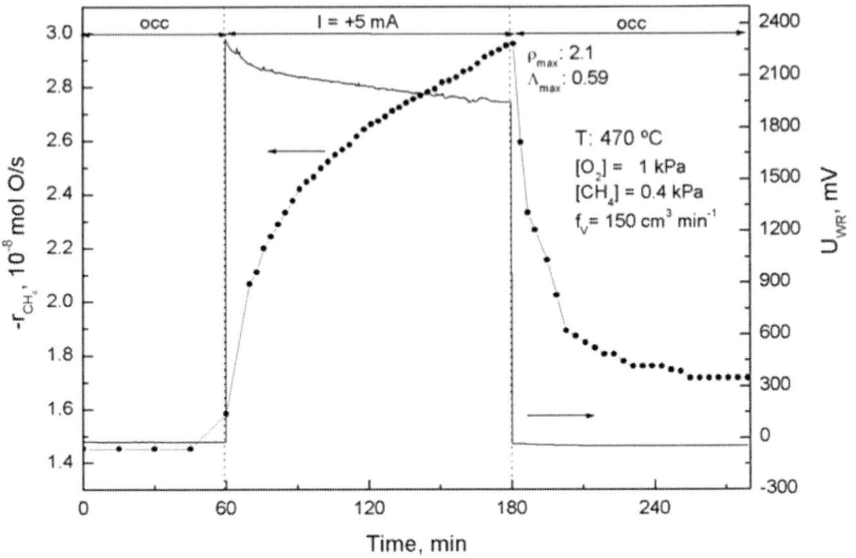

Figure 3. Transient response of the catalytic CH_4 consumption rate and of the catalyst potential upon application of I = +5 mA. Catalyst: impregnated Pd/CeO_2/YSZ/Au.

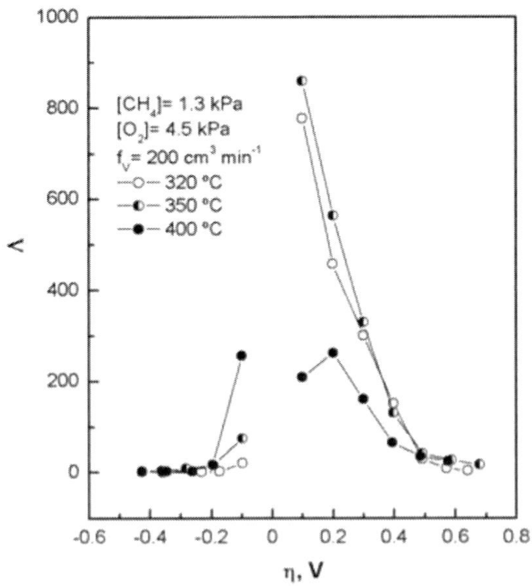

Figure 4. Effect of the applied overpotential (η) and temperature on the Faradaic efficiency (Λ). Catalyst: impregnated Pd/YSZ/Au.

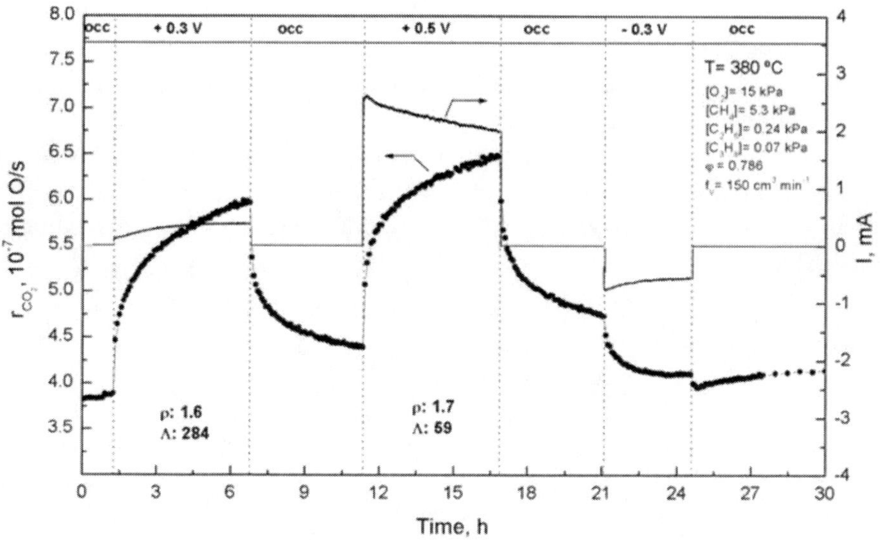

Figure 5. Effect of step changes in the catalyst potential on the CO_2 formation rate and the current through the catalyst (Reprinted with permission from Elservier [36]).

It was also found that the slight reduction of PdO achieved by the pre-treatment improved the catalytic activity towards methane oxidation in comparison to either fully oxidized or reduced catalysts.

Thus, Pd impregnated directly on YSZ was suitable for electropromotion at low temperatures and enhancements in the reaction rate of up to 120% and Faradaic efficiencies of up to 800 were obtained. As an example, the effect of the applied overpotential (η) and temperature on the Faradaic efficiency (Λ) for the catalyst-impregnated Pd/YSZ/Au is shown in Figure 4.

After studying and optimizing the electrochemical promotion of methane oxidation on Pd-based catalyst-electrodes deposited on YSZ, the electrochemically promoted combustion of natural gas over impregnated palladium catalyst-electrodes was investigated. It was found that positive overpotentials, leading to O^{2-} supply to the catalyst, caused a marked increase in the CO_2 formation rate at low temperatures.

The transient effect of constant overpotentials at 380 °C on both the overall CO_2 formation rate and the current through the electrochemical cell is shown in Figure 5.

An electrophobic behavior with respect to the CO_2 formation rate was detected. In this case, positive potentials of 0.3 and 0.5 V led to an increase in the reaction rate of 60 and 70%, respectively, and Faradaic efficiencies of 284 and 59, respectively. Interestingly, it can be observed that a steady-state for the catalytic rate was not attained even after prolonged polarization times. This can be explained by the progressive palladium oxidation during the anodic polarization.

Besides, it can be observed that once the polarization ceased, the value of the catalytic rate returned to that obtained under open circuit conditions. This finding highlights the reversibility of the NEMCA effect on the combustion of natural gas.

CONCLUSION

Methane represents a challenge for research in terms of its reuse and maximum exploitation, for example as an energy source. For this reason, the methane combustion process is being widely investigated. The catalytic process, when compared to traditional homogenous combustion, presents some advantages such as the possibility, in principle, of obtaining complete oxidation of the fuel at low temperatures. In this sense, the electrochemical promotion of catalysis phenomenon may improve the catalytic performance of this process.

This phenomenon is based on the control, by applied current or potential, of the work function due to the electrochemical pumping of ions between the solid electrolyte and the catalytic surface deposited on it. Studies on the electrochemical promotion of methane combustion have shown that noble metals, mainly palladium, used as catalysts/electrodes improve the catalytic activity of this reaction with enhancements in the rate ratios up to 70.

Recently, our group has developed suitable Pd catalyst-electrodes, prepared by an impregnation technique or by the addition of a ceria or porous YSZ interlayer, for the electrochemically promoted oxidation of methane and natural gas combustion. Enhancements in the reaction rate up to 120% and Faradaic efficiencies up to 800 were obtained at temperatures as low as 320 °C by the application of anodic catalyst overpotentials. Moreover, natural gas combustion was electrochemically promoted over impregnated Pd catalyst-electrodes deposited on dense YSZ. In this case, rate enhancements higher than 250% for CO_2 formation were obtained upon positive polarizations at temperatures as low as 340 °C.

REFERENCES

[1] United States Environmental Protection Agency (EPA): http://www.epa.gov/climatechange/ ghgemissions/gases.html.

[2] The Natural Gas Supply: Association http://www.naturalgas.org.

[3] Forzatti, P. *Catal Today,* 2003, 83, 3-18.

[4] Thevenin, P. O.; Pocoroba, E.; Pettersson, L. J.; Karhu, H.; Väyrynen, I. J.; Järås, S. G. *J. Catal.,* 2002, 207, 139-149.

[5] Chin, Y.-H.; Resasco, D. E., Catalytic Oxidation of methane on supported palladium under lean conditions: Kinetics, structure and properties. In Catalysis, The Royal Society of Chemistry: 1999; Vol. 14, pp 1-39.

[6] Yarw-Nan, W.; Herman, R. G.; Klier, K. Surf Sci.1992, 279, 33-48.

[7] Ciuparu, D.; Lyubovsky, M. R.; Altman, E.; Pfefferle, L. D.; Datye, A. *Catal. Rev.,* 2002, 44, 593-649.

[8] Gélin, P.; Primet, M. *Appl. Catal. B-Environ.,* 2002, 39, 1-37.

[9] Choudhary, T. V.; Banerjee, S.; Choudhary, V. R. *Appl. Catal. A-Gen.,* 2002, 234, 1-23.

[10] Lee, J. H.; Trimm, D. L. *Fuel Process Technol.,* 1995, 42, 339-359.

[11] Burch, R.; Crittle, D. J.; Hayes, M. *J. Catal. Today,* 1999, 47, 229-234.

[12] Hicks, R. F.; Qi, H.; Young, M. L.; Lee, R. G. *J. Catal.,* 1990, 122, 280-294.

[13] Ribeiro, F. H.; Chow, M.; Dallabetta, R. A. J Catal.1994, 146, 537-544.

[14] Baldwin, T. R.; Burch, R. *Appl. Catal.,* 1990, 66, 359-381.

[15] Specchia, S.; Conti, F.; Specchia, V. *Ind. Eng. Chem. Res.,* 2010, 49, 11101-11111.

[16] Ismagilov, Z. R.; Kerzhentsev, M. A. *Catal. Rev.,* 1990, 32, 51-103.

[17] Pfefferle, L. D.; Pfefferle, W. C. *Catal. Rev.,* 1987, 29, 219-267.

[18] Karavalakis, G.; Durbin, T. D.; Villela, M.; Miller, J. W. *J. Natural. Gas Sci. Eng.,* 2012, 4, 8-16.

[19] Gélin, P.; Urfels, L.; Primet, M.; Tena, E. Catal Today.2003, 83, 45-57.

[20] Stoukides, M.; Vayenas, C. G. *J. Catalysis,* 1981, 70, 137-146.

[21] Vayenas, C. G.; Bebelis, S.; Pliangos, C.; Brosda, S.; Tsiplakidis, D., Electrochemical Activation of Catalysis: Promotion, Electrochemical Promotion and Metal-Support Interactions. Kluwer Academic / Plenum Publisher: New York, 2001.

[22] Tsiakaras, P.; Vayenas, C. G. *J. Catal.,* 1993, 140, 53-70.

[23] Belyaev, V. D.; Gal'vita, V. V.; Gorelov, V. P.; Sobyanin, V. A. *Catal. Lett.,* 1994, 30, 151-158.

[24] Giannikos, A.; Petrolekas, P.; Pliangos, C.; Frenzel, A.; Vayenas, C. G.; Pütter, H. Ionics.1998, 4, 161-169.
[25] Frantzis, A. D.; Bebelis, S.; Vayenas, C. G. *Solid. State Ionics.*, 2000, 136-137, 863-872.
[26] Roche, V.; Siebert, E.; Steil, M. C.; Deloume, J. P.; Roux, C.; Pagnier, T.; Revel, R.; Vernoux, *P. Ionics,* 2008, 14, 235-241.
[27] Roche, V.; Revel, R.; Vernoux, P. Catal Commun.2010, 11, 1076-1080.
[28] Nakos, A.; Souentie, S.; Katsaounis, A. *Appl. Catal. B-Environ.,* 2010, 101, 31-37.
[29] Jiménez-Borja, C.; Dorado, F.; de Lucas-Consuegra, A.; García-Vargas, J. M.; Valverde, J. L. *Catal. Today*, 2009, 146, 326-329.
[30] Jiménez-Borja, C.; Dorado, F.; De L.-Consuegra, A.; G.-Vargas, J. M.; Valverde, J. L. *Fuel Cells,* 2011, 11, 131-139.
[31] Jiménez-Borja, C.; Brosda, S.; Makri, M.; Sapountzi, F.; Dorado, F.; Valverde, J. L.; Vayenas, C. G. *Solid State Ionics,* 2012, 225, 376-381.
[32] Jiménez-Borja, C.; Brosda, S.; Matei, F.; Makri, M.; Delgado, B.; Sapountzi, F.; Ciuparu, D.; Dorado, F.; Valverde, J. L.; Vayenas, C. G. *Appl. Catal. B-Environ.,* 2012, 128, 48-54.
[33] Matei, F.; Ciuparu, D.; Jiménez-Borja, C.; Dorado, F.; Valverde, J. L.; Brosda, S. *Appl. Catal. B- Environ.,* 2012, 127, 18-27.
[34] Jiménez-Borja, C.; Delgado, B.; Díaz-Díaz, L. F.; Valverde, J. L.; Dorado, F. *Electrochem. Commun.,* 2012, 23, 9-12.
[35] Matei, F.; Jiménez-Borja, C.; Canales-Vázquez, J.; Brosda, S.; Dorado, F.; Valverde, J. L.; Ciuparu, D. *Appl. Catal. B: Environ.,* 2013, 132, 80-89.
[36] Jiménez-Borja, C.; Delgado, B.; Dorado, F.; Valverde, J. L. *Chem. Eng. J.,* 2013, 225, 315-322.

In: Combustion
Editor: Joseph M. Grier

ISBN: 978-1-62948-967-4

Chapter 7

NOVEL NITRATE-FREE ACETATE-H_2O_2 COMBUSTION SYNTHESIS

Narendar Nasani and Duncan P. Fagg*

Nanotechnology Research Division, Centre for Mechanical Technology and Automation, Department of Mechanical Engineering, University of Aveiro, Aveiro, Portugal

ABSTRACT

One of the key challenges in materials science research is the synthesis of ceramic materials with the desired phase purity, morphology, particle size and expected properties for defined applications, while controlling cost. The selection of the synthesis method is vital to regulate the composition, structure, and morphology of a preferred material. Ceramic oxide materials are typically prepared by solid-state reactions, for example simple mechanical mixing of ceramic oxide powders, or by precipitation from solution and following decomposition. Nonetheless, the solid-state route can often lead to the appearance of non-homogeneity in the final products and low phase purity that can deteriorate the desired properties.

A novel nitrate-free acetate-H_2O_2 combustion synthesis method has recently been developed that has been demonstrated for the successful preparation of Ni-$BaZr_{0.85}Y_{0.15}O_{3-\delta}$ cermet anodes and Ba(Ce,Zr,Y)$O_{3-\delta}$

* Corresponding author: Duncan P. Fagg. E-mail: duncan@ua.pt; Fax: +351-234-370953; Tel.: +351-234-370830.

electrolytes for Proton Ceramic fuel cell (PCFC) applications. Metal acetates (M-($OCOCH_3$)x yH2O) and 30% H_2O_2 are employed as starting precursors for this combustion reaction, using microwave heating. The new route contrasts dramatically to traditional nitrate-based combustion processes in that the metal precursors now constitute the fuel rather than the oxidant in the combustion reaction. Another significant distinction of the novel route is that the acetate precursor solution is neutral in contrast to typical nitrate-based combustion routes that involve acidic solutions. This factor has been shown to be highly important for the successful preparation of cermet anodes such as Ni-$BaZr_{0.85}Y_{0.15}O_{3-\delta}$ that contain alkaline earth metals. The novel combustion route produces powders with a fine crystallite size in the nanometric scale. The presence of an intermediate peroxy (-OOH) bond has been identified and is suggested to aid the combustion reaction by harnessing the required oxidant in the form of labile peroxide complexes. This soft chemical method offers homogeneous distribution and stabilization of the metal ions in the solution state and can be applied for the formation of further multi-element ceramic oxide materials, offering economical advantages over more classical nitrate-based combustion routes, as well as significant environmental benefits due to the avoidance of releasing NO_x gases.

Keywords: Combustion, ceramic oxides, proton ceramic fuel cells

1. INTRODUCTION

Combustion synthesis (CS) is an intriguing approach to produce a comprehensive range of advanced materials including powders of ceramic oxides, intermetallics, metal alloys and composite materials that can be applicable in solid oxide fuel cells (SOFC), high temperature semi conductors, gas sensors, aerospace and turbine blade materials, hydrogen storage and catalytic reactions [1]. The combustion technique was discovered by Merzhanov and Borovinskaya in 1972 (in the beginning of 1970's) in the former Soviet Union and the progress of this technique has shown its uniqueness in terms of the production of novel materials with desired properties. Combustion is a type of reaction that occurs between a fuel and an oxidant by the production of a massive amount of heat (exothermic reaction, ΔH = -ve). These types of combustion reactions can associate metals and non-metals to produce a wide variety of novel advanced materials for specific application with unique properties. Combustion is also known as self-propagating high temperature synthesis (SHS) [2]. The well known commonly used oxidants are water soluble metal nitrates, while the most usual fuels are glycine, urea and hydrazine related compounds. The stoichiometric ratio of the precursors are vital in the combustion process [3, 4]. The appropriate amount of fuel and oxidant can be calculated according to propellant chemistry [5, 6].

All solution combustion reactions are redox type reactions involving concurrent oxidation i.e. addition of oxygen or electronegative element (non-metal) and a reduction processes i.e. addition of hydrogen or electropositive element (metal). The combustion occurs in the gas-phase (flame type), solid-gas and explosive type of reaction. In fact, the combustion reaction can occur either in a linear or volume combustion manner. Linear combustion occurs slowly by burning the gel from top to bottom and requires a longer time to ignite whereas voluminous combustion occurs by burning the whole reaction mixture with a flame. A self propagating combustion reaction should evolve more heat than the heat required for initiating the combustion [7]. The combustion technique has been explored to use in materials science firstly by Merzhanov and then further improved by KC Patil and ST Aruna. The most widely used combustion reactions in materials science are the traditional nitrate-glycine or urea combustion syntheses [8]. However the nitrate precursors, used in these processes are harmful to the environment, releasing copious amount of NO_x gases. This factor led the current authors to introduce a new type of combustion reaction that is nitrate-free and more environmental friendly, by considering metal acetate as the fuel and 30% H_2O_2 as the oxidant. The new nitrate-free combustion route is very rapid, cost effective, self-propagating, and environmental friendly, while its importance and flexibility has already been demonstrated in ceramic material production for fuel cell applications [9, 10].

2. Novel Nitrate-Free Acetate-H_2O_2 Combustion Synthesis

Nowadays, numerous synthesis routes are available in research for the synthesis of ceramic oxide materials. Some examples include solid-state, glycine-nitrate combustion, sol-gel and pechini methods etc. [11–14]. Each of the aforementioned methods has their own advantages and disadvantages. Despite its simplicity, several researchers have reported that the solid state route often requires multiple grinding and annealing steps to obtain pure ceramic phases, commonly leading to inhomogeneities in the final materials.

For this reason, the solid state synthesis is not suggestible for the synthesis of pure ceramic oxides, especially when these materials are highly refractory. In contrast, solution-based soft chemical synthesis processes show some advantages, such as offering a higher degree of homogeneity by the uniform mixing of each component at the atomic-level. In addition, the oxidation state of the metal cations can be stabilized in the solution state. Nonetheless, all the available soft chemical synthesis routes, glycine-nitrate combustion, sol-gel, pechini etc., conventionally require metal nitrate precursors leading to the production of polluting NO_x gases. An additional drawback is that many of these solution-based synthesis routes often require expensive chemical precursors.

For this reason a novel acetate-H_2O_2 combustion route has been developed to synthesize homogenous ceramic oxide materials from low cost acetate precursors and without harmful NOx emissions. As such, the new combustion route is suggested as an attractive alternative to potentially replace typical nitrate-based combustion for the synthesis of numerous ceramic oxide materials. A major difference between the traditional glycine-nitrate and the novel acetate-H_2O_2 combustion route is that the metallic precursors represent the fuel in the new process (metal acetates), whereas they represent the oxidant (metal nitrates) in traditional nitrate-based combustion processes.

Additionally, an important advantage of the new acetate-H_2O_2 combustion is that the pH of solution formed by acetate-H_2O_2 precursors is neutral and, therefore, does not require the addition of any further base to increase the pH of the solution. In contrast, the nitrate-glycine combustion process forms metal chelates with the corresponding metal cations with amino and carboxylic group present in glycine, forming an acidic resultant solution [8].

Consequently, in the traditional nitrate-glycine process additional amounts of base (NH_4OH) are commonly added, to attempt to counter this factor and increase the pH of the final solution. Unfortunately, these additions have the negative consequence of further increasing undesired NOx emissions. Stoichiometric amounts of the metal acetates (fuel) and 30 % H_2O_2 (oxidant) are calculated according to propellant chemistry, expressed in equation 1 as.

$$\Phi e = \frac{\sum(Valency\ of\ element\ in\ fuel)\times(Stoichiometry)}{(-1)\sum(Valency\ of\ element\ in\ oxidant)\times(Stoichiometry)} \quad (1)$$

The propellant chemistry, therefore, indicates the fuel/oxidant molar ratio (F/O) required for a stoichiometric mixture ($\Phi e = 1$) by summing the product of the combustion valency of each element and their stoichiometry in the fuel component(s) and dividing by the equivalent sum for the oxidizer component(s). According to the combustion valencies, oxygen (O, -2) is an oxidizing element; carbon (C, +4), hydrogen (H, +1), and all metal cations are reducing elements and nitrogen (N, 0) is neutral element [1, 5]. Thus, in the novel combustion process, H_2O_2, with an overall combustion valence of -2, is an oxidant for the combustion reaction, whereas the metal acetates, which have overall positive combustion valences, correspond to fuels (e.g. Ni $(OCOCH_3)_2$ with an overall combustion valence of 16).

Thus, the novel acetate-H_2O_2 combustion route is in distinct contrast to traditional nitrate-based combustion reactions in which the metal nitrates correspond to the oxidant, with negative overall combustion valences, and glycine or urea etc., with positive combustion valences, correspond to the fuels. The progression of combustion reaction depends on its fuel to oxidant ratio, where $\Phi e = 1$, is considered as stoichiometric, $\Phi e > 1$ is fuel rich and $\Phi e < 1$ is fuel lean.

In the acetate-H_2O_2 combustion process described below the precursor gels were subjected to microwave heating, with constant rotation, in a domestic 2.45 GHz, 800W microwave oven and set at the maximum power. The microwave assisted combustion technique has previously been suggested to offer a homogenous reaction throughout the whole volume of the gel, in contrast with simple hot-plate reactions that typically initiate combustion at the base of the gel in contact with the hotplate [4, 15].

2.1. Synthesis of Multi Element Ceramic Oxides

To demonstrate the new, simple and environmentally friendly acetate-H_2O_2 combustion synthesis route for the synthesis of multi-element ceramic oxide materials, it has been applied for the preparation of $Ba(Ce,Zr,Y)O_{3-\delta}$ proton ceramic fuel cell electrolyte membranes. $Ba(CH_3COO)_2$ (>99.9% pure, Sigma Aldrich), Zr(IV) $(CH_3COO)_x$ $(OH)_y$ (99.9% pure, Sigma Aldrich), Ce $(CH_3COO)_3$ $.1.5H_2O$ (99.9% pure, Alfa Aesar), $Y(CH_3COO)_3$ $.4H_2O$ (99.9% pure, Alfa Aesar) and 30% H_2O_2 (Riedelde Haen, 30% by weight) were used as starting materials for the combustion synthesis, Figure 1. Stoichiometric amounts of the metal acetates were dissolved in distilled water with constant stirring at 25 °C to obtain a clear, transparent solution. To the solution 30 % H_2O_2 was added slowly to achieve a fuel to oxidant ratio of unity as per the propellant chemistry, eq.1. An orange-brown color solution with bubbles was observed for cerium containing compositions as shown in Figure 2.

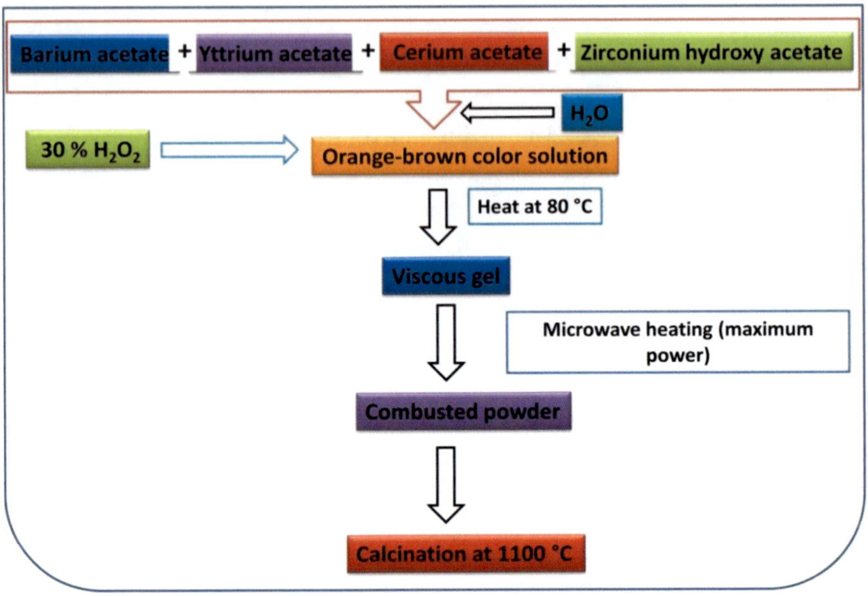

Figure 1. A schematic representation of acetate–H_2O_2 combustion synthesis for BCZY ceramic oxides.

The solution was heated on a hot plate at 80 °C with constant stirring to form a viscous gel. During heating the orange-brown color converts to pale yellow. The gel is then subjected to microwave heating, under rotation, in a domestic 2.45 GHz, 800W microwave oven set at the maximum power.

After few minutes, the dried gel burnt with a flame in a self-propagating combustion manner, releasing plentiful fumes, to form a black powder (for cerium containing compounds). The combustion powder was calcined at 1100 °C for 6h in order to obtain the pure perovskite phase. The progression of the new acetate-H_2O_2 combustion route is summarized in Figure 2.

2.1.1. The Mechanism of Acetate-H_2O_2 Combustion Synthesis and the Role of (H_2O_2)

Generally, H_2O_2 serves as an oxidizing agent with the tendency to raise the oxidation state of the precursors in any chemical reaction. It also can offer some interesting additional effects on the mechanisms of reaction and the morphology of the resultant products.

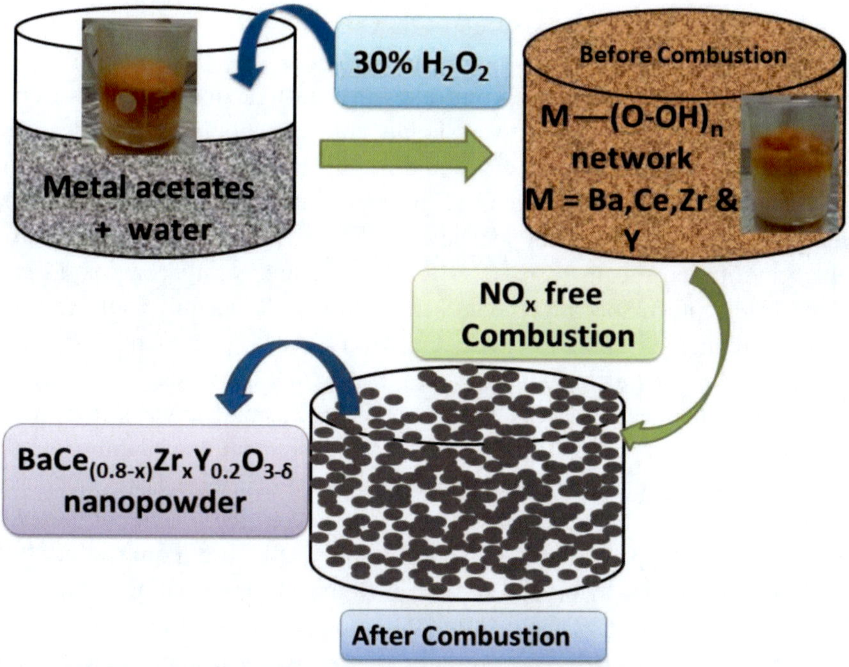

Figure 2. Progression of nitrate free acetate-H_2O_2 combustion reaction.

For example, the creation of nanocrystallites has often been shown to be promoted during the crystallization process, by the presence of intermediate peroxo (M-O-O-M) ligands that prevent the formation of compact M-O-M networks [16-18].

Profiting from these features, several researchers have successfully used H_2O_2 in the preparation of nanoscale ceramic materials such as CeO_2 [16], tetragonal crystalline ZrO_2 [17] and doped ceria, $Ce_{0.8}Sm_{0.2}O_{(1.9-x)}$ [18]. The role of hydrogen peroxide in the new acetate-H_2O_2 combustion synthesis of BCZY powders is principally that of an oxidant in the propellant chemistry. However, additional mechanistic information may be provided by comparison of colour changes to literature data on H_2O_2 preparations of ceria-containing materials. The observance of an orange precipitate in the precursor solution is commonly stated to be characteristic of Ce (+4) hydroxyl peroxo species, Figure 2, while the yellow colour of the gel is typically suggested to be characteristic of the presence of CeO_2 [16].

These observations would suggest that the presence of H_2O_2 in these precursor solutions is leading to a twofold action, i) to oxidize Ce^{3+} to the more easily hydrolysable Ce^{4+} oxidation state

$$Ce^{3+} + H_2O_2 + 2H^+ (aq.) \rightarrow Ce^{4+} + 2H_2O \quad (2)$$

and ii) to form labile peroxide complexes that aid reaction in comparison to simple hydroxide precursors. One should note that both these mechanisms involve the storage of oxygen that could be subsequently liberated during the combustion reaction. For this reason, the presence of peroxide complexes has been investigated in the novel acetate-H_2O_2 combustion process using Fourier transform infrared spectroscopy (FT-IR). The presence of hydroperoxyl ligand –OOH in the precursor gel of $BaCe_{0.4}Zr_{0.4}Y_{0.2}O_{3-\delta}$ is unequivocally revealed by the stretching vibration observed at nearly 1400 cm^{-1} (represented, in Figure 3 by dotted line and arrow), thereby, confirming the participation of peroxide complexes in the acetate-H_2O_2 combustion mechanism. The weak intensity of the –OOH band is probably associated with aforementioned lability of this bond. In this respect, note that Pappeccena et al. had used H_2O_2 in the co-precipitation synthesis of $Ce_{0.8}Sm_{0.2}0_{1.9-x}$, however, these authors failed to identify the –OOH peak due to its short lifetime [18]. After combustion, the –OOH peak is shown to be absent and only the –OH stretching vibration at 3400 cm^{-1} and the –C=O stretching vibration at nearly 1450 cm^{-1} can be identified, Figure 3 [10]. The presence of the –C=O can be related to the formation of some $BaCO_3$ during combustion.

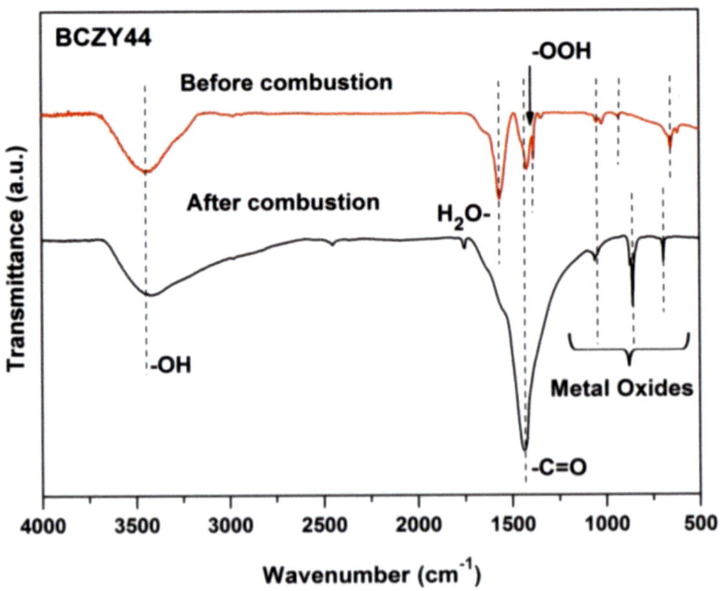

Figure 3. FT-IR spectra of BCZY44 powder before and after combustion.

Thus, it appears that the presence of H_2O_2 in these precursor solutions may store oxygen both in the form of labile peroxide complexes and also by oxidation of cerium (III) to the cerium (IV) oxidation state. Both these phenomenon are, therefore, typical effects that are expected on addition of H_2O_2 to such precursor solutions. Upon combustion, this stored oxygen would be liberated and is suggested to provide the required oxidant for the acetate-H_2O_2 combustion reaction.

2.1.2. Characterization of BCZY Ceramic Oxides

X-ray diffraction analysis (XRD) confirms the pure perovskite phase formation of $BaCe_{0.8-x}Zr_xY_{0.2}O_{3-\delta}$ (BCZY) ceramic oxide powders from acetate-H_2O_2 combustion synthesis, calcined at 1250 °C, Figure 4.

For the comparison, a traditional glycine/nitrate synthesis, performed at a fuel/oxidant ratio of unity with an identical microwave assisted combustion technique as described for the acetate process, also succeeded to achieve the formation of pure perovskite phases, Figure 5.

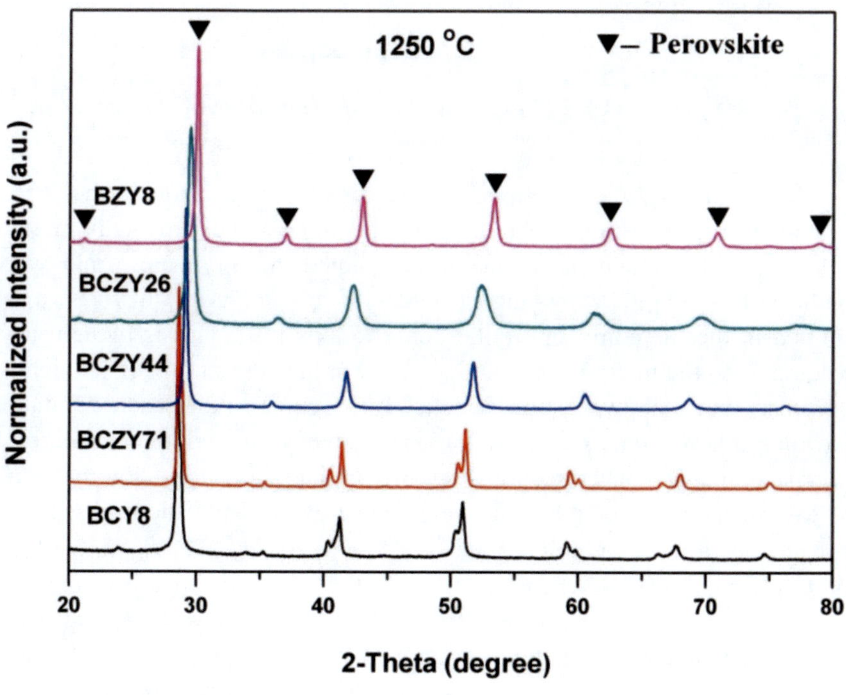

Figure 4. XRD patterns of $BaCe_{8-x}Zr_xY_{0.2}O_{3-\delta}$ (BCZY) ($x = 0$, 0.1, 0.4, 0.6 and 0.8) powders calcined at 1250 °C.

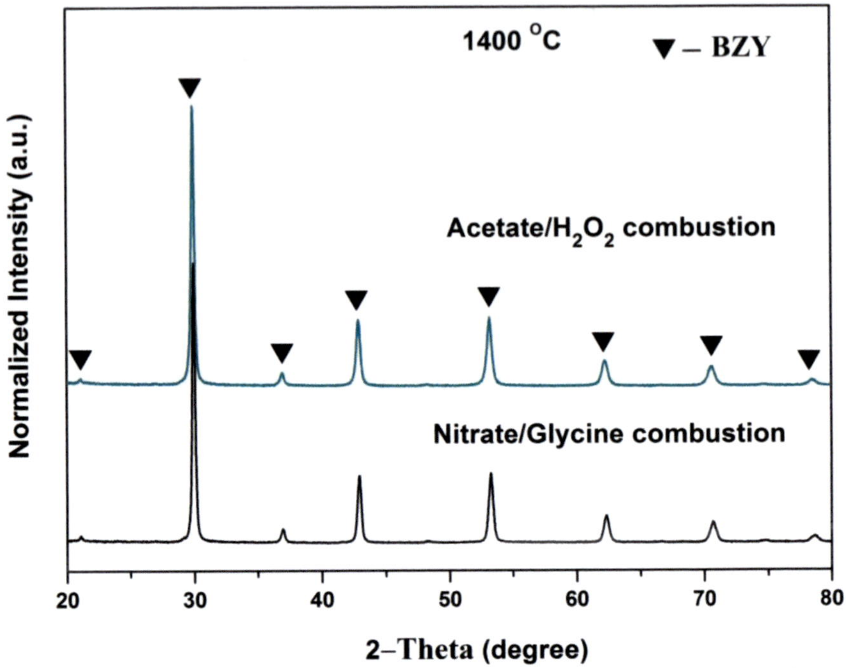

Figure 5. XRD patterns of BZY8 synthesized by both new acetate-H_2O_2 and nitrate-glycine combustion powder calcined at 1400 °C.

This highlights the potential for the direct replacement of the typical nitrate-based combustion techniques by the more environmentally friendly acetate-H_2O_2 technique. It should be also noted that both these microwave combustion synthesis show higher levels of phase purity than previously reported in the literature for similar compositions [19]. This is thought to be attributable to the microwave technique used in the current process, which has previously been shown to be beneficial for combustion reactions due to providing a homogenous reaction throughout the whole volume of the gel. In contrast simple hot-plate reactions typically initiate the combustion reaction at the base of the gel in contact with the hotplate, with combustion subsequently propagating from this ignition point [4, 15]. All $BaCe_{0.8-x}Zr_xY_{0.2}O_{3-\delta}$ (BCZY) compositions show distinct peak shifts for the perovskite phase in the XRD results with changing composition, Figure 4, in agreement with the larger ionic size of Ce than Zr. The lattice volumes of materials prepared by the novel acetate-H_2O_2 combustion process and previous literature data of materials prepared by a sol gel method were shown to be comparable, highlighting the accuracy and effectiveness of the new combustion method in the formation of stoichiometrically accurate quaternary oxides [10]. The morphology of as synthesized and calcined $BaZr_{0.8}Y_{0.2}O_{3-\delta}$ ceramic powders was observed by Scanning Electron Microscopy (SEM), Figure 6. After combustion, the morphology of the powders presents very small crystallite sizes in the nanometric range that are agglomerated into micron-sized particles.

The average crystallite size of as synthesized and calcined powders (1250° C), were in the range of 10-20 nm and 30-60 nm, respectively, as determined by using the Scherrer formula; values that are comparable to those reported for identical materials formed by the traditional nitrate-glycine combustion process. On sintering at 1500 °C, the micrographs of BCZY80, BCZY71 and BCZY44 pellets appear to be fully dense whereas BCZY26 and BCZY08 are significantly less dense, Figure 7.

The proton conductivity of $BaCe_{0.8-x}Zr_xY_{0.2}O_{3-\delta}$ electrolytes was shown to correspond to literature data [10]. Thus, this new synthesis route appears to offer materials of comparable morphology, phase purity and electrical performance to that obtainable by more traditional nitrate-combustion routes, while utilizing cheaper acetate metal precursors and liberating no harmful NO_x gases. Thus, the novel acetate-H_2O_2 combustion route can be recommended for the preparation of many other ceramic oxide materials, offering an alternative to potentially replace classical nitrate-based processes.

Figure 6. (Continued)

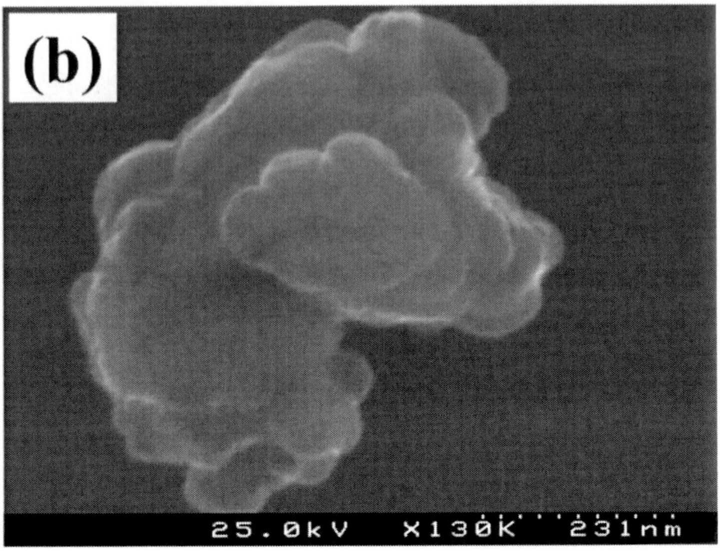

Figure 6. Scanning Electron Micrographs of (a) as synthesized and (b) calcined at 1100 °C BCZY08 nanopowder.

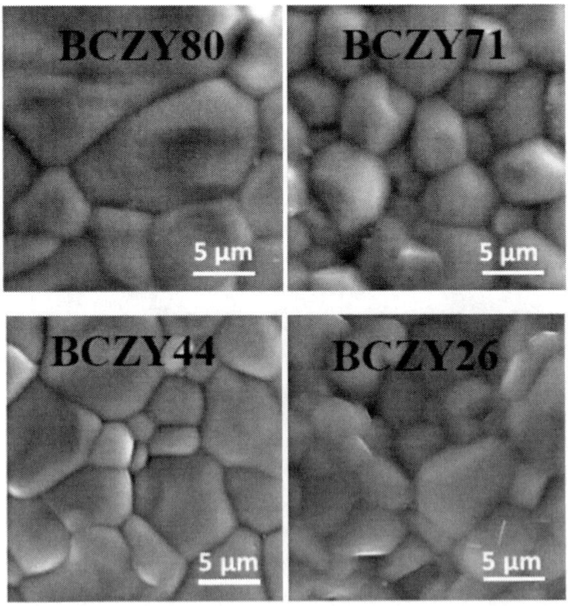

Figure 7. Scanning Electron Micrographs of surface of BCZY pellets sintered at 1500 °C for 8 h.

2.2. Synthesis of Ni-BZY Cermet Anode Materials

The nitrate free acetate-H_2O_2 microwave induced combustion route can also be demonstrated for the synthesis of Ni-$BaZr_{0.85}Y_{0.15}O_{3-\delta}$ (Ni-BZY) cermet anodes for proton ceramic fuel cells (PCFCs).

In this work $BaZr_{0.85}Y_{0.15}O_{3-\delta}$ nanopowder was pre-prepared by mechanosynthesis using a planetary ball mill (Retsch PM200) with constant planetary rotation of 650 rpm using 125 cm^3 tetragonal zirconia vials (Retsch) and balls (Tosoh Co.) as described in the literature [11]. A stoichiometric amount of nickel acetate (Aldrich, 99% pure) was dissolved in distilled water, and the pre-prepared BZY was dispersed using an ultrasonic bath.

An appropriate amount of 30% H_2O_2 (Riedel-de Haen, 30% by weight) was then slowly added, fixing the fuel to oxidant ratio of Ni-acetate/H_2O_2 to 1:1, according to the propellant chemistry, eq.1. The solution was heated on a hot plate at 80 °C until a viscous gel formed as shown in Figure 8.

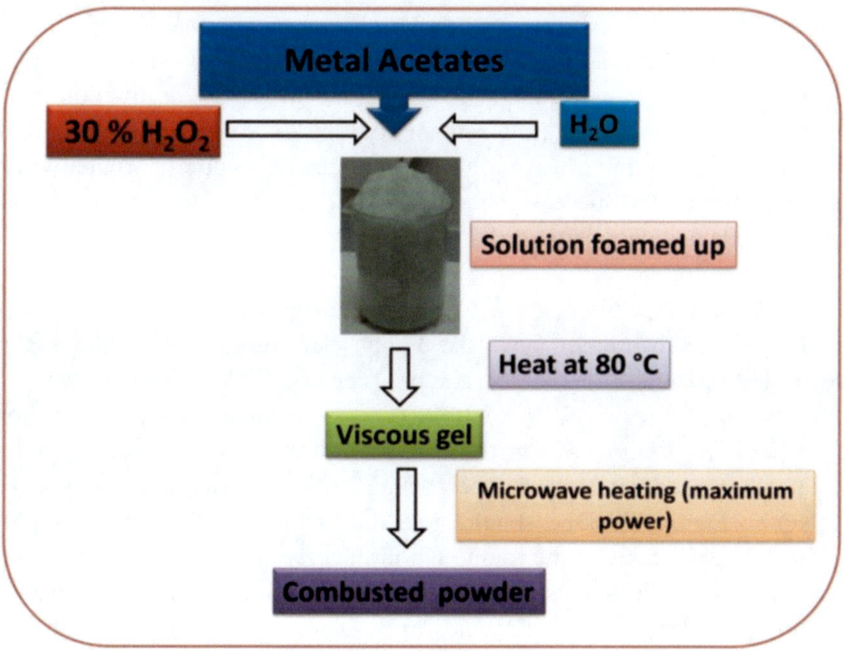

Figure 8. Illustration of acetate-H_2O_2 combustion synthesis steps for cermet anode material preparation.

The viscous gel was then heated and simultaneously rotated in a domestic 2.45 GHz, 800W microwave oven set at the maximum power, in the same way as the previous example. After a few minutes, the dried gel burnt with a flame in a self-propagating combustion manner, releasing plentiful fumes until the gels were completely burnt to form a black powder.

Previously, several researchers have used one-step nitrate-based combustion processes for the preparation of Ni-BZY cermet anodes [20]. However, the nitrate precursor based processes can partially decompose the BZY ceramic oxide and commonly forms impurity phases in these materials, such as BaY_2NiO_5 or Y_2O_3 [9, 21-24]. Such deviations of BZY from the nominal stoichiometry have been shown to be significantly deplete levels of proton conductivity obtained in both the grain and grain boundary [25].

The following section highlights the benefits of this nitrate-free combustion method in comparison to the more common nitrate/glycine combustion route usually adopted in the literature. Both methods use the same microwave combustion route and fuel/oxidant ratio, differing only by the choice of initial precursors; acetate-H_2O_2 cf. nitrate/ glycine. The results will show that typical nitrate/fuel combustion routes cannot be used for Ni-phase dispersion in pre-prepared BZY powders, due to partial destruction of the perovskite phase in the acidic precursor solution. In contrast, the new nitrate-free, acetate-H_2O_2 microwave-assisted combustion synthesis route will be shown to be a very successful method to prepare pure Ni-$BaZr_{0.85}Y_{0.15}O_{3-\delta}$ cermet anode powders [9].

2.2.3. Characterization of Ni-BZY Cermet Anodes

The X-ray diffraction analysis, Figure 9a, shows that in the case of acetate-H_2O_2 combustion method, the pre-prepared BZY phase is shown to be stable in the precursor gel. In contrast, the pre-prepared BZY phase is shown to be partially decomposed in the common nitrate/glycine precursor gel, as revealed by a decrease of the perovskite peak intensity and formation of $Ba(NO_3)_2$, Figure 9b. One should note that acidity of the precursor gel is different in each case. In the traditional nitrate-glycine route the precursor gel is acidic, whereas in the new acetate-H_2O_2 combustion route the precursor gel is neutral. It is this factor that appears to be the cause of the degradation of the pre-prepared BZY phase in the nitrate-glycine case. The acetate-H_2O_2 route avoids such degradation by providing a neutral medium for powder dispersion. After combustion, both the acetate-H_2O_2 and nitrate/glycine processes show the presence of the perovskite phase, NiO and small amounts of Ni-metal due to the high heats of reaction, Figures 9a, 9b. Nonetheless, in the nitrate/glycine route, a significant amount of $BaCO_3$ can also be observed in the XRD pattern, Figure 9b, which is attributed to carbonation of the $Ba(NO_3)_2$ that was present in the precursor gel. In contrast, the acetate-H_2O_2 combustion method continues to successfully avoid BZY phase decomposition [9].

Figure 10 provides the XRD patterns of powders prepared by both the acetate-H_2O_2 and nitrate combustion method after calcination at 1400 °C for 5h in air and reduction in 10% H_2/N_2 at 700 °C for 10 hrs. The powders prepared by acetate-H_2O_2 combustion method exhibit only two phases, that of the perovskite (BZY) and Ni metal. In contrast, the XRD patterns of powders prepared by the conventional nitrate method, calcined at the same temperature, reveal an impurity phase of BaY_2NiO_5 (PDF 00-041-0463) in addition to the perovskite and Ni phases, Figure 10.

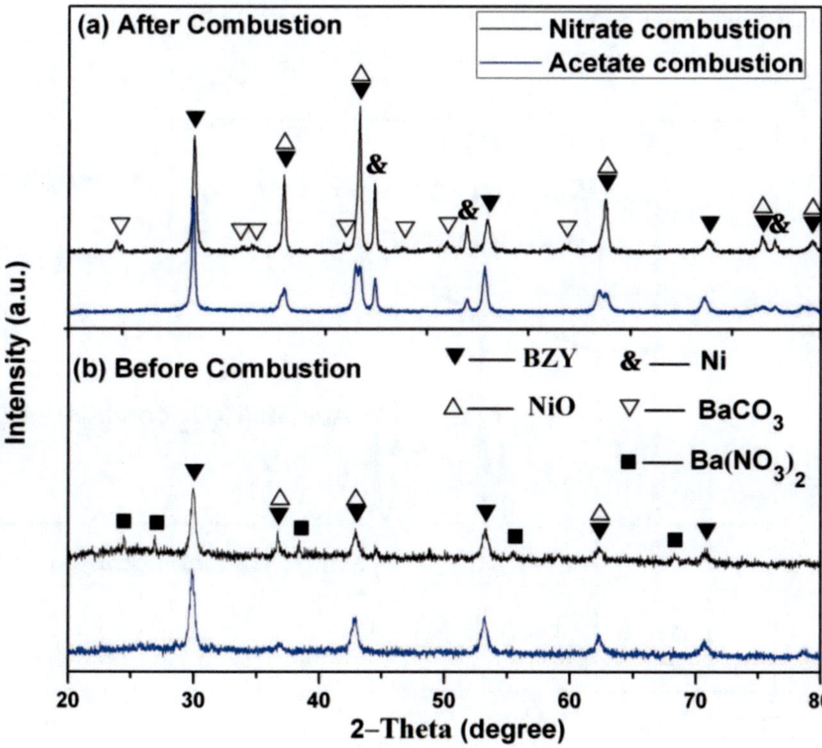

Figure 9. (a) XRD patterns of NiO-BZY cermet anodes (a) after and (b) before combustion by acetate/H_2O_2 and nitrate/glycine combustion method.

Thus, in the nitrate-based case, the removal of Ba from the perovskite phase and formation of $Ba(NO_3)_2$ suffered in the initial precursor solution, has subsequently led to the presence of BaY_2NiO_5 in the final product. As significant Ba-deficiency in the perovskite material is typically shown to be accompanied by loss of Y_2O_3 [25], all the elements required to form the BaY_2NiO_5 phase (Y_2O_3, $BaCO_3$ and NiO) are likely to be available in the nitrate/glycine combusted powder. In contrast, the BZY phase was preserved in the acetate-H_2O_2 combustion gel and, consequently, no BaY_2NiO_5 impurity phase was observed, Figure 10. To underscore this fact, BZY powder was mixed with NiO manually in a pestle and mortar, fired at 1400 °C for 5hrs and reduced under the same conditions [9]. No reaction of BZY and NiO was observed, supporting the conclusion that the degradation of the BZY phase is the source of the final BaY_2NiO_5 impurity in this material and that the origin of this degradation occurs right at the onset of the reaction due to the acidic nature of the nitrate-glycine combustion precursor gel.

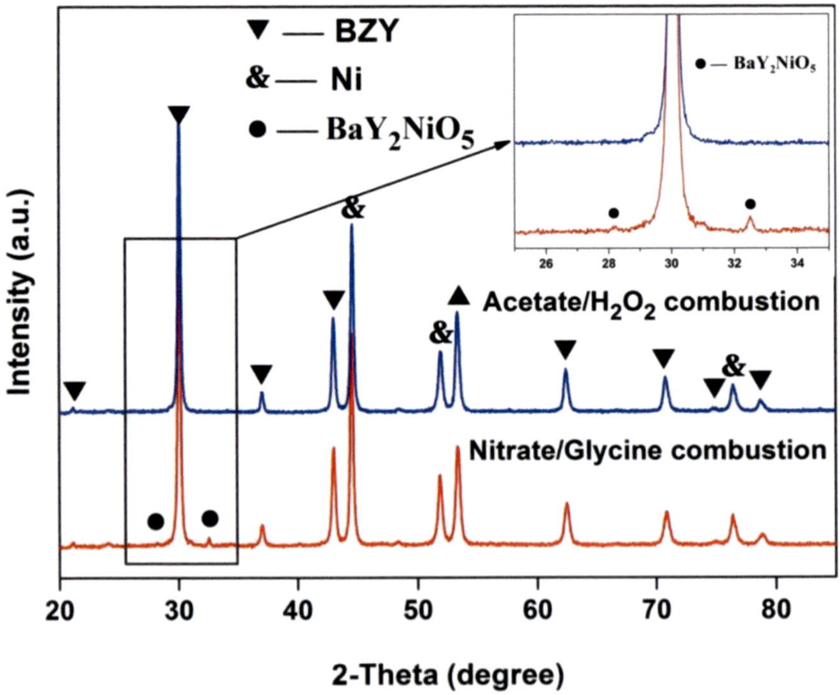

Figure 10. XRD patterns for Ni-BZY composite powders (reduced in 10% H_2/N_2 at 700 °C for 10h, prepared by both nitrate-glycine and acetate-H_2O_2 combustion method).

The BZY lattice parameter* of the nitrate/glycine prepared material, 4.2067(1)Å, is significantly lower than that formed by the novel nitrate-free process and the base BZY material, 4.2107(2)Å and 4.2129(3) Å, respectively, an indication of superior retention of the desired stoichiometry in acetate-H_2O_2 combustion process (*as determined by Reitveld refinement analysis using Fullprof program, employing Ni metal as internal standard) [9].

The morphology of as synthesized 40 vol% NiO-BZY powder produced from the acetate-H_2O_2 combustion method was observed by Scanning electron microscopy, Figure 11. A complete homogeneous distribution of nanoscale NiO and BZY particles was observed, which was further confirmed by SEM-EDS elemental analysis. The microstructure of NiO-BZY anode pellets before reduction is shown in Figure 12a.

Large and small grains correspond to NiO and BZY respectively and suitable percolation pathways appear to exist throughout the NiO and BZY phases. Figure 12b shows the microstructure of the anode after reduction at 700 °C for 6h in a 10% H_2/N_2 atmosphere, while Figure 12c highlights the fact that there are no significant differences in microstructure of Ni-BZY prepared by the acetate-H_2O_2 combustion or the traditional nitrate methods.

Figure 11. SEM micrograph of as synthesized NiO-BZY anode powder by the acetate-H_2O_2 combustion route.

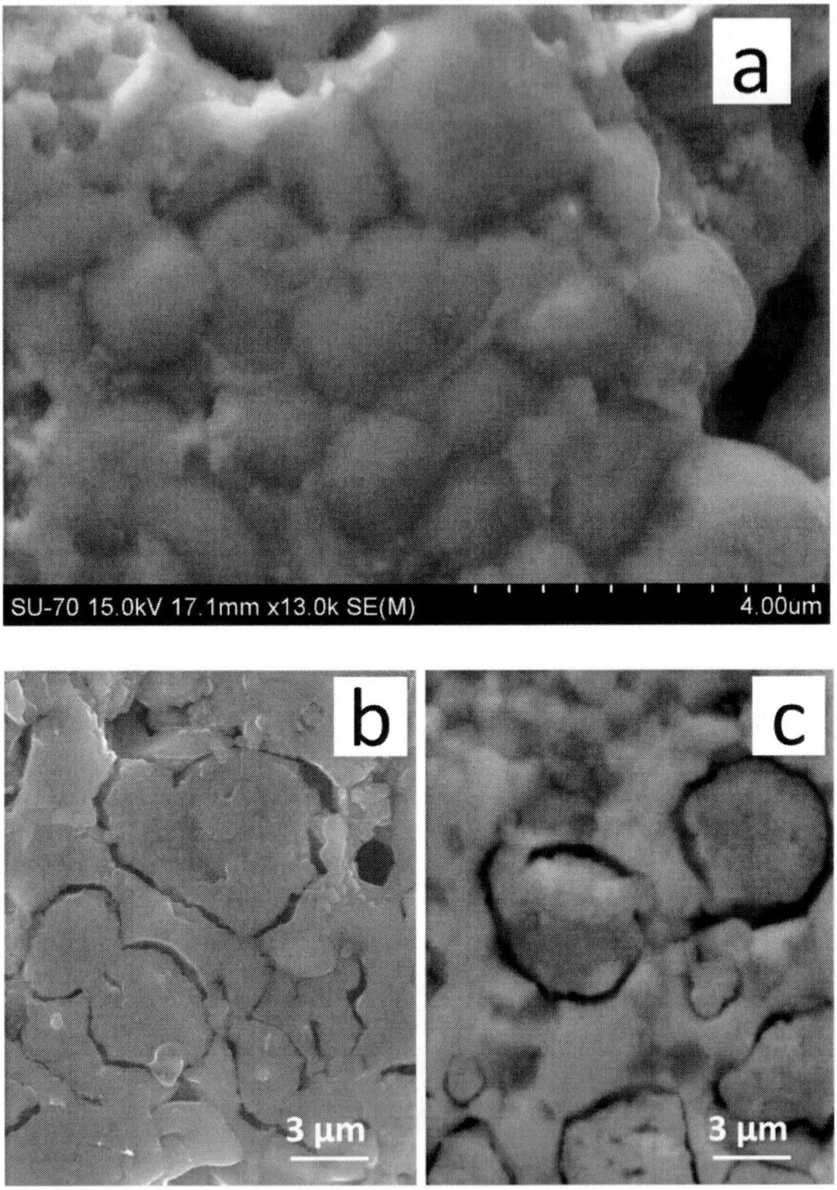

Figure 12. SEM micrographs of 40 vol% Ni-BZY anode pellets formed by the acetate-H_2O_2 combustion route (a) before and (b) after reduction compared to that of (c) the traditional nitrate method after reduction.

Clear percolation pathways for Ni and BZY grains are maintained after reduction, Figures 12b,c and this is further highlighted by Figure 13, which provides SEM-EDS elemental mapping of the formed Ni-BZY anode. The importance of phase purity in Ni-BZY anodes were studied by electrochemical impedance spectroscopy for proton ceramic fuel cell applications. The results demonstrates that anode formed by acetate-H_2O_2 combustion method has superior polarisation behavior, with the anode formed by nitrate route showing the higher polarisation resistance [9].

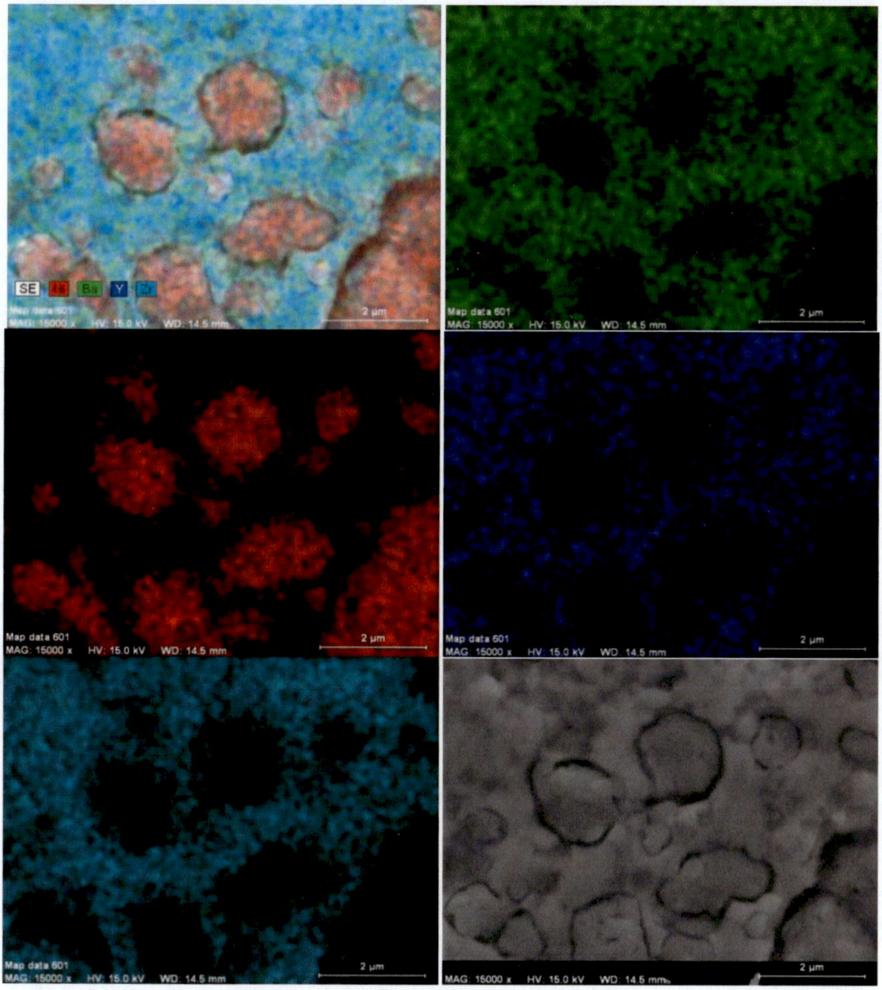

Figure 13. SEM-EDS elemental analysis of Ni-BZY anode pellets formed by the acetate-H_2O_2 combustion route.

3. Significant Features of Acetate-H_2O_2 Combustion Reaction

This chapter has provided some examples of the new acetate-H_2O_2 combustion method for the preparation of both multi component ceramic oxides and for the dispersion of metallic phases in cermet materials. These demonstrations have exhibited several advantages of the new acetate-H_2O_2 route over that of traditional nitrate-based combustion synthesis processes in the literature. These advantages include

i Reduced cost due to cheaper acetate precursors.
ii Effective and environmental friendly process with no NO_x emissions.
iii The formation of a neutral precursor solution, without the need for addition of extra base. This once again avoids NOx emissions and in the case of formation of Ba-containing cermet materials was also shown to be obligatory to obtain phase purity.

Moreover, the new acetate route can mirror the traditional advantages provided by nitrate-based combustion routes.

i Metal ions oxidation state can be stabilized in solution state
ii Homogeneous mixing of reactants
iii Rapid synthesis process
iv Ability to produce pure ceramic oxide powders
v The preparation of nanocrystallites

It can be concluded, therefore, that the novel acetate-H_2O_2 process offers very attractive alternative that can potentially replace classical nitrate-based processes, and is strongly recommended for the preparation of many other ceramic oxide materials in the future.

Acknowledgments

The authors acknowledge financial aid from the FCT, FEDER and COMPETE, PTDC/CTM/100412/2008, PTDC/CTM/105424/2008, and Narendar N thankful to FCT for doctoral research grant SFRH/BD/80949/ 2011.

REFERENCES

[1] Patil, K. C., Hegde, M. S., Rattan, T., Aruna, S. T. *Chemistry of Nanocrystalline Oxide Materials: Combustion Synthesis, Properties and Applications*; 2008th ed.; World Scientific, 2008; p. 345.

[2] Patil, K. C. Advanced ceramics: Combustion synthesis and properties. *Bull. Mater. Sci.* 1993, 16, 533.

[3] Aruna, S. T., Mukasyan, A. S. Combustion synthesis and nanomaterials. *Curr. Opin. Solid State Mater. Sci.* 2008, 12, 44.

[4] Mohebbi, H., Ebadzadeh, T., Hesari, F. a. Synthesis of nano-crystalline (Ni/NiO)–YSZ by microwave-assisted combustion synthesis method: The influence of pH of precursor solution. *J. Power Sources* 2008, 178, 64.

[5] Jain, S. R., Adiga, K. C., Pai Verneker, V. R. A new approach to thermochemical calculations of condensed fuel-oxidizer mixtures. *Combust. Flame* 1981, 40, 71.

[6] Varma, A., Rogachev, A. S., Mukasyan, A. S., Hwang, S. Combustion Synthesis of Advanced Materials: Principles and Applications. *Adv. Chem. Eng.* 1998, 24, 79.

[7] Patil, K. C., Aruna, S. T., Mimani, T. Combustion synthesis: an update. *Curr. Opin. Solid State Mater. Sci.* 2002, 6, 507.

[8] Thomas, L. E., Exarhos, G. J. combustion synthesis of oxide ceramic powders. *Mater. Lett.* 1990, 10, 6.

[9] Narendar, N., Mather, G. C., Dias, P. A. N., Fagg, D. P. The importance of phase purity in Ni–BaZr0.85Y0.15O3−δ cermet anodes – novel nitrate-free combustion route and electrochemical study. *RSC Adv.* 2013, 3, 859.

[10] Nasani, N., Dias, P. A. N., Saraiva, J. A., Fagg, D. P. Synthesis and conductivity of Ba(Ce,Zr,Y)O3−δ electrolytes for PCFCs by new nitrate-free combustion method. *Int. J. Hydrogen Energy* 2013, 38, 8461.

[11] Antunes, I., Brandão, A., Figueiredo, F. M., Frade, J. R., Gracio, J., Fagg, D. P. Mechanosynthesis of nanopowders of the proton-conducting electrolyte material Ba(Zr, Y)O3−δ. *J. Solid State Chem.* 2009, 182, 2149.

[12] Chevallier, L., Zunic, M., Esposito, V., Di Bartolomeo, E., Traversa, E. A wet-chemical route for the preparation of Ni–BaCe0.9Y0.1O3−δ cermet anodes for IT-SOFCs. *Solid State Ionics* 2009, 180, 715.

[13] Deganello, F., Marcì, G., Deganello, G. Citrate–nitrate auto-combustion synthesis of perovskite-type nanopowders: A systematic approach. *J. Eur. Ceram. Soc.* 2009, 29, 439.

[14] Mangalaraja, R. V., Mouzon, J., Hedström, P., Camurri, C. P., Ananthakumar, S., Odén, M. Microwave assisted combustion synthesis of nanocrystalline yttria and its powder characteristics. *Powder Technol.* 2009, 191, 309.

[15] Baghurst, A. R., Chippindale, A. M., Mingos, D. M. P. Microwave syntheses for Superconducting ceramics. *Nature* 1988, 332, 311.

[16] Scholes, F., Hughes, A., Hardin, S. Influence of hydrogen peroxide in the preparation of nanocrystalline ceria. *Chem. Mater.* 2007, 19, 2321.

[17] Panova, T. I., Popov, V. P., Glushkova, V. B., Domanskii, a. V. Preparation of nanodisperse solid solutions based on ZrO2 and HfO2 from hydroperoxides. *Glas. Phys. Chem.* 2007, 33, 652.

[18] Pappacena, A., Porreca, P., Boaro, M., de Leitenburg, C., Trovarelli, A. Effect of process modification and presence of H2O2 in the synthesis of samaria-doped ceria powders for fuel cell applications. *Int. J. Hydrogen Energy* 2012, 37, 1698.

[19] Chien, R. R., Tu, C.-S., Schmidt, V. H., Lee, S.-C., Huang, C.-C. Synthesis and characterization of proton-conducting Ba(Zr0.8−xCex Y0.2)O2.9 ceramics. *Solid State Ionics* 2010, 181, 1251.

[20] Bi, L., Fabbri, E., Sun, Z., Traversa, E. BaZr0.8Y0.2O3−δ-NiO Composite Anodic Powders for Proton-Conducting SOFCs Prepared by a Combustion Method. *J. Electrochem. Soc.* 2011, 158, B797.

[21] Tong, J., Clark, D., Bernau, L., Sanders, M., O'Hayre, R. Solid-state reactive sintering mechanism for large-grained yttrium-doped barium zirconate proton conducting ceramics. *J. Mater. Chem.* 2010, 20, 6333.

[22] Tong, J., Clark, D., Bernau, L., Subramaniyan, A., O'Hayre, R. Proton-conducting yttrium-doped barium cerate ceramics synthesized by a cost-effective solid-state reactive sintering method. *Solid State Ionics* 2010, 181, 1486.

[23] Tong, J., Clark, D., Hoban, M., O'Hayre, R. Cost-effective solid-state reactive sintering method for high conductivity proton conducting yttrium-doped barium zirconium ceramics. *Solid State Ionics* 2010, 181, 496.

[24] Bi, L., Fabbri, E., Sun, Z., Traversa, E. Sinteractive anodic powders improve densification and electrochemical properties of BaZr0.8Y0. 2O3−δ electrolyte films for anode-supported solid oxide fuel cells. *Energy Environ. Sci.* 2011, 4, 1352.

[25] Yamazaki, Y., Hernandez-Sanchez, R., Haile, S. M. Cation non-stoichiometry in yttrium-doped barium zirconate: phase behavior, microstructure, and proton conductivity. *J. Mater. Chem.* 2010, 20, 8158.

In: Combustion
Editor: Joseph M. Grier
ISBN: 978-1-62948-967-4

Chapter 8

SOLUTION COMBUSTION METHOD: A CONVENIENT APPROACH FOR PREPARING Ni-PROMOTED Mo AND Mo-W SULPHIDE HYDROTREATING CATALYSTS

Sergio L. González-Cortés*[*], *Serbia M. Rodulfo-Baechler and Freddy E. Imbert
Laboratorio de Cinética y Catálisis, Departamento de Química, Facultad de Ciencias, Universidad de Los Andes, Mérida, Venezuela

ABSTRACT

The development of advanced catalyst synthesis methods to meet the stringent environmental regulations in the production of ultraclean transportation fuels represents a major challenge for catalyst suppliers. In this chapter, the positive impact of solution combustion synthesis (SCS)-based preparation approach in the synthesis of alumina-supported NiMoP and NiMoW hydrotreating catalysts has been examined. The results show that the use of urea as fuel can tune the metal oxide-support interaction, altering the distribution of Mo-oxo species onto alumina support. Furthermore, the hydrodesulphurisation (HDS) activity of alumina-supported NiMoP sulphide catalysts achieved a maximum at Φ ratios between 0.6 and 1.2. This maximum HDS activity is attributed to a higher

[*] E-mail address: goncor@ula.ve and slgoncor@gmail.com.

surface density of polymolybdate relative to molybdate species in the catalyst precursor and the effective interaction of nickel with the edges of MoS_2 that facilitates the formation of NiMoS structure. On the other hand, the sequential deposition of NiMo-urea solution on metal oxide-modified alumina surface produces a reduction of the catalyst bed temperature rate and an increase of the decomposition period of the redox mixture. The addition of Ni and Mo to WO_x-modified alumina assists the segregation of MoO_x and WO_x species on the alumina surface and promotes their reducibility. Furthermore, the NiMoW catalyst prepared by impregnation-solution based combustion method showed higher hydrotreating activity than a commercial NiMo catalyst.

1. Introduction

1.1. Hydrotreating (HDT) Process in the Crude Oil Refinery

Oil refining is a mature industry whose development has been closely linked to the use of catalytic processes. A refinery employs different physical and chemical processes such as distillation, extraction, reforming, hydrogenation, cracking and blending to separate and then convert the crude oil fractions to higher value products [1].

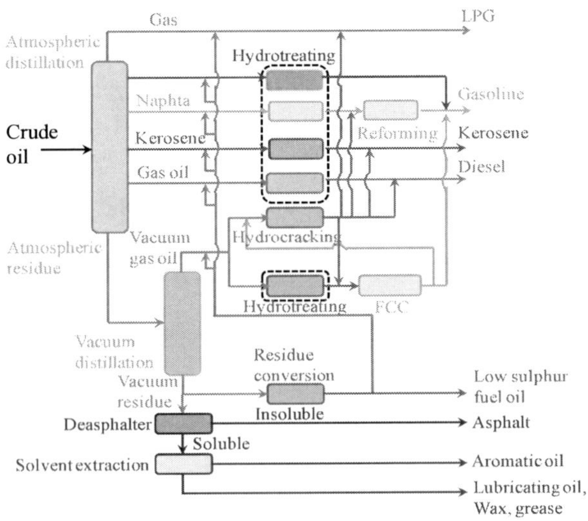

Figure I. Diagram of a typical crude oil refinery. (Note that FFC was changed for FCC).

A simplified schematic diagram of a typical petroleum refinery is illustrated in Figure I [2], highlighting the processes involved in making liquid transportation fuels, in which heterogeneous catalysts play a key role.

Crude oil into the refinery is first is distilled to separate into various fractions such as gases, liquid (naphtha, kerosene and gas oil) and the heavy residue that remains after vacuum distillation. The gaseous fraction, together with the by-products of cracking process, are sold as liquefied petroleum gas (LPG) in containers for camping or domestic use, or in some countries as transportation fuel.

All fractions other than LPG are upgraded in subsequent processes. The lighter liquid petroleum fraction in the gasoline boiling range is called naphtha (boiling range 40 - 180 °C); this can be processed in several ways, including isomerization, hydrotreating and/or reforming, these catalytic processes involve metal based or bifunctional metal/acid catalysts. The distillate fraction (i.e., kerosene, boiling range 180 - 230 °C, and gas oil, boiling range 230 - 360 °C) which boils in the range needed for making diesel and jet fuel, is typically hydrotreated to remove heteroatoms such as sulfur and nitrogen or in some cases hydrocracked to both remove heteroatoms and to crack some of the distillate into gasoline [3].

The next boiling fraction is vacuum gas oil (VGO) (boiling range 360 - 530 °C), which is often sent directly to a fluid catalytic cracker (FCC), however, hydrotreating of the VGO fraction prior to FCC is becoming more and more common, or hydrocracked to eliminate heteroatoms and to crack some of the distillate into diesel.

The vacuum residue fraction typically contains the highest concentrations of heteroatoms and significant amounts of highly aromatics and polar molecules that readily poison catalysts. Thus, the heaviest residue fraction is most often converted via non-catalytic thermal process known as coking which generates lighter liquid stream and a solid coke by-product. Different components of vacuum residue can be recovered by solvent extraction. However, catalytic processes relying on solid catalysts and high hydrogen pressure are sometimes used to convert the residue to lighter products with low sulphur content.

The most important processes and the catalysts normally employed in the crude oil refining to make liquid transportation fuels are summarized in Table 1 [2].

Note that alumina-supported Co- and/or Ni-promoted Mo and W catalysts play an important role in various refinery processes employed for the production of fuels.

Table 1. Major refinery processes for fuel production using solid catalysts

Catalytic process	Goal of the process	Catalyst formulation
Hydrotreating (HDT)	Removal of hetereoatoms (S, N and O) and hydrogention of poly-armatics to produce fuels	Co (or Ni)-MoS_2/Al_2O_3, Ni-WS_2/Al_2O_3
Reforming	Transformation of naphtha into high quality gasoline generating hydrogen as by-product, which is then employed in the hydrogen-requiring processes	Pt-Re/Al_2O_3,, Pt-Ir/Al_2O_3
Hydrocracking (HC)	Upgrading of vacuum gas oil into transportation fuel with hydrogen in a combination of hydrotreating and cracking reactions using by-functional catalysts.	Ni-Mo, Ni-W, Pr-Pd, acidic supports (alumina, zeolites)
Fluidized catalytic cracking (FCC)	Upgrading of vacuum gas oil into transportation fuels using acidic catalysts.	Zeolite Y
Residue Hydroconversion	Transformation of residue using hydrogen into low-sulfur fuel oil, transportation fuel and feedstock for FCC and hydrocracking.	Thermal and catalytic processes

1.2. Solution Combustion Synthesis (SCS) Method for Preparing Hydrotreating Catalysts

Nowadays, the hydrotreating (HDT) reactions play a major role not only on the reduction of environmental pollution but also on the upgrading process of low grade feedstock, in which a number of variables such as catalyst formulation, operation conditions and feed quality (i.e., concentrations and type of aromatics, S- and N-containing compounds, their reactivity, inhibition effects, etc) need to be considered to produce ultraclean transportation fuel [4]. The preparation of hydrotreating catalysts is normally carried out by dry or wet impregnation, in which an aqueous solution containing Ni (or Co) promoter and Mo (or W) is deposited onto the γ-Al_2O_3 or SiO_2-Al_2O_3 support with appropriate textural properties, suitable mechanical strength and geometrical shape. After the calcination process, the catalyst oxide precursor is then presulfiding to convert into the Ni (or Co)-Mo (or W) sulphide [5].

Alternative catalyst preparation methods (e.g., STARS technology) and catalyst formulation (e.g., NEBULA catalyst) have significantly enhanced the HDT catalyst performance [6]. With reference to novel preparation methods, a self-propagating high temperature synthesis (SHS), called solution combustion synthesis (SCS) consists of using a saturated aqueous solution of the desired metal salts (nitrates are generally preferred because of their oxidizing property and high solubility in water) and a suitable organic fuel as reducing agent (e.g., urea, carbohydrazide, maleic hydrazide, etc.) [7, 8]. The redox mixture is ignited and eventually combusts in a self-sustaining and fast combustion reaction, resulting, usually, in a nanocrystalline oxide material [8, 9]. We have previously showed that urea-matrix combustion method can produce active alumina-supported CoMo, NiMo and NiW catalysts [10] for the hydrodesulphurisation (HDS) of thiophene and also mixed-metal oxides [i.e., Co(Ni,Mg)MoO_4] well-dispersed on alumina surface [11]. Furthermore, this preparation method was employed for synthesizing CoMo catalysts to assess not only the effect of different supports (i.e., Al_2O_3, MgO and MgO–$MgAl_2O_4$) [12], but also the influence of the preparation method [13] over the HDS reaction of various model feedstocks. It should be remarking that all theses catalysts were prepared by SCS method under uncontrolled condition of thermal treatment [14]. The objective of this chapter is to gain insight into the effect of the urea/Ni ratio in the HDS activity of alumina-supported NiMoP sulphide catalysts synthesised by impregnation solution-based combustion method under controlled condition of thermal treatment. The optimal urea/Ni ratio to obtain a maximum HDS activity was then used for preparing the alumina-supported NiMoW sulphide catalyst, whose hydrodesulphurisation (HDS), hydrodenitrogenation (HDN) and hydrodearomatisation (HDA) activities were compared to a commercial alumina-supported NiMo sulphide catalyst to find out the capability of this controlled combustion process to produce active hydrotreating catalysts.

2. SYNTHESIS, EVALUATION AND CHARACTERIZATION OF THE SOLID CATALYSTS

2.1. Preparation and Catalyst Performance Test

An example to illustrate the calculation of fuel/oxidizer ratio (Φ) using nickel nitrate as the oxidizer and urea as the fuel for the synthesis of nickel oxide is given below. It is assumed that H_2O, CO_2 and N_2 are the gaseous products formed in the combustion reaction. The eqn. 1 is the overall combustion reaction.

$$Ni(NO_3)_2.6H_2O + (\frac{5}{3}\Phi)CO(NH_2)_2 + (\frac{5}{2})(\Phi - 1)O_2 \longrightarrow NiO + (\frac{10}{3}\Phi + 6)H_2O + (\frac{5}{3}\Phi)CO_2 + (\frac{5}{3}\Phi + 1)N_2 \quad (1)$$

The optimal stoichiometric composition of the redox mixture is obtained when no molecular oxygen is required (i.e., $\Phi = 1$). When Φ is below unit (Φ

< 1) the redox mixture is under fuel-lean regime and molecular oxygen is produced. On the other hand, when Φ > 1 the redox mixture is under fuel-rich condition, requiring molecular oxygen to fully covert the fuel.

The alumina-supported NiMoP oxide catalyst precursors were prepared by impregnation solution-based combustion using urea as fuel. Initially, alumina-based trilobe extrudates (i.e., 1.5 mm of diameter, 5-10 mm of length and 250 $m^2.g^{-1}$) were impregnated with 10 % H_2O_2 solution containing urea (Acros Organics, 98 %), nickel (II) nitrate hexahydrate (Acros Organics, 99 %), ammonium molybdate (VI) tetrahydrate (Acros Organics) and phosphoric acid (Aldrich , 85 % in water) to obtain a loading of 5 wt. % NiO, 23 wt. % MoO_3 and 1 wt. % P_2O_5. Four formulations were prepared with various urea/Ni ratios (Φ) (i.e., 0; 0.3; 0.6 and 1.2). Then, the impregnated solid was dried at 100 °C and calcined up to 300 °C for 5 h under flowing air.

The alumina-supported NiMoW oxide catalyst precursor was also prepared by impregnation solution-based combustion using urea as fuel. The γ-Al_2O_3 extrudate support was initially impregnated with an aqueous solution of ammonium metatungstate hydrate (Aldrich) by incipient wetness impregnation (IWI). The incipient wetness solid was dried at 100 °C and calcined at 300 °C for 5 h. Subsequently, the WO_x-Al_2O_3 sample was impregnated with an aqueous solution containing nickel (II) nitrate hexahydrate, ammonium molybdate (VI) tetrahydrate and urea (i.e., Φ = 0.60) by sequential impregnation steps. After each IWI step, the wet solid sample was dried at 100 °C and calcined at 300 °C for 5 h. A final loading of 9 wt. % NiO, 15 wt. % MoO_3 and 20 wt. % WO_3 was deposited on alumina.

Finally, the alumina-supported metal oxide catalyst precursors were-presulphided at liquid hour space velocity (LHSV) of 2.4 h^{-1}; H_2/oil volume ratio of 300 N m^3/m^3 and 3.0 MPa of hydrogen using middle distillate spiked with dimethyl disulfide to total sulphur content of 4 wt. % as sulphiding agent. The temperature was increased at 0.5 °C.min^{-1} to 240 °C and kept it for 60 min. The temperature was then further risen at 0.5 °C.min^{-1} to 360 °C and kept at 360 °C for 8 h. After that period, the catalyst bed temperature decreased to 220 °C under sulfiding agent and H_2 flow condition. Then, the hydrogen pressure, the flows of hydrogen and feedstock were set according to the required reaction condition whereas the catalyst bed was increased to reaction temperature at a 0.5 °C.min^{-1} heating rate. The catalysts were evaluated at each reaction condition for periods between 48 and 168 h. The catalytic performance of Al_2O_3-supported NiMoP sulphide catalyst was evaluated on feed 1, whilst the NiMoW sulphide catalyst was assessed on feed 2 and feed 3, whose properties are shown in Table 2.

Table 2. Compositions and physical properties of various feedstocks

Fuel Properties	Feed 1	Feed 2	Feed 3
Sulphur / $\mu g.g^{-1}$	4,180	4,250	5,900
Nitrogen / $\mu g.g^{-1}$	190	1000	80
Total Aromatics / wt. %	42.0	2.0	27.2
Density (g/cm^3)	0.872	0.890	0.840
Boiling point range (°C)	114-412	186-365	139-446

2.1. Catalyst Characterization

The temperature-time (T-t) profiles for the urea-containing samples were carried out at 0.1 MPa and 3 $cm^3(min.g_{cat})^{-1}$ (i.e., weight minute space velocity) under a flow of air on a fixed-bed reactor. Approximately 11 grams of dried catalyst extrudates were loaded into the reactor with a type K thermocouple (Omega Engineering) placed on the center of the catalysts bed to record the dependence of the catalyst bed temperature at seven measurements per minute with the increase of the time. The sample was initially preheated at 100 °C for 10 minutes and then the temperature was lineally increased up to 260-300 °C for NiMoP catalyst and ca. 320 °C for NiMoW catalyst using a heating rate of 1.0 $°C.min^{-1}$.

The Raman spectra of the oxide precursors were recorded in a Jobin Yvon Labram spectrometer with a 632 nm HeNe laser, run in a back-scattered co-focal arrangement. The solid samples were pressed in a microscope slide; with a 30 s scanning time and 2 cm^{-1} resolution. Several points of each catalyst surface were probed to explore homogeneity of the sample and reproducibility of the data. Temperature-programmed oxidation coupled to mass spectrometer (TPO-MS) analysis for the alumina-supported NiMoP sulphide catalysts was carried out on a Micromeritics AutoChem II 2920 instrument. Between 100 and 200 mg of the solid samples was pre-treated before analysis with a 50 $cm^3.min^{-1}$ N_2 flow rate, from room temperature up to 100 °C for 30 min. The sample was then cooled to room temperature and the TPO-MS profiles were obtained by using 20 % O_2 in nitrogen (50 $cm^3.min^{-1}$) and 10 $°C.min^{-1}$ heating rate from room temperature up to 650 °C.

Temperature-programmed reduction (TPR) analysis for the oxide precursors was carried out on a Micromeritics AutoChem II 2920 instrument. Between 110 and 150 mg of the solid samples was pre-treated before analysis with a 50 $cm^3.min^{-1}$ He flow rate, from room temperature up to 150 °C for 30 min. The sample was then cooled to room temperature and the TPR profiles were obtained by using 10 % H_2 in Ar (50 $cm^3.min^{-1}$) and 10 $°C.min^{-1}$ heating rate from room temperature up to 950 °C and hold this temperature for 30 min. The amount of consumed H_2 was quantified by using a thermal conductivity detector whose response was calibrated with different H_2 concentrations.

3. Results and Discussion

3.1. Synthesis and Catalytic Performance of Alumina-Supported NiMoP Catalysts

An aqueous solution containing nickel nitrate, ammonium molybdate, phosphoric acid and variable content of urea as suitable reducing agent was deposited onto Al_2O_3 by impregnation solution-based combustion synthesis. The impregnated sample was uniformly heated in the mode of controlled thermal treatment approach [14]. Figure II shows the temperature–time (T–t) profiles for the synthesis of alumina-supported 5 wt. % NiO-23 wt. % MoO_3-1wt. % P_2O_5 catalyst at different urea/Ni ratios (Φ).

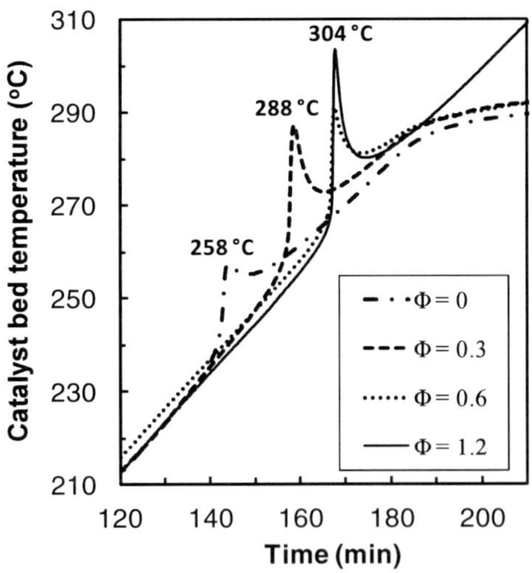

Figure II. Temperature-time profiles for alumina extrudate-supported NiMoP catalyst synthesised by impregnation solution-based combustion method with different urea/Ni ratios (Φ).

Table 3. Influence of the urea/Ni ratio over the ignition temperature, maximum temperature, catalyst bed temperature rate and HDS rate constant for Al_2O_3-supported NiMoP sulphide catalyst. Tests conducted at 340 °C, 5.0 MPa and 400 Nm^3/m^3 (H_2/oil ratio)

Urea/Ni ratio (Φ)	T_{Ign} (°C)	T_{max} (°C)	dT/dt (°C/min)	k_{HDS} (1/g.s.wppm$^{0.5}$)*10^{-4}
0.0	237	258	12	33.0
0.3	256	288	32	37.4
0.6	266	292	46	44.1
1.2	263	304	69	40.8

A single exothermic peak is clearly shown, whose maximum temperature and ignition temperature is dependent on the fuel/oxidizer ratio. In Table 3 is listed the combustion parameters from the T-t profiles for NiMoP catalysts prepared with different Φ and also the HDS rate constants. The NiMoP precursor without urea showed an exothermic peak at lower ignition temperature (i.e., 237 °C) than those observed in urea-containing catalyst precursors (i.e., 256-266 °C), owing to the dependence of the ignition temperature with the chemical composition of the redox mixture. The ignition temperature (T_{ign}) and the peak maximum temperature (T_{max}) did not change markedly at Φ ratio between 0.3 and 1.2, whereas the catalyst bed temperature rate (*dT/dt*) lineally increased with urea content in the catalyst precursor. This trend indicates that the T_{ign} and T_{max}, are associated with the chemical nature of the redox mixture, whilst the *dT/dt* is strongly dependent upon the fuel content. On the other hand, the rise of urea/Ni ratio increased the HDS activity to achieve a maximum HDS rate constant at Φ ratio between 0.6 and 1.2. It is worth remarking that the calcination temperatures between 260 and 300 °C did not affect the catalytic performances of γ-Al_2O_3 supported NiMoP sulphide catalyst, because the catalyst bed temperature was fairly independent of the final calcination temperature (i.e., 292-295 °C).

In order to find out the chemical nature of Mo species deposited on alumina surface after the combustion process, laser Raman (LR) spectroscopy was used to characterize a series of sample prepared with different urea/Ni ratios (Figure III). The Raman spectrum of alumina-supported NiMoP oxide prepared without urea (i.e., Φ = 0) displayed a main band at 950 cm^{-1} and a wide small peak centred at 360 cm^{-1}, attributed to the symmetric and bending modes of the terminal Mo=O bond of octahedrally coordinated Mo species for the hydrated $Mo_7O_{24}^{6-}$ anion [15], (Figure III a). A wide Raman band at 870 cm^{-1} is associated with the antisymmetric mode of the Mo–O–Mo bridge for the polymolybdate and also with the presence of molybdate (MoO_4^{2-}) anion. Similar LR spectra showed the catalyst precursors prepared with Φ ratios of 0.6 and 1.2 (see Figures III b and c). However, the peak intensity of the main Raman bands at 950 and 870 cm^{-1} varied with the fuel composition. Indeed, the intensity of the peak at 950 cm^{-1} compared to the peak at 870 cm^{-1} increased as raising the urea content from 1.20 (Φ = 0) to 2.21 (Φ = 1.2). This finding indicates that the abundance of the polymolybdate anion relative to the single molybdate on alumina surface increased with urea content in the redox mixture after the combustion process, tentatively attributed to the higher catalyst bed temperature rate. However, we cannot rule out a possible effect of urea on the phosphate-Ni-Mo-oxo and even phosphate-alumina interactions that could affect the dispersion of Ni and Mo-oxo species onto alumina support. Nevertheless, an optimal domain size of Mo-oxo clusters and their possible interaction with nickel oxide at Φ ratio between 0.6 and 1.2 are crucial to obtain a maximum HDS activity.

To gain further insight into the effect of urea on the HDS catalyst performance, the alumina supported NiMoP sulphide catalysts synthesised with urea (Φ = 0.6) and without urea (Φ = 0) were analysed by TPO-MS. Figure IV shows the evolution of SO_2 produced upon the oxidation of various alumina supported Ni, Mo and NiMoP sulphide catalysts. It should be mentioned that a similar study for commercial CoMo sulphide catalyst was reported elsewhere [16].

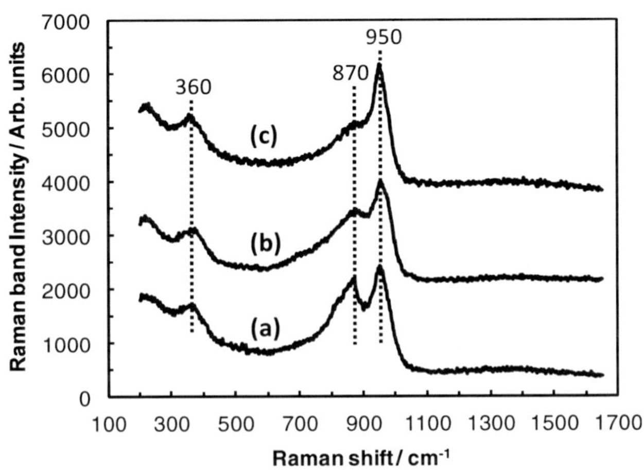

Figure III. Laser Raman spectra of Al_2O_3-supported NiMoP catalyst precursors at various urea/Ni ratios. Φ = 0 (a); Φ = 0.6 (b) and Φ = 1.2 (c).

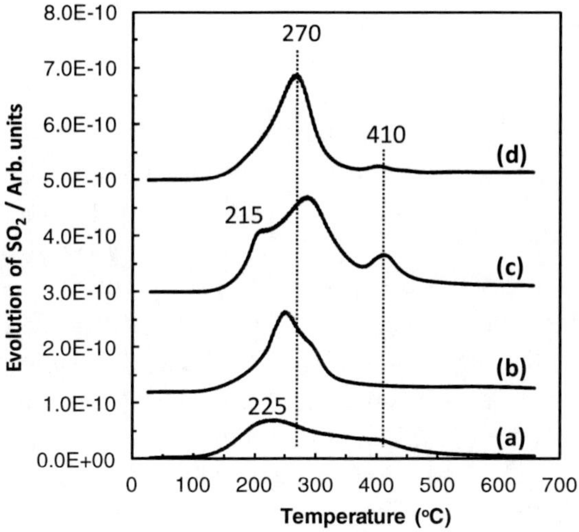

Figure IV. Evolution of SO_2 by temperature-programmed oxidation coupled to MS (TPO-MS) for Al_2O_3-supported NiS_x (a), Al_2O_3-supported MoS_x (b), Al_2O_3-supported NiMoP sulphide ($\Phi = 0$) (c) and Al_2O_3-supported NiMoP sulphide ($\Phi = 0.6$) (d).

The TPO-MS profiles for alumina-supported Ni and Mo sulphides are listed as references, which were synthesized by similar procedure and composition that alumina-supported NiMoP sulphide catalysts. The oxidation of NiS_x occurs into a wide range of temperature with a broad peak centred at 225 °C and a shoulder at about 410 °C (Figures IV a), whilst the TPO-MS profile for alumina-supported MoS_x sample shows a major peak at 270 °C and a shoulder at ca. 285 °C (Figure IV b).

On the other hand, the oxidation of alumina-supported NiMoP sulphide prepared without urea ($\Phi = 0$) shows a main peak at 285 °C characteristics of MoS_x oxidation, the oxidation of NiS_x is associated to the shoulder at 215 °C and a small peak at 410 °C (Figure IV c). A single major peak at 270 °C associated with the oxidation of NiMo sulphide is given in the TPO-MS profile for alumina-supported NiMoP sulphide prepared with urea ($\Phi = 0.6$), Figure IV d. This finding clearly reveals that the use of urea upon the catalyst preparation controls the metal oxide-support interaction and facilitates the formation of NiMoS structure, which enhances the HDS catalyst performance (see Table 3) in agreement with previous reports [17, 18]. Indeed, a urea/Ni ratio between 0.6 and 1.2 improved the formation of polymolybdate relative to the single molybdate, which is reflected on an optimal HDS activity as a consequence of the effective interaction of nickel with the edges of MoS_2 that facilitates the formation of NiMoS structure.

3.2. Synthesis and Catalytic Performance of Alumina-Supported NiMoW Catalysts

The depositions of the redox mixture of NiMo-urea ($\Phi = 0.6$) on WO_x-modified Al_2O_3 were carried out by impregnation solution-based combustion synthesis. The impregnated sample was uniformly heated in the mode of controlled thermal treatment approach [14]. The Temperature-time (T-t) profiles for the impregnated samples are given in Figure V.

The first impregnation step (i.e., NiMo-1) produced a clear exothermic peak with an ignition temperature of 272 °C and a maximum temperature of 294 °C. The T-t profiles for the second (NiMo-2) and third (NiMo-3) impregnations showed ignition temperatures (ca. 272 °C) and maximum temperatures (i.e., 298 and 290 °C) comparable to those obtained in the T-t profile for the first impregnation step. However, the peak intensity significantly decreased for the third NiMo impregnation step. It is worth remarking that the T_{ign} was not affected by the number of impregnations, owing to its direct dependence with the chemical composition of the impregnation solution.

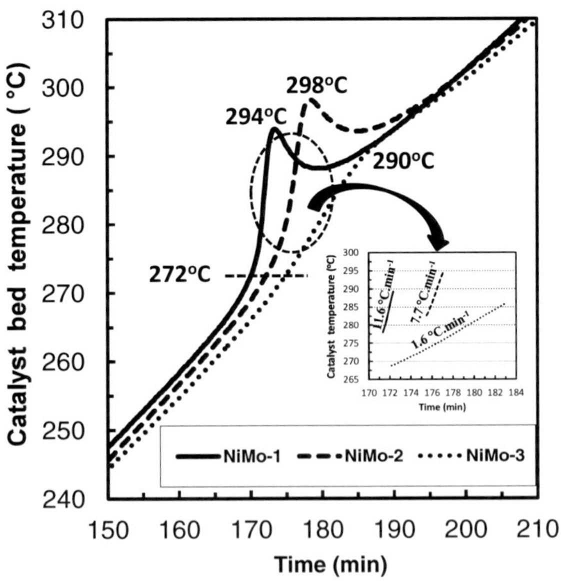

Figure V. Temperature-time profiles for WO_x-Al_2O_3-deposited NiMo-urea solutions synthesized by sequential co-impregnations. Inset Figure, amplification of the data within the time interval of 171-183 minutes.

Table 4. Effect of the sequential impregnations on the decomposition period of the metal precursor, catalyst bed temperature rate, loading of Ni-Mo oxides and pore volume

Sample	*Δt (min)*	*dT/dt (°C/min)*	*NiMo oxides/metal oxides-Al_2O_3 (g/g)*	*[a] NiMo oxides (wt. %)*	*[b] Reduction of PV (%)*
NiMo-1	23	11.6	0.17	9.2	37.5
NiMo-2	41	7.7	0.13	7.9	14.0
NiMo-3	65	1.6	0.11	7.0	11.6

[a] Ni and Mo oxides wt. % relative to the overall catalyst composition, [b] Pore volume (PV) reduction relative to the metal oxide-support pore volume of the former impregnation step.

The catalyst bed temperature rate, on the other hand, decreased from 11.6 to 1.6 °C.min^{-1}, whereas increasing the number of impregnation steps as is given in inset Figure. An opposite trend showed the decomposition period of the metal oxide precursors (i.e., 23, 41 and 65 minutes), Table 4.

The ratio of Ni and Mo oxides to metal oxides-alumina and the relative amount of deposited Ni and Mo oxides decreased with the number of impregnations because of the reduction of pore volume (PV) after each step. Therefore, the diminution of the *dT/dt* and the increase of *Δt* is associated to the lower content of NiMo oxides relative to the metal oxide-support (dilution effect) when increasing the number of impregnations, since the urea/nickel ratio was kept constant. It should be mentioning that the major components of the catalyst (i.e., Al_2O_3, MoO_3 and WO_3) have a high molar heat capacity (between 79 and 73 J.K^{-1}.mol^{-1}) but NiO has a value of 44.3 J.K^{-1}.mol^{-1} [19]. Based on those values, it is expected that the trend observed on the T-t profiles after the impregnation steps are not a consequence of the variations in the overall heat capacity. It is worth highlighting that the ignition temperature of the nickel nitrate-urea (i.e., 245 °C) is significantly lower than that for NiMoP-urea (i.e., 256-263 °C) deposited on alumina and even NiMo-urea (ca. 272 °C) deposited on WO_x-alumina surface as a consequence of the different components of the redox mixture and the support composition.

The TPR profiles of the alumina-supported NiMo and NiMoW oxides are given in Figure VI. In order to compare the different reduction stages, the TPR profiles for alumina-supported Ni, Mo and W oxides are also included. The reduction profile of alumina-supported 9 wt.% NiO (Figure VI a) shows a well-defined hydrogen consumption peak at 305 °C and a broad band at 580 °C, associated with the reduction processes of low crystalline nickel oxide [20] and surface Ni^{2+} ions of a not well-defined Ni–Al spinel [21], respectively.

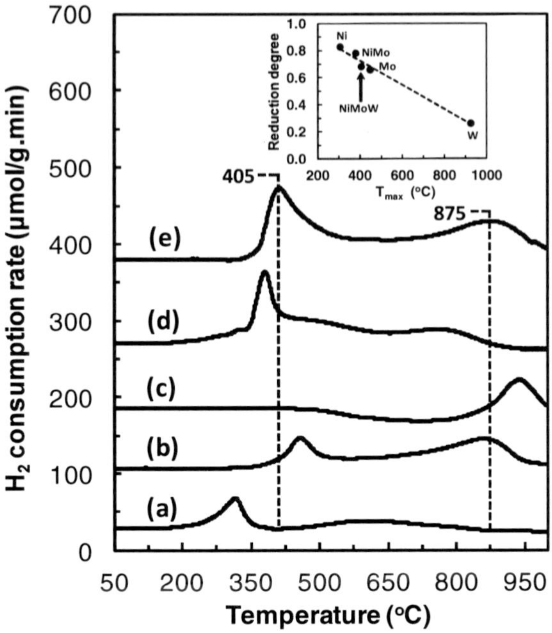

Figure VI. Temperature-programmed reduction of 9 wt. % NiO/Al_2O_3 (a), 15 wt. % MoO_3/Al_2O_3 (b), 20 wt. % WO_3/Al_2O_3 (c), 9 wt. % NiO-15 wt. % MoO_3/Al_2O_3 (d) and 9 wt. % NiO-15 wt. % MoO_3-20 wt. % WO_3/Al_2O_3 (e). Inset Figure: Dependence of the reduction degree with the maximum temperature of the initial reduction peak.

The TPR profiles of 15 wt.% MoO_3/Al_2O_3 sample displays defined reduction processes at low-temperature (*T*L) around 447 °C and a asymmetric broad peak at high-temperature (*T*H) at 857 °C (Figure VI b). The *T*L peak is normally assigned to the partial reduction of polymolybdate probably from Mo(VI) to Mo(IV) and the *T*H reduction peak is associated with the subsequent reduction of polymolybdate-partially reduced species and monomeric tetrahedrally coordinated Mo species [15].

In contrast, the alumina-supported 20 wt. % WO_3 profile (Figure VI c) shows a major single reduction peak at 925 °C attributed to the reduction of WO_x species, whose relatively higher abundance of W=O terminal group compared to Mo=O group rationalise its lower reducibility. Note also that the TPR profiles for alumina supported NiMo and NiMoW catalysts (i.e., Figures VI d and e) show the reduction peaks at intermediate temperatures between Ni- and W-containing samples, as a consequence of the effective promoter effect of nickel on the reducibility of MoO_x and WO_x domains. A linear dependence of the sample reduction degree with the maximum temperature of the first reduction peak is clearly illustrated in the inset Figure, which confirms that reduced nickel catalyses the H_2 dissociation and subsequently the H atoms could spill-over and reduce the Mo- and W-oxo species. The reduction degree is below 1.0 and this would indicate that alumina-supported metal oxides are not fully reduced within the analysis temperature range (50-950 °C).

The assignment of the overall reduction process, the theoretical and experimental moles of H_2 consumed per moles of metals (i.e., Ni, Mo and W) and the maximum reduction temperatures are listed in Table 5. The measured overall H_2 consumptions per moles of metals are comparable to those expected for the partial extent of reduction in Mo-containing samples. In those samples, Mo^{6+} is reduced to Mo^{2+} and Ni^{2+} is completed reduced to Ni^0, whilst W^{6+} appears to be reduced to various oxidation states (i.e., W^{2+}, W^{3+}).

When one compares the experimental (i.e., 0.79) and theoretical (i.e., 1.0) overall H_2 consumptions for alumina-supported W (+6) oxide an uncompleted reduction toward W^{4+} also appears to occurs. Indeed, the co-impregnation of alumina-supported W oxide with Ni and Mo oxides enhanced the reducibility of WO_x species, this finding is also supported by the diminution of the high temperature reduction peak (from 925 °C to 875 °C). Apparently, NiO is fully reduced in alumina-supported NiMo samples, indicating that MoO_x species could hinder the NiO-Al_2O_3 interaction and hence facilitating NiO reduction.

The overall analysis of H_2-TPR results suggest that NiO-MeO_x interaction facilitates the segregation of MoO_x and WO_x species on alumina surface and promotes their reducibility.

Table 5. Maximum reduction temperatures, moles of H_2 consumed per moles of metals and assignment of overall reduction process for alumina-supported Ni-, Mo-, and W-containing samples

Sample	$T_{1\ max}$(°C)	$T_{2\ max}$(°C)	(H_2/Metals) Experimental	(H_2/Metals) Theoretical	Assignment
Ni	305	580	0.83	1.00	$Ni^2+ \rightarrow Ni^0$
Mo	447	857	1.98	2.00	$Mo^{6+} \rightarrow Mo^{2+}$
W	---	925	0.79	1.00	$W^{6+} \rightarrow W^{4+}$
NiMo	379	737	1.51	1.50	$Ni^{2+} \rightarrow Ni^0$ $Mo^{6+} \rightarrow Mo^{2+}$
NiMoW	405	875	1.52	1.67	$Ni^{2+} \rightarrow Ni^0$ $Mo^{6+} \rightarrow Mo^{2+}$ $W^{6+} \rightarrow W^{2+}$

According to the above-discussed results, no evidence for the formation of Mo-O-W link was obtained, despite the relatively high surface density of MeO_x (8.2 Me atoms/nm^2). Indeed, a former report showed that MoO_x and WO_x formed a mixed metal oxide nanostructure on alumina surface at high surface density (i.e., 2-8 Me atoms/nm^2) [22]. The absent of Mo-O-W links could be considered as an additional evidence of the selective NiO-MeO_x interaction relative to MoO_x-WO_x interaction. It is envisaged that the proximity of NiO with MeO_x species could improve the promoter effect of Ni on MoS_2 (or WS_2) after the presulphiding process and hence enhancing the hydrotreating performance.

To find out the capability of the impregnation solution-based combustion method to produce deep hydrotreating catalysts, the apparent reaction orders and rate constants for alumina-supported NiMoW sulphide and commercial alumina-NiMo sulphide catalysts using two different feedstocks (i.e., feed 2 and feed 3) are compared in Table 6. There is a number of interesting points to be mentioned. 1) The apparent reaction order for HDS reactions for feed 2 (i.e., 1.3) is lower than that obtained for feed 3 (i.e., 1.6); an opposite trend was obtained for the HDS rate constants. 2) The HDN reaction data were satisfactorily fitted with an apparent reaction order of one independently of the feed composition. This relatively low value of reaction order compared to the HDS reaction orders (i.e., 1.3 and 1.6) implicates that N-containing molecules would have a higher coverage (or stronger interactions) on the surface of Ni-promoted Mo (and/or W) sulphide than S-containing molecules. This was markedly reflected on higher HDN rate constants relative to HDS activity, particularly for feed 3, which suggests that this feedstock is indeed less reactive (more demanding) than feed 1 based on the reactivity of S and N compounds. 3) The NiMoW catalyst showed HDS rate constants between 72 % (feed 2) and 57 % (feed 3) higher than those of NiMo catalyst, clearly indicating that the alumina-supported NiMoW sulphide catalyst is more active that the commercial alumina-supported NiMo sulphide catalyst for hydride-sulfurization reactions.

However, the HDN rate constants for feed 2 only showed a small increase with the catalyst composition; an even smaller rise was observed for feed 3. This trend is in agreement with the former conclusion that the feed 3 is less reactive than feed 2.

In order to find out the capability of the NiMo and NiMoW catalysts for the hydrogenation of aromatics, the fluorene concentrations present in feed 2 treated at different LHSVs were fitted on a pseudo-first order of reaction with correlation factor (i.e., R^2) above 0.990.

Table 6. Apparent reaction orders and rate constants for HDS and HDN reactions over alumina-supported NiMoW and commercial alumina-NiMo catalyst for model feed (feed 2) and straight run gas oil (feed 3). Reactions conducted at 340 °C, 3.5 MPa and 250 Nm^3/m^3

Catalyst	Feed	HDS apparent order	k_{HDS} [1/g.s.wppm $^{(n-1)}$]*10^{-4}	HDN apparent order	k_{HDN} (1/g.s)*10^{-4}
NiMo	2	1.3	8.29	1.0	70.70
	3	1.6	0.36	1.0	16.41
NiMoW	2	1.3	13.04	1.0	76.71
	3	1.6	0.62	1.0	17.34

The commercial alumina-supported NiMo and alumina-supported NiMoW sulphide catalysts showed hydrodearomatization (HDA) rate constants of $6.05x10^{-4}$ $(g.s)^{-1}$ and $8.87x10^{-4}$ $(g.s)^{-1}$, respectively.

Again, the latter catalyst exhibits a nearly 47 % superior HDA activity relative to the commercial NiMo formulation. This increase follows a similar trend in the improvement observed for the HDS activity (72 %), indicating that the hydrogenation sites appear to be the main cause of the higher HDS activity of NiMoW catalyst.

This is likely due to the presence of the type II Ni-(Mo/W)-S phase that would facilitate the hydrogenation pathway [23, 24]. This trend was not markedly reflected on the HDN activity, indicating that the contribution of the hydrogenation reaction in the overall rate of the HDN pathways is rather limited.

A previous report suggests that the enhanced S-Me bond strength by the proper combination of Ni, Mo and W can significantly improve the HDS performance of NiMoW sulphide catalyst [25]. On the other hand, it is envisaged that the formation of multiple slabs of Mo (and W) sulphide promoted with Ni can also produce a highly active type II Ni-(Mo/W)-S structures (i.e., Ni-Mo-S, Ni-W-S and Ni-Mo-W-S phases). These phases would be equivalent to the type II Co-Mo-S structure [26], whose type II phase has weaker interaction with the support and higher intrinsic activity than type I active phase.

It is expected that N-containing compounds cover a large fraction of the hydrogenation sites of the type II Ni-(Mo/W)-S phase, suppressing hence the hydrogenation of sterically hindered S compounds and aromatics.

CONCLUSION

The influence of the urea/Ni ratio in the HDS activity of alumina-supported NiMoP sulphide catalysts synthesised by impregnation solution-based combustion method has been catalytically examined. The synthesis of alumina-supported NiMoW oxide was carried out using a Φ ratio of 0.6. The HDS, HDN and HDA activities were compared to a commercial alumina-supported NiMo sulphide catalyst. Based on the catalyst preparation, characterization and performance, the following conclusions can be drawn:

1. The urea/Ni ratio can tune metal oxide-support interaction and hence the catalytic properties of hydrotreating catalysts. The ignition temperature and the maximum temperature of the exothermic peak in the *T-t* profiles showed a strong dependence of the chemical nature of the redox mixture deposited on alumina surface. The catalyst bed temperature rate increased lineally with the rise of fuel (i.e., urea) content.
2. The HDS activity of alumina-supported NiMoP sulphide catalysts achieved a maximum at Φ ratio between 0.6 and 1.2. This maximum HDS rate constant is attributed to a higher abundance of polymolybdate clusters relative to single molybdate species and Ni-Mo-oxo interaction in the catalyst precursor, which is reflected on the effective incorporation of nickel onto the edges of MoS_2 that facilitates the formation of NiMoS structure.
3. The sequential deposition of NiMo-urea solution on metal oxides-modified alumina surface produces a reduction of the catalyst bed temperature rate and an increase of the decomposition period of the metal oxide precursors and urea as a consequence of the lower content of NiMo oxides relative to the metal oxide-support (dilution effect) when increasing the number of impregnations. The addition of Ni and Mo to WO_x-modified alumina assists the segregation of MoO_x and WO_x species on the alumina surface and promotes their reducibility.
4. The addition of tungsten to the NiMo catalyst selectively enhances the HDS reaction rates compared to the HDN reactions, which is also reflected on the HDA reaction rates. This finding is rationalized assuming the formation of type II Ni-(Mo/W)-S structures and an electronic factor that improves the S-Me bond strength when a suitable catalyst synthesis approach and a combination of Ni, Mo and W are used.

ACKNOWLEDGMENT

The authors are indebted to a number of co-workers that have taken part in this research program during the last five years.

REFERENCES

[1] J. H. Gary, G.E. Handwerk, M.J. Kaiser, *Petroleum Refining. Technology and Economics*, Fifth Edition, CRC Press, London, 2007.

[2] Chorkendorff, J.W. Niemantsverdriet, *Concepts of Modern Catalysis and Kinetics,* Wiley-VCH, Weinheim, 2003, Chapter 9.

[3] T.G. Kaufmann, A. Kaldor, G.F. Stuntz, M.C. Kerby, L.L. Ansell, *Catal. Today,* 62, 77 (2000).

[4] Stanislaus, A. Marafi, M. S. Rana, *Catal. Today,* 153, 1 (2010).

[5] J. Eijsbouts: in K.P. de Jong, (Ed.), *Synthesis of Solid Catalysts,* Wiley-VCH, Weinheim, 2009, p. 301.

[6] S. Eijsbouts, S.W. Mayo, K. Fujita, *Appl. Catal. A: Gen.* 322, 58 (2007).

[7] J.J. Kingsley, K.C. Patil, *Mater Lett.* 6, 427 (1988).

[8] K.C. Patil, M.S. Hedge, R. Rattan, S.T. Aruna, *Nanocristalline Oxide Materials. Combustion Synthesis, Properties and Applications,* World Scientific, London, 2008.

[9] S. Specchia E. Finocchio, G. Busca, V. Specchia, in: M. Lackner, F. Winter, A.K. Agarwal (Eds.), *Handbook of Combustion*, VHC-Wiley, Weinheim, 2010, pp. 439-472.

[10] S.L. González-Cortés, S.M.A. Rodulfo-Baechler, T. Xiao, M.L.H. Green, *Catal. Lett.* 111, 57 (2006).

[11] S.L. González-Cortés, T. Xiao, T. Lin, M.L.H. Green, *Appl. Catal. A: Gen.* 302, 264 (2006).

[12] L. Wu, D. Jiao, L. Chen, J. Wang, F. Cao, *Adv. Mater. Res.* 132, 45 (2010).

[13] D. Jiao, Y. Ma, F. Cao, *Particuology* 4, 468 (2012).

[14] S.L. González-Cortés, F.E. Imbert, *Appl. Catal. A:Gen.* 452, 117 (2013).

[15] S.L. González-Cortés, T. Xiao, P.M.F.J. Costa, B. Fontal, M.L.H. Green, *Appl. Catal. A:Gen.* 270, 209 (2004).

[16] C. Glasson, C. Geantet, M. Lacroix, F. Labruyere, P. Dufresne, *J. Catal.* 212, 76 (2002).

[17] H. Topsøe, *Appl. Catal. A:Gen.* 322, 3 (2007).

[18] S.L. González-Cortés, *React. Kinet. Catal. Lett.* 97, 131 (2009).

[19] J. Leitner, P. Chuchvalec, Sedmidubsky, A. Strejc, P. Abrman, *Thermochimica Acta* 395, 27 (2003).

[20] S.L. González-Cortés, I. Aray, S.M.A. Rodulfo-Baechler, C.A. Lugo, H.L. Del Castillo, A. Loaiza-Gil, F.E. Imbert, H. Figueroa, W. Pernía, A. Rodríguez, O. Delgado, R. Casanova, J. Mendialdua, F. Rueda, *J. Mater. Sci.* 42, 6532 (2007).

[21] L. Zhang, J. Lin, Y. Chen, *J. Chem. Soc., Faraday Trans.,* 88, 497 (1992).

[22] H. Nair, M.J. Liszka, J.E. Gatt, C.D. Baertsch, *J. Phys. Chem. C,* 112, 1612 (2008).

[23] W. Lai, W. Song, L. Pang, Z. Wu, N. Zheng, J. Li, J. Zheng, X. Yi, *J. Catal.* 303, 80 (2013).

[24] N. Koizumi, Y. Hamabe, S. Yoshida, M. Yamada, *Appl. Catal. A:Gen.* 383, 79 (2010).

[25] C. Thomazeau, C. Geantet, M. Lacroix, M. Danot, V. Harlé, P. Raybaud, *Appl. Catal. A: Gen.* 322, 92 (2007).

[26] B. Hinnemann, J.K. Nørskov, H. Topsøe, *J. Phys. Chem. B,* 109, 2245 (2005).

INDEX

C

D

E

F

G

H

I

R

S

X

Y

Z